Grothendieck Construction of Bipermutative-Indexed Categories

The Grothendieck construction provides an explicit link between indexed categories and opfibrations. It is a fundamental concept in category theory and related fields with far-reaching applications. Bipermutative categories are categorifications of rings. They play a central role in algebraic K-theory and infinite loop space theory.

This monograph is a detailed study of the Grothendieck construction over a bipermutative category in the context of categorically enriched multicategories, with new and important applications to inverse K-theory and pseudo symmetric E_∞-algebras. After carefully recalling preliminaries in enriched categories, bipermutative categories, and enriched multicategories, we show that the Grothendieck construction over a small tight bipermutative category is a pseudo symmetric Cat-multifunctor and generally not a Cat-multifunctor in the symmetric sense.

Pseudo symmetry of Cat-multifunctors is a new concept we introduce in this work.

The following features make it accessible as a graduate text or reference for experts:

- Complete definitions and proofs.
- Self-contained background. Parts of Chapters 1–3, 7, 9, and 10 contain background material from the research literature.
- Extensive cross-references.
- Connections between chapters. Each chapter has its own introduction discussing not only the topics of that chapter but also its connection with other chapters.
- Open questions. Appendix A contains open questions that arise from the material in the text and are suitable for graduate students.

This book is suitable for graduate students and researchers with an interest in category theory, algebraic K-theory, homotopy theory, and related fields. The presentation is thorough and self-contained, with complete details and background material for non-expert readers.

Donald Yau is a Professor of Mathematics at The Ohio State University at Newark, specializing in homotopy theory and algebraic K-theory. He obtained his PhD at MIT under the direction of Haynes Miller and held a post-doctoral position at the University of Illinois at Urbana-Champaign. He is the author of over fifty research articles and ten books.

Grothendieck Construction of Bipermutative-Indexed Categories

Donald Yau

CRC Press
Taylor & Francis Group
Boca Raton London New York

CRC Press is an imprint of the
Taylor & Francis Group, an **informa** business
A CHAPMAN & HALL BOOK

First edition published 2024
by CRC Press
6000 Broken Sound Parkway NW, Suite 300, Boca Raton, FL 33487-2742

and by CRC Press
4 Park Square, Milton Park, Abingdon, Oxon, OX14 4RN

CRC Press is an imprint of Taylor & Francis Group, LLC

ISBN: 978-1-032-58404-1 (hbk)
ISBN: 978-1-032-58725-7 (pbk)
ISBN: 978-1-003-45123-5 (ebk)

DOI: 10.1201/9781003451235

Typeset in Latin Modern font
by KnowledgeWorks Global Ltd.

Publisher's note: This book has been prepared from camera-ready copy provided by the authors.

The author dedicates this book to Jacqueline.

Contents

Preface

The Grothendieck construction is a fundamental link between indexed categories and opfibrations. With Cat denoting the category of small categories, for each small category C, the Grothendieck construction associates to each pseudo functor $F : \mathsf{C} \longrightarrow \mathsf{Cat}$ a category $\int_{\mathsf{C}} F$ together with an opfibration over C:

$$\int_{\mathsf{C}} F \xrightarrow{U_F} \mathsf{C}.$$

Extended to strong transformations and modifications, this construction is a 2-equivalence between 2-categories [JY21, 10.6.16]

$$[\mathsf{C}, \mathsf{Cat}] \xrightarrow[\sim]{\int_{\mathsf{C}}} \mathsf{Fib}^{\mathsf{op}}(\mathsf{C}). \tag{0.1}$$

There are many variations and extensions of the Grothendieck construction in the literature. See, for example, [BW19] for an enriched version, [MV20] for a monoidal version, [Shu08] for a version involving double categories, and [Her04] where F is a multifunctor.

Purpose and Main Results

This monograph is a detailed study of the Grothendieck construction over a bipermutative category $(\mathcal{D}, \oplus, \otimes)$ in the context of Cat-enriched multicategories, with applications to inverse K-theory and pseudo symmetric E_∞-algebras. Next we describe the main results of this monograph.

Characterization of Cat-Multiequivalences

A key ingredient of the proof of the 2-equivalence $\int_{\mathsf{C}}$ in (0.1) is a characterization of a 2-equivalence as a 2-functor that is

- essentially surjective on objects and
- an isomorphism on each hom category.

See [JY21, 7.5.8] for a detailed proof. Just as a multicategory is a multiple-input extension of a category, a Cat-multicategory is a multiple-input generalization of a 2-category. From this view point, our first main result extends the characterization of 2-equivalences to Cat-*multiequivalences*, which are equivalences in the 2-category Cat-Multicat of small Cat-multicategories, Cat-multifunctors, and Cat-multinatural transformations. The Multiequivalence Theorem 3.69 says that a Cat-multifunctor is a Cat-multiequivalence if and only if it is

- essentially surjective on objects and
- an isomorphism on each multimorphism category.

The non-symmetric variant, where there is no symmetric group action, is also true. Moreover, this result recovers the characterization of 2-equivalences by restricting to 1-ary multimorphisms; see Theorem 3.70.

Pseudo Symmetric Cat-Multifunctors

We introduce a weaker variant of a Cat-multifunctor, which is called a *pseudo symmetric* Cat-*multifunctor*, that preserves the symmetric group action up to coherent natural isomorphisms

$$\begin{array}{ccc} \mathsf{M}(\langle c\rangle\,;c') & \xrightarrow{F} & \mathsf{N}(F\langle c\rangle\,;Fc') \\ {\scriptstyle\sigma}\downarrow & \overset{F_\sigma}{\underset{\cong}{\Longrightarrow}} & \downarrow{\scriptstyle\sigma} \\ \mathsf{M}(\langle c\rangle\sigma\,;c') & \xrightarrow[F]{} & \mathsf{N}(F\langle c\rangle\sigma\,;Fc') \end{array} \tag{0.2}$$

called *pseudo symmetry isomorphisms* (4.2). As we will discuss shortly, pseudo symmetric Cat-multifunctors are important in the context of the Grothendieck construction over a bipermutative category and inverse K-theory. With an appropriate notion of *pseudo symmetric* Cat-multinatural transformations, we prove in Theorem 4.34 that there is a 2-category $\mathsf{Cat\text{-}Multicat}^{\mathsf{ps}}$ with

- small Cat-multicategories as objects,
- pseudo symmetric Cat-multifunctors as 1-cells, and
- pseudo symmetric Cat-multinatural transformations as 2-cells.

In this context, we extend Theorem 3.69 to characterize pseudo symmetric Cat-multiequivalences; see Theorem 4.64.

Cat-Multicategory of Bipermutative-Indexed Categories

A *bipermutative category* $(\mathcal{D}, \oplus, \otimes)$ is equipped with two permutative—that is, strict symmetric monoidal—structures, $\oplus$ and $\otimes$, that interact via factorization natural transformations

$$\begin{array}{ccc} (A\otimes C)\oplus(B\otimes C) & \xrightarrow{\partial^l_{A,B,C}} & (A\oplus B)\otimes C \\ (A\otimes B)\oplus(A\otimes C) & \xrightarrow{\partial^r_{A,B,C}} & A\otimes(B\oplus C) \end{array} \tag{0.3}$$

and satisfy appropriate axioms (Definition 2.41). Bipermutative categories arise naturally in multiplicative infinite loop space theory and algebraic K-theory spectra; see [EM06] and [JY∞, Ch. 11]. In this monograph, bipermutative categories are mostly small and *tight*, which means that ∂^l and ∂^r in (0.3) are natural isomorphisms.

In the original Grothendieck construction $\int_{\mathsf{C}}$ in (0.1) with C a small category, the domain $[\mathsf{C}, \mathsf{Cat}]$ is a 2-category. For a small tight bipermutative category $(\mathcal{D}, \oplus, \otimes)$, we prove in Theorem 5.71 that there is a Cat-multicategory

$$\mathcal{D}\mathsf{Cat}$$

with symmetric monoidal functors

$$(\mathcal{D}, \oplus) \longrightarrow (\mathsf{Cat}, \times)$$

as objects. The multimorphism categories of $\mathcal{D}\mathsf{Cat}$ (Definitions 5.39 and 5.52) make full use of the bipermutative structure of $\mathcal{D}$. We emphasize that the Cat-multicategory structure on $\mathcal{D}\mathsf{Cat}$ is *not* induced by a monoidal structure (Explanation 5.64).

Pseudo Symmetric Cat-Multifunctorial Grothendieck Construction

We prove in Theorem 6.110 that the Grothendieck construction over a small tight bipermutative category $(\mathcal{D}, \oplus, \otimes)$ is a *pseudo symmetric* Cat-multifunctor

$$\mathcal{D}\mathsf{Cat} \xrightarrow{\int_{\mathcal{D}}} \mathsf{PermCat}^{\mathsf{sg}}. \tag{0.4}$$

The codomain $\mathsf{PermCat}^{\mathsf{sg}}$ is a Cat-multicategory with small permutative categories as objects (Theorem 6.31). Moreover, if the multiplicative braiding $\beta^{\otimes}$ in $\mathcal{D}$ is the identity natural transformation, then $\int_{\mathcal{D}}$ is a Cat-multifunctor, which means that it strictly preserves the symmetric group action. A weaker converse is true. If $\int_{\mathcal{D}}$ is a Cat-multifunctor, then

$$x \otimes y = y \otimes x$$

for all objects $x, y \in \mathcal{D}$. An important implication of this theorem is that inverse K-theory is a pseudo symmetric Cat-multifunctor but *not* a Cat-multifunctor in the symmetric sense.

Permutative Opfibrations from the Grothendieck Construction

For a symmetric monoidal functor

$$(\mathcal{D}, \oplus) \xrightarrow{X} (\mathsf{Cat}, \times),$$

the object assignment of $\int_{\mathcal{D}}$ in (0.4) produces a small permutative category $\int_{\mathcal{D}} X$. On the other hand, the Grothendieck construction $\int_{\mathsf{C}}$ in (0.1) produces not only the category $\int_{\mathsf{C}} F$ but also an opfibration $\int_{\mathsf{C}} F \longrightarrow \mathsf{C}$ over C. To take into account the projection

$$\int_{\mathcal{D}} X \longrightarrow \mathcal{D}$$

to $\mathcal{D}$, in Theorem 7.51 we prove that there is a non-symmetric Cat-multicategory $\mathsf{PFib}^{\mathcal{D}}$ with *small permutative opfibrations over* $\mathcal{D}$ as objects. Similar to $\mathcal{D}\mathsf{Cat}$, the objects in $\mathsf{PFib}^{\mathcal{D}}$ only involve the additive structure $(\mathcal{D}, \oplus)$, but the multimorphism categories in $\mathsf{PFib}^{\mathcal{D}}$ make full use of the bipermutative structure of $\mathcal{D}$. On the other hand, unlike $\mathcal{D}\mathsf{Cat}$, the non-symmetric Cat-multicategory $\mathsf{PFib}^{\mathcal{D}}$ is rarely a Cat-multicategory in the symmetric sense (Lemmas 7.44 and 7.48).

To extend the 2-equivalence $\int_{\mathsf{C}}$ in (0.1) to the bipermutative context, first we construct a non-symmetric Cat-multifunctor (7.53)

$$\mathsf{PFib}^{\mathcal{D}} \xrightarrow{\mathsf{U}} \mathsf{PermCat}^{\mathsf{sg}}$$

that forgets the projection down to $\mathcal{D}$. In Theorem 7.64 we prove that $\int_{\mathcal{D}}$ in (0.4) lifts as a non-symmetric Cat-multifunctor to $\mathsf{PFib}^{\mathcal{D}}$ as in the following commutative diagram.

$$\begin{array}{ccc} & \overset{\int_{\mathcal{D}}}{\nearrow} & \mathsf{PFib}^{\mathcal{D}} \\ & & \downarrow \mathsf{U} \\ \mathcal{D}\mathsf{Cat} & \xrightarrow{\int_{\mathcal{D}}} & \mathsf{PermCat}^{\mathsf{sg}} \end{array} \tag{0.5}$$

Utilizing the characterization Theorem 3.69, in Theorem 8.76 we prove that the lifted Grothendieck construction

$$\mathcal{D}\mathsf{Cat} \xrightarrow[\sim]{\int_{\mathcal{D}}} \mathsf{PFib}^{\mathcal{D}} \tag{0.6}$$

is a non-symmetric Cat-*multiequivalence*. Below is a summary.

- $\mathcal{D}$. Small tight bipermutative category (Definition 2.41).
- $\mathcal{D}\mathsf{Cat}$. Cat-multicategory (Theorem 5.71).
- $\mathsf{PermCat}^{\mathsf{sg}}$. Cat-multicategory (Theorem 6.31).
- $\mathsf{PFib}^{\mathcal{D}}$. Non-symmetric Cat-multicategory (Theorem 7.51).
- $\int_{\mathcal{D}} : \mathcal{D}\mathsf{Cat} \longrightarrow \mathsf{PermCat}^{\mathsf{sg}}$. Pseudo symmetric Cat-multifunctor (Theorem 6.110).
- $\mathsf{U} : \mathsf{PFib}^{\mathcal{D}} \longrightarrow \mathsf{PermCat}^{\mathsf{sg}}$. Non-symmetric Cat-multifunctor (Corollary 7.52).
- $\int_{\mathcal{D}} : \mathcal{D}\mathsf{Cat} \xrightarrow{\sim} \mathsf{PFib}^{\mathcal{D}}$. Non-symmetric Cat-multiequivalence (Theorem 8.76).

Pseudo Symmetric Cat-Multifunctorial Inverse K-Theory

The theorems described above have major implications in inverse K-theory. In [Man10] Mandell constructed a functor

$$\Gamma\text{-}\mathsf{Cat} \xrightarrow{\ \mathcal{P}\ } \mathsf{PermCat}^{\mathsf{sg}}, \tag{0.7}$$

from the category of Γ-categories to the category of small permutative categories, and proved that $\mathcal{P}$ is a homotopy inverse of Segal K-theory. It is proved in [JY22b] that the inverse K-theory functor $\mathcal{P}$ is a non-symmetric Cat-multifunctor. We emphasize that $\mathcal{P}$ is *not* a Cat-multifunctor in the symmetric sense because it does not strictly preserve the symmetric group action.

By construction $\mathcal{P}$ is a composite $\int_{\mathcal{A}} \circ A$ of two functors. We prove in Theorem 9.120 that the first functor A extends to a Cat-multifunctor

$$\Gamma\text{-}\mathsf{Cat} \xrightarrow{\ A\ } \mathcal{A}\mathsf{Cat}. \tag{0.8}$$

The codomain $\mathcal{A}\mathsf{Cat}$ is the Cat-multicategory $\mathcal{D}\mathsf{Cat}$ discussed above for the small tight bipermutative category $\mathcal{A}$ in Example 2.63. Moreover, by (0.4) the Grothendieck construction $\int_{\mathcal{A}}$ is a pseudo symmetric Cat-multifunctor. As a result, inverse K-theory $\mathcal{P}$

$$\begin{array}{ccc} \Gamma\text{-}\mathsf{Cat} & \xrightarrow{\ \mathcal{P}\ } & \mathsf{PermCat}^{\mathsf{sg}} \\ {\scriptstyle A}\searrow & & \nearrow{\scriptstyle \int_{\mathcal{A}}} \\ & \mathcal{A}\mathsf{Cat} & \end{array} \tag{0.9}$$

is a pseudo symmetric Cat-multifunctor but *not* a Cat-multifunctor; see Theorem 10.12. Combining (0.5), (0.6), and (0.9) yields the further factorization of $\mathcal{P}$

$$\begin{array}{ccc} \Gamma\text{-}\mathsf{Cat} & \xrightarrow{\ \mathcal{P}\ } & \mathsf{PermCat}^{\mathsf{sg}} \\ {\scriptstyle A}\downarrow & \nearrow{\scriptstyle \int_{\mathcal{A}}} & \uparrow{\scriptstyle \mathsf{U}} \\ \mathcal{A}\mathsf{Cat} & \xrightarrow[\sim]{\ \int_{\mathcal{A}}\ } & \mathsf{PFib}^{\mathcal{A}} \end{array} \tag{0.10}$$

into

1. a Cat-multifunctor A,
2. a non-symmetric Cat-multiequivalence $\int_{\mathcal{A}}$, and
3. a non-symmetric Cat-multifunctor U.

Below is a summary.

- $\mathcal{A}$. Small tight bipermutative category (Example 2.63).
- $\mathcal{A}\mathsf{Cat}$. Cat-multicategory (Example 5.74).
- $\mathsf{PFib}^{\mathcal{A}}$. Non-symmetric Cat-multicategory (Example 7.55).

- Γ-Cat. Cat-multicategory (Theorem 9.23).
- $\int_{\mathcal{A}} : \mathcal{A}\mathsf{Cat} \longrightarrow \mathsf{PermCat}^{\mathsf{sg}}$. Pseudo symmetric Cat-multifunctor (Example 6.114).
- $\mathsf{U} : \mathsf{PFib}^{\mathcal{A}} \longrightarrow \mathsf{PermCat}^{\mathsf{sg}}$. Non-symmetric Cat-multifunctor (Example 7.55).
- $\int_{\mathcal{A}} : \mathcal{A}\mathsf{Cat} \xrightarrow{\sim} \mathsf{PFib}^{\mathcal{A}}$. Non-symmetric Cat-multiequivalence (Example 8.77).
- $A : \Gamma\text{-}\mathsf{Cat} \longrightarrow \mathcal{A}\mathsf{Cat}$. Cat-multifunctor (Theorem 9.120).
- $\mathcal{P} : \Gamma\text{-}\mathsf{Cat} \longrightarrow \mathsf{PermCat}^{\mathsf{sg}}$. Pseudo symmetric Cat-multifunctor (Theorem 10.12).

Application: Preservation of Pseudo Symmetric Algebras

Since inverse K-theory $\mathcal{P}$ is not a Cat-multifunctor, it does not preserve algebraic structures parametrized by Cat-multifunctors in general. However, the pseudo symmetric Cat-multifunctoriality of $\mathcal{P}$ (0.9) implies that it preserves algebraic structures parametrized by

- non-symmetric Cat-multifunctors (Corollary 10.15) and
- pseudo symmetric Cat-multifunctors (Corollary 10.19).

As a special case of the latter, we observe that $\mathcal{P}$ preserves *pseudo symmetric E_∞-algebras* (Corollary 10.54), which are pseudo symmetric Cat-multifunctors from the categorical Barratt-Eccles operad $E\Sigma_*$. Pseudo symmetric E_∞-algebras in Γ-Cat and $\mathsf{PermCat}^{\mathsf{sg}}$ are described explicitly in Section 10.5.

Chapter Summaries

This monograph consists of the following three parts.

- Part I: Bipermutative Categories, Enriched
  Multicategories, and Pseudo Symmetry
- Part II: Grothendieck Multiequivalence from
  Bipermutative-Indexed Categories
  to Permutative Opfibrations
- Part III: Pseudo Symmetric Enriched Multifunctorial
  Inverse K-Theory

Below is a brief summary of each chapter. In the main text, each chapter starts with a more thorough introduction.

Chapter 1: Preliminaries on Enriched Categories and 2-Categories

To keep the prerequisite to an absolute minimum, in this preliminary chapter we review basic concepts related to monoidal categories, enriched categories, and 2-categories. We recall the symmetric monoidal closed structure on a diagram category $\mathcal{D}$-V when the domain category $\mathcal{D}$ and the codomain category V have the relevant structure. Due to its importance in later chapters, the special case $\mathsf{V} = \mathsf{Cat}$ is explained further. More detailed discussion of the material in this chapter can be found in [JY21, JY∞, Yau∞I].

Chapter 2: Symmetric Bimonoidal and Bipermutative Categories

This chapter has two main purposes. First we review the closely related notions of *symmetric bimonoidal categories* and *bipermutative categories*. We discuss several important examples of small tight bipermutative categories, including the indexing categories $\mathcal{F}$ and $\mathcal{A}$ for, respectively, Γ-categories and inverse K-theory. The second purpose of this chapter is to review Laplaza's Coherence Theorem 2.90 for symmetric bimonoidal categories. Tight bipermutative categories, which have invertible factorization morphisms (0.3), are special cases of tight symmetric bimonoidal categories. Thus Theorem 2.90 also applies to tight bipermutative categories. More detailed discussion of the material in this chapter can be found in [Yau∞I, Yau∞II].

Chapter 3: Enriched Multicategories and Multiequivalences

This chapter is about enriched multicategories and contains the first main theorem of this monograph. First we recall the 2-category V-$\mathsf{Multicat}$ of small V-multicategories, V-multifunctors, and V-multinatural transformations for a symmetric monoidal category V. The most relevant case for this monograph is $\mathsf{V} = \mathsf{Cat}$. More detailed discussion of enriched multicategories can be found in [JY21, JY∞, Yau16]. The second half of this chapter is devoted to the proof of Theorem 3.69, which characterizes (adjoint) Cat-multiequivalences locally. From this theorem we recover the analogous local characterization of 2-equivalences (Theorem 3.70).

Chapter 4: Pseudo Symmetry

This chapter introduces the notion of *pseudo symmetric* Cat-*multifunctors*. They preserve the symmetric group action up to coherent natural isomorphisms (0.2) that satisfy four coherence axioms. They are required to strictly preserve the colored units and the composition. In later chapters we show that

- the Grothendieck construction $\int_{\mathcal{D}}$ for each small tight bipermutative category $\mathcal{D}$ and

- inverse K-theory $\mathcal{P}$

are pseudo symmetric Cat-multifunctors. Further lax variants of Cat-multifunctors are discussed in Explanation 4.35 and Question A.1. Theorem 4.34 proves the existence of a 2-category $\mathsf{Cat\text{-}Multicat^{ps}}$ with small Cat-multicategories as objects, pseudo symmetric Cat-multifunctors as 1-cells, and pseudo symmetric Cat-multinatural transformations as 2-cells. Theorem 4.64 provides a local characterization of (adjoint) equivalences in the 2-category $\mathsf{Cat\text{-}Multicat^{ps}}$. This concludes Part I.

Chapter 5: Enriched Multicategories of Indexed Categories

Part II is about the Grothendieck construction $\int_{\mathcal{D}}$ over a small tight bipermutative category $\mathcal{D}$. In Section 5.1 we first discuss in detail the Cat-multicategory $\mathcal{D}$-Cat, which is also denoted by $\mathcal{D}^{\otimes}$-Cat, of $\mathcal{D}$-indexed categories for a small permutative category $(\mathcal{D}, \otimes)$. The rest of this chapter discusses a different Cat-multicategory $\mathcal{D}\mathsf{Cat}$, which serves as the domain of $\int_{\mathcal{D}}$, for a small tight bipermutative category $(\mathcal{D}, \oplus, \otimes)$. The objects of $\mathcal{D}\mathsf{Cat}$ are symmetric monoidal functors $(\mathcal{D}, \oplus) \longrightarrow \mathsf{Cat}$, instead of just $\mathcal{D}$-indexed categories. Moreover, the multimorphism categories of $\mathcal{D}\mathsf{Cat}$ involve the full extent of the bipermutative structure of $\mathcal{D}$ and Laplaza's Coherence Theorem 2.90. Furthermore, while the Cat-multicategory structure on $\mathcal{D}^{\otimes}$-Cat is induced by its symmetric monoidal closed structure, the Cat-multicategory structure on $\mathcal{D}\mathsf{Cat}$ is *not* induced by a monoidal structure in general (Explanation 5.64).

Chapter 6: The Grothendieck Construction is a Pseudo Symmetric Cat-Multifunctor

Section 6.1 discusses the Cat-multicategories PermCat and its strong variant $\mathsf{PermCat^{sg}}$, which serves as the codomain of $\int_{\mathcal{D}}$ when the projection down to $\mathcal{D}$ is not taken into account. The Cat-multicategory $\mathsf{PermCat^{sg}}$ is also the codomain of inverse K-theory. In each of PermCat and $\mathsf{PermCat^{sg}}$, the objects are small permutative categories, and the n-ary 2-cells are n-linear natural transformations. The n-ary 1-cells in PermCat and $\mathsf{PermCat^{sg}}$ are, respectively, n-linear functors and their strong variants. The rest of this chapter is devoted to the proof of Theorem 6.110, which proves that the Grothendieck construction $\int_{\mathcal{D}}$ in (0.4) is a pseudo symmetric Cat-multifunctor in the sense of Chapter 4. It is, furthermore, a Cat-multifunctor if the multiplicative braiding $\beta^{\otimes}$ in $\mathcal{D}$ is the identity. Conversely, if $\int_{\mathcal{D}}$ is a Cat-multifunctor, then $x \otimes y = y \otimes x$ for all objects $x, y \in \mathcal{D}$.

Chapter 7: Permutative Opfibrations from Bipermutative-Indexed Categories

This chapter studies the Grothendieck construction $\int_{\mathcal{D}}$ on a small tight bipermutative category $\mathcal{D}$ with the projection down to $\mathcal{D}$ taken into account. Theorem 7.51 proves the existence of the non-symmetric Cat-multicategory

$\mathsf{PFib}^{\mathcal{D}}$ with small permutative opfibrations over $\mathcal{D}$ as objects. Its multimorphism categories are more subtle and involve the bipermutative structure on $\mathcal{D}$ and Laplaza's Coherence Theorem 2.90. The non-symmetry of $\mathsf{PFib}^{\mathcal{D}}$ in general is once again related to the multiplicative braiding $\beta^{\otimes}$ in $\mathcal{D}$. It is explained in Lemmas 7.44 and 7.48. The second main result of this chapter is Theorem 7.64, which lifts the pseudo symmetric Cat-multifunctor $\int_{\mathcal{D}}$ to a non-symmetric Cat-multifunctor with codomain $\mathsf{PFib}^{\mathcal{D}}$ as in (0.5).

Chapter 8: The Grothendieck Construction is a Cat-Multiequivalence

The main result of this chapter is Theorem 8.76, which proves that the lifted Grothendieck construction $\int_{\mathcal{D}}$ in (0.6) is a non-symmetric Cat-*multiequivalence*. This proof uses the local characterization for (adjoint) equivalences in Theorem 3.69. In other words, instead of explicitly constructing an adjoint inverse of $\int_{\mathcal{D}}$, we show that $\int_{\mathcal{D}}$ is (i) essentially surjective on objects and (ii) an isomorphism on each multimorphism category. This concludes Part II.

Chapter 9: The Cat-Multifunctor A

Part III is about

- the pseudo symmetric Cat-multifunctoriality of inverse K-theory $\mathcal{P}$ and
- its ramification in terms of preservation of pseudo symmetric algebraic structures.

Inverse K-theory $\mathcal{P} = \int_{\mathcal{A}} \circ A$ involves two constructions:

- A in (0.8) and
- the Grothendieck construction $\int_{\mathcal{A}}$ for the small tight bipermutative category $\mathcal{A}$ in Example 2.63.

This chapter discusses the first construction A. For the domain of $\mathcal{P}$ and A, Sections 9.1 and 9.2 discuss the symmetric monoidal closed category Γ-Cat of Γ-categories and its Cat-multicategory structure. Then we discuss the underlying functor of A and its extension to a Cat-multifunctor.

Chapter 10: Inverse K-Theory is a Pseudo Symmetric Cat-Multifunctor

This chapter contains our main Theorem 10.12 on inverse K-theory. It constructs

- the pseudo symmetric Cat-multifunctorial inverse K-theory $\mathcal{P} = \int_{\mathcal{A}} \circ A$ as in (0.9) and
- a further factorization of $\mathcal{P}$ as in (0.10) that involves the non-symmetric Cat-multiequivalence $\int_{\mathcal{A}} : \mathcal{A}\mathsf{Cat} \xrightarrow{\sim} \mathsf{PFib}^{\mathcal{A}}$.

In Section 10.3 we completely unravel the pseudo symmetric Cat-multifunctorial structure of $\mathcal{P}$. In the second half of this section, we discuss related work in the literature [Elm∞, JY22b] about the multifunctoriality of $\mathcal{P}$. To illustrate the utility of Theorem 10.12, the rest of this chapter discusses the fact that $\mathcal{P}$ preserves pseudo symmetric E_∞-algebras. As preparation in Section 10.4 we discuss the categorical Barratt-Eccles operad $E\Sigma_*$, its Coherence Theorem 10.39, and E_∞-algebras in Γ-Cat. In Section 10.5 we define *pseudo symmetric* E_∞*-algebras* as pseudo symmetric Cat-multifunctors from $E\Sigma_*$. Inverse K-theory sends pseudo symmetric E_∞-algebras in Γ-Cat to those in $\mathsf{PermCat}^{\mathsf{sg}}$ (Corollary 10.54). Then we completely unravel these structures in Γ-Cat and $\mathsf{PermCat}^{\mathsf{sg}}$. This concludes Part III.

Appendix A: Open Questions

In this appendix, we discuss a number of open questions related to the topics of this monograph. They provide further motivation for the main text.

Appendix B: List of Main Facts

In this appendix, we summarize key facts in each chapter.

Audience

This monograph is aimed at graduate students and researchers with an interest in category theory, algebraic K-theory, and homotopy theory. Our highly detailed exposition is designed to make this monograph accessible to a wide audience.

Related Literature

The following references provide background and further discussion.

2-Dimensional Categories: [JY21].

Monoidal Categories and Enriched Multicategories: [JY∞, Yau16].

Bimonoidal and Bipermutative Categories: [Yau∞I, Yau∞II].

Grothendieck Construction: [BW19, Her04, JY21, MV20, Shu08].

Algebraic K-Theory Spectra: [BO20, JY∞, JY23, JY22a, JY22c, BF78, May78, Qui73, Seg74, Tho95, Wal85].

Multifunctorial K-Theory: [EM06, EM09, JY∞].

Inverse K-Theory: [GJO17, Man10, JY22b].

Author

Donald Yau is a Professor of Mathematics at The Ohio State University at Newark, specializing in homotopy theory and algebraic K-theory. He obtained his PhD at MIT under the direction of Haynes Miller and held a post-doctoral position at the University of Illinois at Urbana-Champaign. He is the author of over fifty research articles and ten books, including *Colored Operads* and *2-Dimensional Categories*.

List of Notations

Standard Notations

$\mathsf{Ob}(\mathsf{C})$, $\mathsf{Ob}\,\mathsf{C}$: objects in a category C
$\mathsf{C}(X,Y)$, $\mathsf{C}(X;Y)$: set of morphisms $X \longrightarrow Y$
1_X: identity morphism
$g \circ f$, gf: composition of morphisms
$\cong$, $\stackrel{\cong}{\longrightarrow}$: an isomorphism
$F : \mathsf{C} \longrightarrow \mathsf{D}$: a functor
Id_{C}, 1_{C}: identity functor
$\mathbf{1}$: terminal category
$(\mathsf{Set}, \times, *)$: category of sets and functions
θ_X: a component of a natural transformation θ
1_F: identity natural transformation
$\phi\theta$: vertical composition of natural transformations
$\theta' * \theta$: horizontal composition of natural transformations
(L, R, ϕ), $L \dashv R$: an adjunction
η, ε: unit and counit of an adjunction
$\varnothing$, $\varnothing^{\mathsf{C}}$: an initial object
$\coprod$, $\amalg$: a coproduct
$\prod$, $\sqcap$, $\times$: a product
Σ_n: symmetric group on n letters

Chapter 1

Chapter 2

Chapter 3

Chapter 4

Chapter 5

Chapter 6

$\langle x \circ_i x'_i \rangle$, $\langle x \rangle \circ_i x'_i$: $\langle x \rangle$ with its i-th entry replaced by x'_i, 136
$\langle x \circ_i x'_i \circ_\ell x'_\ell \rangle$: $\langle x \circ_i x'_i \rangle$ with its ℓ-th entry replaced by x'_ℓ, 136
$(F, \langle F^2_j \rangle_{j=1}^n)$, $(F, \langle F^2_j \rangle)$: an n-linear functor, 136
F^2_j: j-th linearity constraint of F, 136
$\mathsf{PermCat}(\langle \mathsf{C} \rangle ; \mathsf{D})$: category of n-linear functors and natural transformations, 140
$\mathsf{PermCat}^{\mathsf{sg}}(\langle \mathsf{C} \rangle ; \mathsf{D})$: strong variant of $\mathsf{PermCat}(\langle \mathsf{C} \rangle ; \mathsf{D})$, 140
$\mathsf{PermCat}(\langle \rangle ; \mathsf{D})$: D, 140
$\mathsf{PermCat}^{\mathsf{sg}}(\langle \rangle ; \mathsf{D})$: D, 140
$\mathsf{PermCat}$: Cat-multicategory of small permutative categories, 143
$\mathsf{PermCat}^{\mathsf{sg}}$: strong variant of $\mathsf{PermCat}$, 143
$\int_{\mathcal{D}} X$: Grothendieck construction of a functor $X : \mathcal{D} \longrightarrow \mathsf{Cat}$, 144
$(0, X^0 *)$: monoidal unit of $\int_{\mathcal{D}} X$, 145
$\square$: monoidal product of $\int_{\mathcal{D}} X$, 145
$\beta^{\square}$: braiding of $\int_{\mathcal{D}} X$, 146
$\int_{\mathcal{D}} \phi$: Grothendieck construction of an additive natural transformation, 149
$(\int_{\mathcal{D}} \phi)^2_i$: i-th linearity constraint of $\int_{\mathcal{D}} \phi$, 151
$\int_{\mathcal{D}} \Phi$: Grothendieck construction of an additive modification, 156
$(\int_{\mathcal{D}})_\sigma$: a pseudo symmetry isomorphism of $\int_{\mathcal{D}}$, 164
$(\int_{\mathcal{D}})_{\sigma,\phi}$, $(\int_{\mathcal{D}})_{\sigma,\phi,\langle (a_j, x_j) \rangle}$: components of $(\int_{\mathcal{D}})_\sigma$, 165
$\int_{\mathcal{D}}$: pseudo symmetric Cat-multifunctor $\mathcal{D}\mathsf{Cat} \longrightarrow \mathsf{PermCat}^{\mathsf{sg}}$, 174

Chapter 7

$\langle y ; f \rangle$: a fore-lift with object y and morphism f, 180
$\overline{f} : y \longrightarrow y_f$: a lift of a fore-lift $\langle y ; f \rangle$, 180
$\langle g, h; f \rangle$: a fore-raise, 181
$\widetilde{f}$: a raise of a fore-raise $\langle g, h; f \rangle$, 181
$P^{\mathsf{op}} : \mathsf{E}^{\mathsf{op}} \longrightarrow \mathsf{B}^{\mathsf{op}}$: opposite functor of $P : \mathsf{E} \longrightarrow \mathsf{B}$, 182
$\langle P \rangle \longrightarrow P_0$: an opcartesian n-linear functor, 185
$\mathsf{PFib}^{\mathcal{D}}(\langle P \rangle ; P_0)$: a subcategory of $\mathsf{PermCat}^{\mathsf{sg}}(\langle \mathsf{C} \rangle ; \mathsf{C}_0)$, 192
$\mathsf{PFib}^{\mathcal{D}}$: non-symmetric Cat-multicategory of small permutative opfibrations over $\mathcal{D}$, 194
$\otimes^{\mathsf{op}}$: opposite monoidal product, 198
U: non-symmetric Cat-multifunctor $\mathsf{PFib}^{\mathcal{D}} \longrightarrow \mathsf{PermCat}^{\mathsf{sg}}$, 202
U_X: first-factor projection $\int_{\mathcal{D}} X \longrightarrow \mathcal{D}$, 204
$\langle \int_{\mathcal{D}} X \rangle$: $\langle \int_{\mathcal{D}} X_j \rangle_{j=1}^n$, 205
$\langle U_X \rangle$: $\langle U_{X_j} \rangle_{j=1}^n$, 205
$\int_{\mathcal{D}}$: non-symmetric Cat-multifunctor $\mathcal{D}\mathsf{Cat} \longrightarrow \mathsf{PFib}^{\mathcal{D}}$, 208

Chapter 8

$P^{-1}(a)$: preimage subcategory, 213
$F_2\langle x \rangle$: second entry in $F\langle (a_j, x_j) \rangle$, 224
$F_2\langle p \rangle$: second entry in $F\langle (1_{a_j}, p_j) \rangle$, 225

Chapter 9

$(\mathsf{C}, *)$, (C, ι): a pointed category C with basepoint $\iota : \mathbf{1} \longrightarrow \mathsf{C}$, 243
Cat_*: category of small pointed categories and pointed functors, 244
E: smash unit $\mathbf{1} \coprod \mathbf{1}$, 244

Chapter 10

Part I

Bipermutative Categories, Enriched Multicategories, and Pseudo Symmetry

1

Preliminaries on Enriched Categories and 2-Categories

To prepare for later chapters, in this chapter we review some relevant definitions and theorems about monoidal categories (Section 1.1), enriched categories (Section 1.2), 2-categories (Section 1.3), and the Day convolution for functors between symmetric monoidal categories (Section 1.4). Below is a summary of some of the definitions and results in this chapter.

- Monoidal categories (braided, symmetric, closed). Definition 1.3 (1.8, 1.11, 1.14).
- Diagram categories. Definition 1.15.
- Monoidal functors and natural transformations. Definitions 1.17 and 1.21.
- Enriched categories, functors, and natural transformations. Definitions 1.25, 1.28, and 1.30.
- Tensor products of enriched categories. Definition 1.31.
- 2-categories, 2-functors, and 2-natural transformations. Definitions 1.32, 1.36, and 1.37.
- Modifications. Definition 1.41.
- Adjoint equivalences in a 2-category. Definition 1.46.
- 2-equivalences. Definition 1.49.
- Day convolution. Definition 1.50.
- Symmetric monoidal closed diagram categories. Theorems 1.63 and 1.65.
- Permutative-indexed categories. Explanation 1.66.

A detailed discussion of enriched categories and the Day convolution can be found in [JY∞]. All the necessary concepts of 2-dimensional category theory, including pasting diagrams, for this monograph are discussed in detail in [JY21].

DOI: 10.1201/9781003451235-1

Connection with Subsequent Chapters

The notion of a symmetric monoidal category in Section 1.1 is central to this work. One of our main objectives is to study the Grothendieck construction from $\mathcal{D}\mathsf{Cat}$, whose objects are additive symmetric monoidal functors $\mathcal{D}^{\oplus} \longrightarrow \mathsf{Cat}$, with $(\mathcal{D}, \oplus, \otimes)$ a small tight bipermutative category (Definition 2.41). In a bipermutative category, each of $\oplus$ and $\otimes$ is a permutative structure, which means a strict symmetric monoidal structure. The codomains of the Grothendieck constructions in Theorems 6.110 and 7.64 also involve small permutative categories. The codomain of inverse K-theory (Theorem 10.12) has small permutative categories as objects.

From Chapter 3 onward we discuss enriched multicategories, especially Cat-multicategories, which are multicategories enriched in the symmetric monoidal category $(\mathsf{Cat}, \times, \mathbf{1})$ of small categories with the Cartesian product. Sections 1.2 and 1.3 on enriched categories and 2-categories serve as motivation for enriched multicategories and the 2-categories that they form. The language of 2-categories also provides a convenient framework to define (adjoint) Cat-multiequivalences for Cat-multicategories (Definition 3.42). One of our main observations in this work is that the Grothendieck construction on bipermutative-indexed categories is a non-symmetric Cat-multiequivalence (Theorem 8.76).

The Day convolution in Section 1.4 provides a symmetric monoidal structure, hence also a Cat-multicategory structure, on $\mathcal{D}$-indexed categories (Theorem 5.1). This, in turn, allows us to define the Cat-multicategory $\mathcal{D}\mathsf{Cat}$ (Definition 5.61). However, $\mathcal{D}\mathsf{Cat}$ is *not* induced by a symmetric monoidal structure due to the non-existence of a potential monoidal unit in general (Explanation 5.64).

1.1 Monoidal Categories

In this section we review the definitions of monoidal categories, monoidal functors, and monoidal natural transformations. For further discussion of monoidal categories, the reader is referred to [JS93, ML98, JY∞, Yau∞I, Yau∞II].

Definition 1.1. A *Grothendieck universe* is a set $\mathcal{U}$ with the following properties.

1. If $x \in \mathcal{U}$ and $y \in x$, then $y \in \mathcal{U}$.

2. If $x \in \mathcal{U}$, then $\mathsf{P}(x) \in \mathcal{U}$, where $\mathsf{P}(x)$ is the set of subsets of x.

3. If $J \in \mathcal{U}$ and $x_j \in \mathcal{U}$ for each $j \in J$, then the union $\bigcup_{j\in J} x_j \in \mathcal{U}$.

4. $\mathbb{N} \in \mathcal{U}$, where $\mathbb{N}$ is the set of finite ordinals. ◇

Convention 1.2 (Grothendieck Universe). We assume Grothendieck's A *of Universes*: every set belongs to some universe. We fix a universe $\mathcal{U}$ and an element in $\mathcal{U}$ a *set*. A subset of $\mathcal{U}$ is called a *class*. Any categorical stru is called *small* if its objects form a set. Whenever necessary, we tacitly re $\mathcal{U}$ by a larger universe $\mathcal{U}'$ in which $\mathcal{U}$ is a set. For more discussion of unive see [JY21, Section 1.1], [ML98, I.6], and [AGV72, ML69].

Definition 1.3. A *monoidal category* is a sextuple

$$(\mathsf{C}, \otimes, \mathbb{1}, \alpha, \lambda, \rho)$$

consisting of the following data and axioms.

- C is a category.
- $\otimes : \mathsf{C} \times \mathsf{C} \longrightarrow \mathsf{C}$ is a functor, which is called the *monoidal product.*
- $\mathbb{1} \in \mathsf{C}$ is an object, which is called the *monoidal unit.*
- α, λ, and ρ are natural isomorphisms

$$(X \otimes Y) \otimes Z \xrightarrow[\cong]{\alpha_{X,Y,Z}} X \otimes (Y \otimes Z)$$

$$\mathbb{1} \otimes X \xrightarrow[\cong]{\lambda_X} X \xleftarrow[\cong]{\rho_X} X \otimes \mathbb{1}$$

for objects $X, Y, Z \in \mathsf{C}$. They are called, respectively, the *assoc isomorphism*, the *left unit isomorphism*, and the *right unit isomor*

The following middle unity and pentagon diagrams are required to co for objects $W, X, Y, Z \in \mathsf{C}$, where all $\otimes$ symbols are omitted to save sp

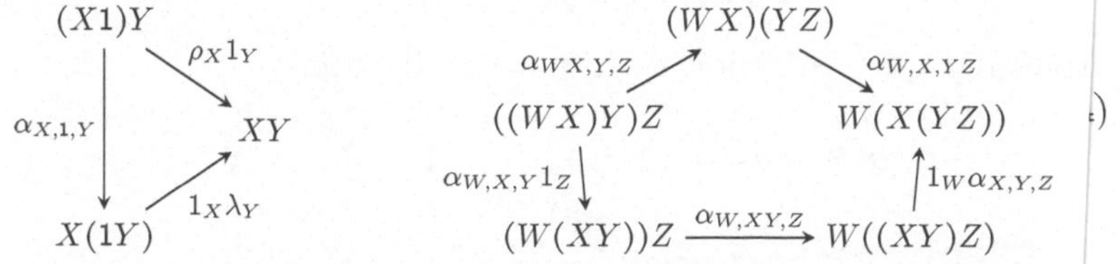

This finishes the definition of a monoidal category. Moreover, we de the following.

- A monoidal category is *strict* if α, λ, and ρ are identity natural sfor- mations.
- A (monoidal) category is also called a *(monoidal) 1-category.* ◇

In each monoidal category, the morphism equality holds:

$$\lambda_{\mathbb{1}} = \rho_{\mathbb{1}} : \mathbb{1} \otimes \mathbb{1} \longrightarrow \mathbb{1}. \tag{1.5}$$

e following left and right unity diagrams commute [Kel74].

$$\begin{array}{ccc} \otimes Y & \xrightarrow{\alpha_{1,X,Y}} & 1 \otimes (X \otimes Y) \\ & & \downarrow \lambda_{X\otimes Y} \\ & = & X \otimes Y \end{array} \qquad \begin{array}{ccc} (X \otimes Y) \otimes 1 & \xrightarrow{\alpha_{X,Y,1}} & X \otimes (Y \otimes 1) \\ \rho_{X\otimes Y} \downarrow & & \downarrow 1_X \otimes \rho_Y \\ X \otimes Y & = & X \otimes Y \end{array} \tag{1.6}$$

ese unity properties in the more general context of bicategories
d in [JY21, Section 2.2].

1.7. A *monoid* in a monoidal category C is a triple (X, μ, η)
f the following data and axioms.

an object.

$X \longrightarrow X$ is a morphism, which is called the *multiplication.*

X is a morphism, which is called the *unit.*

ng associativity and unity diagrams are required to commute.

$$\begin{array}{ccc} X) \otimes X & \xrightarrow{\alpha} & X \otimes (X \otimes X) \\ \downarrow {}_{X} & & \downarrow 1_X \otimes \mu \\ & & X \otimes X \\ & & \downarrow \mu \\ X \otimes X & \xrightarrow{\mu} & X \end{array} \qquad \begin{array}{ccc} 1 \otimes X & \xrightarrow{\lambda_X} & X \\ \eta \otimes 1_X \downarrow & & \| \\ X \otimes X & \xrightarrow{\mu} & X \\ 1_X \otimes \eta \uparrow & & \| \\ X \otimes 1 & \xrightarrow{\rho_X} & X \end{array}$$

A m of monoids

$$(X, \mu^X, \eta^X) \xrightarrow{f} (Y, \mu^Y, \eta^Y)$$

is a ism $f : X \longrightarrow Y$ in C such that the diagrams

$$\begin{array}{ccc} X \otimes X & \xrightarrow{f \otimes f} & Y \otimes Y \\ \mu^X \downarrow & & \downarrow \mu^Y \\ X & \xrightarrow{f} & Y \end{array} \qquad \begin{array}{ccc} 1 & \xrightarrow{\eta^X} & X \\ \| & & \downarrow f \\ 1 & \xrightarrow{\eta^Y} & Y \end{array}$$

comm ◇

Defin n 1.8. A *braided monoidal category* is a pair (C, ξ) consisting of the
follow data and axioms.

- $\mathsf{C} =, \otimes, 1, \alpha, \lambda, \rho)$ is a monoidal category.
- ξ is natural isomorphism

$$X \otimes Y \xrightarrow[\cong]{\xi_{X,Y}} Y \otimes X$$

for $Y \in \mathsf{C}$, which is called the *braiding.*

The hexagon diagrams

$$
\begin{array}{ccc}
(Y\otimes X)\otimes Z & \xrightarrow{\alpha} & Y\otimes(X\otimes Z)\\
{\scriptstyle \xi^{\otimes}_{X,Y}\otimes 1_Z}\nearrow & & \searrow{\scriptstyle 1_Y\otimes\xi^{\otimes}_{X,Z}}\\
(X\otimes Y)\otimes Z & & Y\otimes(Z\otimes X)\\
{\scriptstyle \alpha}\searrow & & \nearrow{\scriptstyle \alpha}\\
X\otimes(Y\otimes Z) & \xrightarrow{\xi^{\otimes}_{X,Y\otimes Z}} & (Y\otimes Z)\otimes X
\end{array}
\qquad
\begin{array}{ccc}
X\otimes(Z\otimes Y) & \xrightarrow{\alpha^{-1}} & (X\otimes Z)\otimes Y\\
{\scriptstyle 1_X\otimes\xi^{\otimes}_{Y,Z}}\nearrow & & \searrow{\scriptstyle \xi^{\otimes}_{X,Z}\otimes 1_Y}\\
X\otimes(Y\otimes Z) & & (Z\otimes X)\otimes Y\\
{\scriptstyle \alpha^{-1}}\searrow & & \nearrow{\scriptstyle \alpha^{-1}}\\
(X\otimes Y)\otimes Z & \xrightarrow{\xi_{X\otimes Y,Z}} & Z\otimes(X\otimes Y)
\end{array}
\tag{1.9}
$$

are required to commute for objects $X, Y, Z \in \mathsf{C}$. ◇

The following two unity diagrams commute for each object X in a braided monoidal category; see [Yau∞II, 1.3.21] for the proof.

$$
\begin{array}{ccc}
X\otimes 1 & \xrightarrow{\xi_{X,1}} & 1\otimes X\\
{\scriptstyle \rho_X}\downarrow & & \downarrow{\scriptstyle \lambda_X}\\
X & = & X
\end{array}
\qquad
\begin{array}{ccc}
1\otimes X & \xrightarrow{\xi_{1,X}} & X\otimes 1\\
{\scriptstyle \lambda_X}\downarrow & & \downarrow{\scriptstyle \rho_X}\\
X & = & X
\end{array}
\tag{1.10}
$$

Definition 1.11. A *symmetric monoidal category* is a pair (C, ξ) consisting of the following data and axioms.

- $\mathsf{C} = (\mathsf{C}, \otimes, 1, \alpha, \lambda, \rho)$ is a monoidal category.
- ξ is a natural isomorphism

$$X \otimes Y \xrightarrow[\cong]{\xi_{X,Y}} Y \otimes X$$

for $X, Y \in \mathsf{C}$, which is called the *symmetry isomorphism* or the *braiding*.

The following symmetry and hexagon diagrams are required to commute for objects $X, Y, Z \in \mathsf{C}$.

$$
\begin{array}{ccc}
X\otimes Y & \xrightarrow{1_{X\otimes Y}} & X\otimes Y\\
{\scriptstyle \xi_{X,Y}}\searrow & & \nearrow{\scriptstyle \xi_{Y,X}}\\
& Y\otimes X &
\end{array}
\qquad
\begin{array}{ccc}
(Y\otimes X)\otimes Z & \xrightarrow{\alpha} & Y\otimes(X\otimes Z)\\
{\scriptstyle \xi_{X,Y}\otimes 1_Z}\nearrow & & \searrow{\scriptstyle 1_Y\otimes\xi_{X,Z}}\\
(X\otimes Y)\otimes Z & & Y\otimes(Z\otimes X)\\
{\scriptstyle \alpha}\searrow & & \nearrow{\scriptstyle \alpha}\\
X\otimes(Y\otimes Z) & \xrightarrow{\xi_{X,Y\otimes Z}} & (Y\otimes Z)\otimes X
\end{array}
\tag{1.12}
$$

A *permutative category* is a strict symmetric monoidal category. For a permutative category, we sometimes denote its monoidal product by $\oplus$ and its monoidal unit by e. ◇

Remark 1.13. The symmetry axiom, $\xi_{Y,X}\xi_{X,Y} = 1$, implies that the left and right hexagons (1.9) are equivalent. Therefore, a symmetric monoidal category is precisely a braided monoidal category that satisfies the symmetry axiom. As a consequence, in each symmetric monoidal category the unity diagrams (1.10) commute. ◇

Definition 1.14. A symmetric monoidal category $(\mathsf{C}, \otimes, 1, \xi)$ is *closed* if, for each object $X \in \mathsf{C}$, the functor $- \otimes X : \mathsf{C} \longrightarrow \mathsf{C}$ admits a right adjoint, which is called an *internal hom.* A right adjoint of $- \otimes X$ is denoted by $\mathsf{Hom}(X, -)$ or $[X, -]$. ◇

Definition 1.15 (Diagrams)**.** For a small category $\mathcal{D}$ and a category C, the *diagram category* $\mathcal{D}$-C has

- functors $\mathcal{D} \longrightarrow \mathsf{C}$ as objects,
- natural transformations between such functors as morphisms,
- identity natural transformations as identities, and
- vertical composition of natural transformations as composition.

Moreover, we define the following.

- A functor $\mathcal{D} \longrightarrow \mathsf{C}$ is also called a *$\mathcal{D}$-diagram in* C.
- For the category Cat of small categories and functors, a functor $\mathcal{D} \longrightarrow \mathsf{Cat}$ is called a *$\mathcal{D}$-indexed category.* ◇

Example 1.16 (Small Categories)**.** $(\mathsf{Cat}, \times, \mathbf{1}, [,])$ is a symmetric monoidal closed category, with Cat the category of small categories and functors.

- The monoidal product is the Cartesian product $\times$.
- The monoidal unit $\mathbf{1}$ is the terminal category with only one object $*$ and its identity morphism 1_*.
- The closed structure $[,]$ is given by diagram categories (Definition 1.15).

The category Cat is complete and cocomplete [Yau20, Section 1.4]. ◇

Monoidal Functors and Natural Transformations

Definition 1.17. Suppose C and D are monoidal categories. A *monoidal functor*

$$(F, F^2, F^0) : \mathsf{C} \longrightarrow \mathsf{D}$$

consists of the following data and axioms.

- $F : \mathsf{C} \longrightarrow \mathsf{D}$ is a functor.
- $F^0 : 1 \longrightarrow F1$ is a morphism in D, which is called the *unit constraint.*
- F^2 is a natural transformation

$$FX \otimes FY \xrightarrow{F^2_{X,Y}} F(X \otimes Y) \quad \text{for} \quad X, Y \in \mathsf{C},$$

which is called the *monoidal constraint.*

The following unity and associativity diagrams are required to commute for objects $X, Y, Z \in \mathsf{C}$.

$$\begin{array}{ccc} 1 \otimes FX & \xrightarrow{\lambda_{FX}} & FX \\ {\scriptstyle F^0 \otimes 1_{FX}}\downarrow & & \uparrow{\scriptstyle F\lambda_X} \\ F1 \otimes FX & \xrightarrow{F^2} & F(1 \otimes X) \end{array} \qquad \begin{array}{ccc} FX \otimes 1 & \xrightarrow{\rho_{FX}} & FX \\ {\scriptstyle 1_{FX} \otimes F^0}\downarrow & & \uparrow{\scriptstyle F\rho_X} \\ FX \otimes F1 & \xrightarrow{F^2} & F(X \otimes 1) \end{array} \tag{1.18}$$

$$\begin{array}{ccc} (FX \otimes FY) \otimes FZ & \xrightarrow{\alpha} & FX \otimes (FY \otimes FZ) \\ {\scriptstyle F^2 \otimes 1_{FZ}}\downarrow & & \downarrow{\scriptstyle 1_{FX} \otimes F^2} \\ F(X \otimes Y) \otimes FZ & & FX \otimes F(Y \otimes Z) \\ {\scriptstyle F^2}\downarrow & & \downarrow{\scriptstyle F^2} \\ F((X \otimes Y) \otimes Z) & \xrightarrow{F\alpha} & F(X \otimes (Y \otimes Z)) \end{array} \tag{1.19}$$

A monoidal functor is said to be

- *strong* if F^0 and F^2 are isomorphisms;
- *strictly unital* if F^0 is the identity morphism; and
- *strict* if F^0 and F^2 are identities.

For braided monoidal categories C and D, a monoidal functor (F, F^2, F^0) as above is *braided* if the diagram

$$\begin{array}{ccc} FX \otimes FY & \overset{\xi_{FX,FY}}{\underset{\cong}{\longrightarrow}} & FY \otimes FX \\ {\scriptstyle F^2}\downarrow & & \downarrow{\scriptstyle F^2} \\ F(X \otimes Y) & \overset{F\xi_{X,Y}}{\underset{\cong}{\longrightarrow}} & F(Y \otimes X) \end{array} \tag{1.20}$$

commutes for all objects $X, Y \in \mathsf{C}$. If C and D are symmetric monoidal categories, then a braided monoidal functor as above is called a *symmetric monoidal functor.* ◇

Definition 1.21. For monoidal functors

$$(F, F^2, F^0) \quad \text{and} \quad (G, G^2, G^0) : \mathsf{C} \longrightarrow \mathsf{D}$$

between monoidal categories C and D, a *monoidal natural transformation* $\theta : F \longrightarrow G$ is a natural transformation of the underlying functors such that the unit constraint and monoidal constraint diagrams

$$\begin{array}{ccc} 1 & \xrightarrow{F^0} & F1 \\ & {\scriptstyle G^0}\searrow & \downarrow{\scriptstyle \theta_1} \\ & & G1 \end{array} \qquad \begin{array}{ccc} FX \otimes FY & \xrightarrow{\theta_X \otimes \theta_Y} & GX \otimes GY \\ {\scriptstyle F^2}\downarrow & & \downarrow{\scriptstyle G^2} \\ F(X \otimes Y) & \xrightarrow{\theta_{X \otimes Y}} & G(X \otimes Y) \end{array} \tag{1.22}$$

commute for all objects $X, Y \in \mathsf{C}$. ◇

Such a (monoidal) natural transformation is also denoted by the following 2-cell notation.

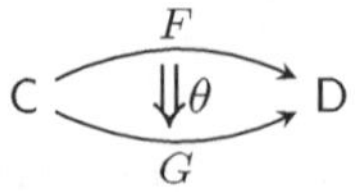

(1.23)

Pasting diagrams involving 2-cells are discussed in [JY21, Ch. 3].

Convention 1.24 (Left Normalized Bracketing)**.** Unless otherwise specified, an iterated monoidal product is *left normalized* with the left half of each pair of parentheses on the far left. ◇

For example, we denote

$$a \otimes b \otimes c \otimes d = \big((a \otimes b) \otimes c\big) \otimes d.$$

We omit parentheses for iterated monoidal products and tacitly insert the necessary associativity and unit isomorphisms. This is valid because each monoidal category C admits a strong monoidal adjoint equivalence $\mathsf{C} \longrightarrow \mathsf{C}_{\mathsf{st}}$ with C_{st} strict monoidal [ML98, XI.3.1]. The braided and symmetric versions can be found in [Yau22, 21.3.1 and 21.6.1]. In each case, the strict diagrams commute if and only if their preimages in C commute.

1.2 Enriched Categories

In this section we briefly review the definitions of enriched categories, enriched functors, enriched natural transformations, and their tensor products. More detailed discussion can be found in [JY21, Section 1.3], [JY∞, Ch. 1], and [Kel05].

Definition 1.25. For a monoidal category $(\mathsf{V}, \otimes, \mathbb{1}, \alpha, \lambda, \rho)$ (Definition 1.3), a V-*category* C, which is also called a *category enriched in* V, consists of the following data and axioms.

- It is equipped with a class $\mathsf{Ob}(\mathsf{C})$ of *objects*.
- For each pair of objects X, Y in C, it is equipped with a *hom object* $\mathsf{C}(X,Y) \in \mathsf{V}$ with domain X and codomain Y.
- For objects X, Y, Z in C, it is equipped with a morphism

  $$\mathsf{C}(Y,Z) \otimes \mathsf{C}(X,Y) \xrightarrow{m_{X,Y,Z}} \mathsf{C}(X,Z) \quad \text{in} \quad \mathsf{V},$$

  called the *composition*.
- For each object X in C, it is equipped with a morphism

  $$\mathbb{1} \xrightarrow{i_X} \mathsf{C}(X,X) \quad \text{in} \quad \mathsf{V},$$

  called the *identity* of X.

The following *associativity diagram* and *unity diagram* are required to commute for objects W, X, Y, Z in C.

$$
\begin{array}{ccc}
\big(\mathsf{C}(Y,Z) \otimes \mathsf{C}(X,Y)\big) \otimes \mathsf{C}(W,X) & \xrightarrow{\alpha} & \mathsf{C}(Y,Z) \otimes \big(\mathsf{C}(X,Y) \otimes \mathsf{C}(W,X)\big) \\
& & \downarrow{\scriptstyle 1\otimes m} \\
\downarrow{\scriptstyle m\otimes 1} & & \mathsf{C}(Y,Z) \otimes \mathsf{C}(W,Y) \\
& & \downarrow{\scriptstyle m} \\
\mathsf{C}(X,Z) \otimes \mathsf{C}(W,X) & \xrightarrow{\quad m \quad} & \mathsf{C}(W,Z)
\end{array}
\tag{1.26}
$$

$$
\begin{array}{ccccc}
1 \otimes \mathsf{C}(X,Y) & \xrightarrow{\lambda} & \mathsf{C}(X,Y) & \xleftarrow{\rho} & \mathsf{C}(X,Y) \otimes 1 \\
\downarrow{\scriptstyle i_Y \otimes 1} & & \| & & \downarrow{\scriptstyle 1 \otimes i_X} \\
\mathsf{C}(Y,Y) \otimes \mathsf{C}(X,Y) & \xrightarrow{m} & \mathsf{C}(X,Y) & \xleftarrow{m} & \mathsf{C}(X,Y) \otimes \mathsf{C}(X,X)
\end{array}
\tag{1.27}
$$

This finishes the definition of a V-category. Moreover, a V-category C is *small* if $\mathsf{Ob}(\mathsf{C})$ is a set. ◇

Definition 1.28. For V-categories C and D, a V-*functor* $F : \mathsf{C} \longrightarrow \mathsf{D}$ consists of the following data and axioms.

- It is equipped with an object assignment $F : \mathsf{Ob}(\mathsf{C}) \longrightarrow \mathsf{Ob}(\mathsf{D})$.
- For each pair of objects X, Y in C, it is equipped with a component morphism

$$\mathsf{C}(X,Y) \xrightarrow{F_{X,Y}} \mathsf{D}\big(FX, FY\big) \quad \text{in} \quad \mathsf{V}.$$

The following two diagrams are required to commute for objects X, Y, Z in C.

$$
\begin{array}{ccc}
\mathsf{C}(Y,Z) \otimes \mathsf{C}(X,Y) & \xrightarrow{\quad m \quad} & \mathsf{C}(X,Z) \\
\downarrow{\scriptstyle F\otimes F} & & \downarrow{\scriptstyle F} \\
\mathsf{D}(FY,FZ) \otimes \mathsf{D}(FX,FY) & \xrightarrow{m} & \mathsf{D}(FX,FZ)
\end{array}
\qquad
\begin{array}{ccc}
1 & \xrightarrow{i_X} & \mathsf{C}(X,X) \\
\| & & \downarrow{\scriptstyle F} \\
1 & \xrightarrow{i_{FX}} & \mathsf{D}(FX,FX)
\end{array}
\tag{1.29}
$$

Moreover, identity V-functors and composition of V-functors are defined separately on objects and component morphisms. ◇

Definition 1.30. For V-functors $F, G : \mathsf{C} \longrightarrow \mathsf{D}$ between V-categories C and D, a V-*natural transformation* $\theta : F \longrightarrow G$ consists of, for each object X in C, a morphism

$$1 \xrightarrow{\theta_X} \mathsf{D}(FX, GX) \quad \text{in} \quad \mathsf{V},$$

called the X-*component* of θ, such that the following naturality diagram commutes for objects X, Y in C.

$$\begin{array}{ccc}
 & 1 \otimes \mathsf{C}(X,Y) \xrightarrow{\theta_Y \otimes F} \mathsf{D}(FY,GY) \otimes \mathsf{D}(FX,FY) & \\
{}^{\lambda^{-1}}\nearrow\cong & & \searrow m \\
\mathsf{C}(X,Y) & & \mathsf{D}(FX,GY) \\
{}_{\rho^{-1}}\searrow\cong & & \nearrow m \\
 & \mathsf{C}(X,Y) \otimes 1 \xrightarrow{G \otimes \theta_X} \mathsf{D}(GX,GY) \otimes \mathsf{D}(FX,GX) &
\end{array}$$

This finishes the definition of a V-natural transformation. Moreover, the following notions are defined componentwise:

- identity V-natural transformations,
- vertical composition of V-natural transformations, and
- horizontal composition of V-natural transformations. ◇

The following definition is from [JY∞, Section 1.3].

Definition 1.31. For V-categories C and D with $(\mathsf{V}, \otimes, 1, \xi)$ a braided monoidal category (Definition 1.8), the *tensor product* $\mathsf{C} \otimes \mathsf{D}$ is the V-category defined by the following data.

Objects $\mathsf{Ob}(\mathsf{C} \otimes \mathsf{D}) = \mathsf{Ob}(\mathsf{C}) \times \mathsf{Ob}(\mathsf{D})$, with objects denoted $X \otimes Y$ for $X \in \mathsf{C}$ and $Y \in \mathsf{D}$.

Hom Objects For objects $X \otimes Y$ and $X' \otimes Y'$, the hom object is given by

$$(\mathsf{C} \otimes \mathsf{D})(X \otimes Y, X' \otimes Y') = \mathsf{C}(X, X') \otimes \mathsf{D}(Y, Y').$$

Composition For objects $X \otimes Y$, $X' \otimes Y'$, and $X'' \otimes Y''$, the composition is given by the following composite in V, with ξ_{mid} interchanging the middle two factors using the associativity isomorphism and braiding in V.

$$\begin{array}{ccc}
\big(\mathsf{C}(X',X'') \otimes \mathsf{D}(Y',Y'')\big) \otimes \big(\mathsf{C}(X,X') \otimes \mathsf{D}(Y,Y')\big) & & \mathsf{C}(X,X'') \otimes \mathsf{D}(Y,Y'') \\
{}_{\xi_{\mathrm{mid}}}\searrow & & \nearrow_{m \otimes m} \\
 & \big(\mathsf{C}(X',X'') \otimes \mathsf{C}(X,X')\big) \otimes \big(\mathsf{D}(Y',Y'') \otimes \mathsf{D}(Y,Y')\big) &
\end{array}$$

Identities The identity of an object $X \otimes Y$ is the following composite.

$$1 \xrightarrow[\cong]{\lambda^{-1}} 1 \otimes 1 \xrightarrow{i_X \otimes i_Y} \mathsf{C}(X,X) \otimes \mathsf{D}(Y,Y)$$

This finishes the definition of the V-category $\mathsf{C} \otimes \mathsf{D}$. Moreover, the tensor product $\otimes$ is extended to V-functors and V-natural transformations componentwise. ◇

1.3 2-Categories

In this section we review the definitions of 2-categories, 2-functors, 2-natural transformations, modifications, (adjoint) equivalences in 2-categories, and (adjoint) 2-equivalences between 2-categories. The reader is referred to [JY21] for an introduction to 2-categories.

Definition 1.32. A *2-category* B consists of the following data and axioms.

Objects It is equipped with a class B_0 of *objects*.

1-Cells For each pair of objects $X, Y \in \mathsf{B}_0$, it is equipped with a class $\mathsf{B}_1(X,Y)$ of *1-cells* from X to Y. Such a 1-cell is denoted $X \longrightarrow Y$.

2-Cells For 1-cells $f, f' \in \mathsf{B}_1(X,Y)$, it is equipped with a set $\mathsf{B}_2(f,f')$ of *2-cells* from f to f'. Such a 2-cell is denoted $f \longrightarrow f'$.

Identities It is equipped with

- an *identity 1-cell* $1_X \in \mathsf{B}_1(X,X)$ for each object X and
- an *identity 2-cell* $1_f \in \mathsf{B}_2(f,f)$ for each 1-cell $f \in \mathsf{B}_1(X,Y)$.

Compositions X, Y, and Z denote objects in B in the following compositions.

- For 1-cells $f, f', f'' \in \mathsf{B}_1(X,Y)$, it is equipped with an assignment
$$\mathsf{B}_2(f',f'') \times \mathsf{B}_2(f,f') \xrightarrow{\;v\;} \mathsf{B}_2(f,f''), \qquad v(\alpha',\alpha) = \alpha'\alpha$$
called the *vertical composition* of 2-cells.
- It is equipped with an assignment
$$\mathsf{B}_1(Y,Z) \times \mathsf{B}_1(X,Y) \xrightarrow{\;c_1\;} \mathsf{B}_1(X,Z), \qquad c_1(g,f) = gf$$
called the *horizontal composition* of 1-cells.
- For 1-cells $f, f' \in \mathsf{B}_1(X,Y)$ and $g, g' \in \mathsf{B}_1(Y,Z)$, it is equipped with an assignment
$$\mathsf{B}_2(g,g') \times \mathsf{B}_2(f,f') \xrightarrow{\;c_2\;} \mathsf{B}_2(gf,g'f'), \quad c_2(\beta,\alpha) = \beta * \alpha$$
called the *horizontal composition* of 2-cells.

The data above are required to satisfy the following four axioms:

1. Vertical composition is associative and unital for identity 2-cells.
2. Horizontal composition preserves identity 2-cells and vertical composition.

3. Horizontal composition of 1-cells is associative and unital for identity 1-cells.

4. Horizontal composition of 2-cells is associative and unital for identity 2-cells of identity 1-cells.

This finishes the definition of a 2-category.

Moreover, we define the following.

- For objects X and Y in a 2-category B, the *hom category* $\mathsf{B}(X,Y)$ is the category defined by the following data.
  - Its objects are 1-cells from X to Y.
  - Morphisms from a 1-cell $f : X \longrightarrow Y$ to a 1-cell $f' : X \longrightarrow Y$ are 2-cells $f \longrightarrow f'$.
  - Its composition is vertical composition of 2-cells.
  - Its identities are identity 2-cells.

- A 2-category is *locally small* if each hom category is a small category.

- A 2-category is *small* if it is locally small and has a set of objects.

- The *underlying 1-category* of a 2-category is the category defined by the following data.
  - It has the same class of objects.
  - Its morphisms are 1-cells.
  - Its composition is horizontal composition of 1-cells.
  - Its identities are identity 1-cells.

The 2-cell notation (1.23) also applies to 2-cells in a 2-category. ◇

Example 1.33 (Small Categories)**.** Cat in Example 1.16 is the underlying 1-category of a 2-category, also denoted by Cat, where the 2-cells are natural transformations. Horizontal, respectively vertical, composition of 2-cells is defined as that of natural transformations. ◇

Example 1.34 (Small Enriched Categories)**.** Each monoidal category V has an associated 2-category $\mathsf{Cat}^{\mathsf{V}}$ defined by the following data.

- Its objects are small V-categories.

- Its 1-cells are V-functors.

- Its 2-cells are V-natural transformations.

The 2-category Cat in Example 1.33 is $\mathsf{Cat}^{\mathsf{V}}$ for $\mathsf{V} = (\mathsf{Set}, \times, *)$. ◇

A Cat-category—that is, a category enriched in $(\mathsf{Cat}, \times, \mathbf{1})$—is a 2-category. The converse is also true up to a set-theoretic condition.

Proposition 1.35. *A locally small 2-category is precisely a* Cat*-category.*

2-Functors

Up to the set-theoretic condition in Proposition 1.35, a 2-functor is a Cat-functor between Cat-categories. We organize the precise definition as follows.

Definition 1.36. For 2-categories B and C, a *2-functor* $F : \mathsf{B} \longrightarrow \mathsf{C}$ consists of the following data and axioms.

- It is equipped with an *object assignment*

$$\mathsf{B}_0 \xrightarrow{F_0} \mathsf{C}_0.$$

- It is equipped with a *1-cell assignment*

$$\mathsf{B}_1(X,Y) \xrightarrow{F_1} \mathsf{C}_1(F_0X, F_0Y)$$

  for each pair of objects X, Y in B.

- It is equipped with a *2-cell assignment*

$$\mathsf{B}_2(f,f') \xrightarrow{F_2} \mathsf{C}_2(F_1f, F_1f')$$

  for each pair of objects X, Y and 1-cells $f, f' \in \mathsf{B}_1(X,Y)$.

The data above are required to satisfy the following three axioms, with F_0, F_1, and F_2 all abbreviated to F:

1. The object and 1-cell assignments of F form a functor between the underlying 1-categories of B and C.
2. For each pair of objects X, Y in B, the 1-cell and 2-cell assignments of F form a functor
$$\mathsf{B}(X,Y) \xrightarrow{F} \mathsf{C}(FX, FY)$$
between hom categories.
3. F preserves horizontal composition of 2-cells.

This finishes the definition of a 2-functor.

Moreover, given a 2-functor $G : \mathsf{C} \longrightarrow \mathsf{D}$, the *composite 2-functor*

$$\mathsf{B} \xrightarrow{GF} \mathsf{D}$$

is defined by separately composing the object assignments, the 1-cell assignments, and the 2-cell assignments. ◇

2-Natural Transformations

Up to the set-theoretic condition in Proposition 1.35, a 2-natural transformation is a Cat-natural transformation between Cat-functors. We organize the precise definition as follows.

Definition 1.37. For 2-functors $F, G : \mathsf{B} \longrightarrow \mathsf{C}$ between 2-categories B and C, a *2-natural transformation* $\phi : F \longrightarrow G$ consists of, for each object X in B, a *component 1-cell*

$$FX \xrightarrow{\phi_X} GX \quad \text{in} \quad \mathsf{C}$$

such that the following two axioms hold:

1-Cell Naturality For each 1-cell $f : X \longrightarrow Y$ in B, the following two composite 1-cells in $\mathsf{C}(FX, GY)$ are equal.

$$\begin{array}{ccc} FX & \xrightarrow{\phi_X} & GX \\ {\scriptstyle Ff}\downarrow & & \downarrow{\scriptstyle Gf} \\ FY & \xrightarrow{\phi_Y} & GY \end{array} \tag{1.38}$$

2-Cell Naturality For each 2-cell $\theta : f \longrightarrow g$ in $\mathsf{B}(X, Y)$, the following two whiskered 2-cells are equal.

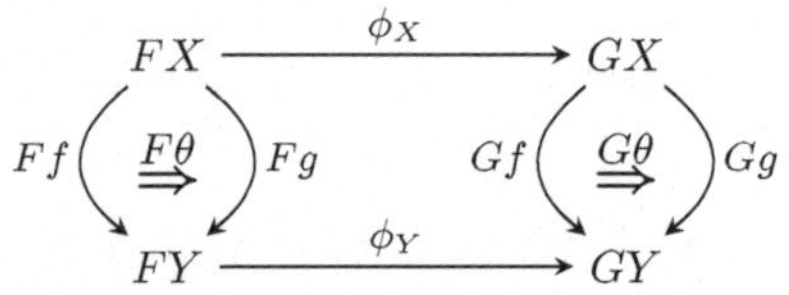

(1.39)

This means that the following 2-cell equality holds in $\mathsf{C}(FX, GY)$.

$$G\theta * 1_{\phi_X} = 1_{\phi_Y} * F\theta$$

This finishes the definition of a 2-natural transformation. The 2-cell notation (1.23) applies to 2-natural transformations. Moreover, a *2-natural isomorphism* is a 2-natural transformation with each component 1-cell an isomorphism in the underlying 1-category. ◇

Definition 1.40. Suppose $\phi : F \longrightarrow G$ and $\varphi : G \longrightarrow H$ are 2-natural transformations for 2-functors $F, G, H : \mathsf{B} \longrightarrow \mathsf{C}$ between 2-categories B and C. The *horizontal composite* 2-natural transformation

$$F \xrightarrow{\varphi\phi} H$$

is defined by, for each object X in B, the horizontal composite component 1-cell

$$FX \xrightarrow{\phi_X} GX \xrightarrow{\varphi_X} HX$$

in $\mathsf{C}(FX, HX)$. ◇

Modifications

Definition 1.41. Suppose $\phi, \varphi : F \longrightarrow G$ are 2-natural transformations for 2-functors $F, G : \mathsf{B} \longrightarrow \mathsf{C}$ between 2-categories B and C. A *modification* $\Phi : \phi \longrightarrow \varphi$ consists of, for each object X in B, a *component 2-cell*

$$\phi_X \xrightarrow{\Phi_X} \varphi_X \quad \text{in} \quad \mathsf{C}(FX, GX)$$

such that the following two whiskered 2-cells in $\mathsf{C}(FX, GY)$ are equal for each 1-cell $f : X \longrightarrow Y$ in B.

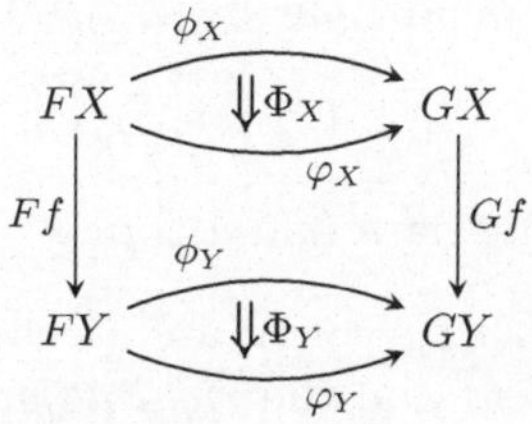

(1.42)

This means the equality

$$1_{Gf} * \Phi_X = \Phi_Y * 1_{Ff}.$$

This is called the *modification axiom.* ◇

Definition 1.43. Suppose $F, G, H : \mathsf{B} \longrightarrow \mathsf{C}$ are 2-functors between 2-categories B and C.

1. Suppose $\phi, \varphi, \psi : F \longrightarrow G$ are 2-natural transformations, and $\Phi : \phi \longrightarrow \varphi$ and $\Psi : \varphi \longrightarrow \psi$ are modifications. The *vertical composite modification*

$$\phi \xrightarrow{\Psi\Phi} \psi \tag{1.44}$$

is defined by, for each object X in B, the vertical composite 2-cell

$$\phi_X \xrightarrow{\Phi_X} \varphi_X \xrightarrow{\Psi_X} \psi_X$$

in $\mathsf{C}(FX, GX)$.

2. Suppose $\Phi' : \phi' \longrightarrow \varphi'$ is a modification for 2-natural transformations $\phi', \varphi' : G \longrightarrow H$. The *horizontal composite modification*

$$\phi'\phi \xrightarrow{\Phi' * \Phi} \varphi'\varphi \tag{1.45}$$

is defined by, for each object X in B, the horizontal composite 2-cell

$$\phi'_X \phi_X \xrightarrow{\Phi'_X * \Phi_X} \varphi'_X \varphi_X$$

in $\mathsf{C}(FX, HX)$. ◇

Adjoint Equivalences

The following definitions are [JY21, 5.1.18, 6.1.1, and 6.2.1] restricted to 2-categories.

Definition 1.46. Suppose B is a 2-category.

1. A 1-cell $f : X \longrightarrow Y$ in B is called an *equivalence* if there exist
   - a 1-cell $h : Y \longrightarrow X$, which is called an *inverse* of f, and
   - 2-cell isomorphisms $\eta : 1_X \xrightarrow{\cong} hf$ and $\varepsilon : fh \xrightarrow{\cong} 1_Y$.
2. An *adjunction* in B is a quadruple
$$(f, h, \eta, \varepsilon)$$
   consisting of the following data and axioms.
   - $f : X \longrightarrow Y$ and $h : Y \longrightarrow X$ are 1-cells, which are called, respectively, the *left adjoint* and the *right adjoint*.
   - $\eta : 1_X \longrightarrow hf$ and $\varepsilon : fh \longrightarrow 1_Y$ are 2-cells, which are called, respectively, the *unit* and the *counit*.

   The following two diagrams of 2-cells, which are called, respectively, the *left triangle identity* and the *right triangle identity*, are required to commute.
$$\begin{array}{ccc} f & \xrightarrow{1_f * \eta} & fhf \\ & \searrow^{1_f} & \downarrow{\scriptstyle \varepsilon * 1_f} \\ & & f \end{array} \qquad \begin{array}{ccc} h & \xrightarrow{\eta * 1_h} & hfh \\ & \searrow^{1_h} & \downarrow{\scriptstyle 1_h * \varepsilon} \\ & & h \end{array} \tag{1.47}$$
3. An adjunction is called an *adjoint equivalence* if the unit and the counit are isomorphisms.

This finishes the definition. ◇

Example 1.48. Consider Definition 1.46.

- If f is an equivalence with (h, η, ε) as in 1, then h is also an equivalence with inverse 1-cell f and 2-cell isomorphisms $(\varepsilon^{-1}, \eta^{-1})$.
- A quadruple $(f, h, \eta, \varepsilon)$ is an adjoint equivalence as in 3 if and only if the quadruple $(h, f, \varepsilon^{-1}, \eta^{-1})$ is an adjoint equivalence.
- If $(f, h, \eta, \varepsilon)$ is an adjoint equivalence, then f and h are equivalences. Conversely, each equivalence is the left adjoint of an adjoint equivalence by [JY21, 6.2.4]. It is also the right adjoint of an adjoint equivalence by the previous item. ◇

Definition 1.49. A 2-functor $F : \mathsf{B} \longrightarrow \mathsf{C}$ is called a *2-equivalence* if there exist

- a 2-functor $G : \mathsf{C} \longrightarrow \mathsf{B}$, which is called an *inverse* of F, and
- 2-natural isomorphisms $\eta : 1_{\mathsf{B}} \xrightarrow{\cong} GF$ and $\varepsilon : FG \xrightarrow{\cong} 1_{\mathsf{C}}$.

Moreover, if the quadruple $(F, G, \eta, \varepsilon)$ satisfies the triangle identities (1.47), then it is called an *adjoint 2-equivalence*. ◇

A functor between categories is an equivalence if and only if it is

- essentially surjective on objects and
- fully faithful on morphism sets.

A similar characterization of 2-equivalences is [JY21, 7.5.8]; see Theorem 3.70. Theorem 3.69 is the extension to Cat-multiequivalences.

1.4 The Day Convolution

In this section we review the Day convolution for functors between symmetric monoidal categories and associated constructions. The special case where the codomain is the symmetric monoidal closed category Cat is discussed in detail in Explanation 1.66. The following discussion is adapted from [JY∞, Ch. 3], where much more detail can be found. The *pointed* Day convolution for Γ-categories, which we need for the inverse K-theory discussion in Chapter 10, is in Definition 9.17.

Definition 1.50. Suppose $(\mathsf{V}, \otimes, \mathbb{1}, \xi, [,])$ is a symmetric monoidal closed category that is complete and cocomplete, and $(\mathcal{D}, \boxdot, e)$ is a small symmetric monoidal category. Suppose $X, Y : \mathcal{D} \longrightarrow \mathsf{V}$ are $\mathcal{D}$-diagrams in V.

- The *Day convolution* of X and Y is the $\mathcal{D}$-diagram in V defined as the coend

$$X \otimes Y = \int^{(a,b) \in \mathcal{D} \times \mathcal{D}} \coprod_{\mathcal{D}(a \boxdot b, -)} (Xa \otimes Yb). \tag{1.51}$$

  For natural transformations $\phi : X \longrightarrow X'$ and $\psi : Y \longrightarrow Y'$ between $\mathcal{D}$-diagrams in V, the *Day convolution* $\phi \otimes \psi$ is the natural transformation

$$X \otimes Y \xrightarrow{\phi \otimes \psi} X' \otimes Y' \tag{1.52}$$

  induced by the morphisms in V

$$Xa \otimes Yb \xrightarrow{\phi_a \otimes \psi_b} X'a \otimes Y'b \quad \text{for} \quad a, b \in \mathcal{D}$$

  via the universal properties of coproducts and coends.

- The *braiding*

$$X \otimes Y \xrightarrow[\cong]{\xi_{X,Y}} Y \otimes X \tag{1.53}$$

is the natural transformation induced by the braiding in V

$$Xa \otimes Yb \xrightarrow[\cong]{\xi_{Xa,Yb}} Yb \otimes Xa \quad \text{for} \quad a, b \in \mathcal{D},$$

the braiding in $\mathcal{D}$, and the universal properties of coproducts and coends.

- The *hom diagram* from X to Y is defined as the end

$$\mathsf{Hom}_{\mathcal{D}}(X,Y) = \int_{(b,c) \in \mathcal{D} \times \mathcal{D}} \Big[\coprod_{\mathcal{D}(-\boxdot b, c)} Xb, Yc \Big]. \tag{1.54}$$

This is extended to natural transformations between $\mathcal{D}$-diagrams in V via the universal properties of coproducts and ends.

- The *unit diagram* J is the $\mathcal{D}$-diagram

$$J = \coprod_{\mathcal{D}(e,-)} \mathbb{1} : \mathcal{D} \longrightarrow \mathsf{V}. \tag{1.55}$$

This finishes the definition. ◇

Explanation 1.56 (Universal Property)**.** The Day convolution $X \otimes Y$ in (1.51) is characterized by the following universal property. For each pair of objects $(a, b) \in \mathcal{D} \times \mathcal{D}$, it is equipped with a structure morphism

$$Xa \otimes Yb \xrightarrow{\zeta_{a,b}} (X \otimes Y)(a \boxdot b) \quad \text{in} \quad \mathsf{V} \tag{1.57}$$

such that the following two conditions hold.

Naturality For each pair of morphisms $f : a \longrightarrow a'$ and $g : b \longrightarrow b'$ in $\mathcal{D}$, the following diagram in V commutes.

$$\begin{array}{ccc} Xa \otimes Yb & \xrightarrow{Xf \otimes Yg} & Xa' \otimes Yb' \\ \downarrow{\scriptstyle \zeta_{a,b}} & & \downarrow{\scriptstyle \zeta_{a',b'}} \\ (X \otimes Y)(a \boxdot b) & \xrightarrow{(X \otimes Y)(f \boxdot g)} & (X \otimes Y)(a' \boxdot b') \end{array} \tag{1.58}$$

Universality The pair $(X \otimes Y, \zeta)$ is initial with respect to the previous condition.

We refer to (1.58) as the *naturality* of ζ. ◇

Explanation 1.59 (Iterated Day Convolution)**.** For $n \geq 2$ suppose $X_1, \ldots, X_n$ are $\mathcal{D}$-diagrams in V. By general coend calculus their iterated Day convolution is the $\mathcal{D}$-diagram given by the coend

$$\bigotimes_{j=1}^{n} X_j \cong \int^{\{a_j\}_{j=1}^n \in \mathcal{D}^n} \coprod_{\mathcal{D}(a_1 \boxdot \cdots \boxdot a_n\, ,\, -)} \Big(\bigotimes_{j=1}^{n} X_j a_j \Big). \tag{1.60}$$

It is characterized by the universal property involving the n-variable variants of the structure morphisms (1.57) and the naturality diagram (1.58). More explicitly, for each n-tuple of objects $\{a_j\}_{j=1}^n$ in $\mathcal{D}$, it is equipped with a structure morphism

$$\bigotimes_{j=1}^n X_j a_j \xrightarrow{\zeta_{a_1,\ldots,a_n}} \big(\bigotimes_{j=1}^n X_j\big)(a_1 \boxdot \cdots \boxdot a_n) \quad \text{in} \quad \mathsf{V} \tag{1.61}$$

such that the following two conditions hold.

Naturality For each n-tuple of morphisms $\{f_j : a_j \longrightarrow a_j'\}_{j=1}^n$ in $\mathcal{D}$, denote by

$$a = a_1 \boxdot \cdots \boxdot a_n, \quad a' = a_1' \boxdot \cdots \boxdot a_n', \quad \text{and} \quad f = f_1 \boxdot \cdots \boxdot f_n.$$

Then the following naturality diagram in V commutes.

$$\begin{array}{ccc} \bigotimes_{j=1}^n X_j a_j & \xrightarrow{\bigotimes_{j=1}^n X_j f_j} & \bigotimes_{j=1}^n X_j a_j' \\ {\scriptstyle \zeta_{a_1,\ldots,a_n}} \big\downarrow & & \big\downarrow {\scriptstyle \zeta_{a_1',\ldots,a_n'}} \\ \big(\bigotimes_{j=1}^n X_j\big)(a) & \xrightarrow{(\bigotimes_{j=1}^n X_j)(f)} & \big(\bigotimes_{j=1}^n X_j\big)(a') \end{array} \tag{1.62}$$

Universality The pair $\big(\bigotimes_{j=1}^n X_j\, ,\, \zeta\big)$ is initial with respect to the previous condition.

In (1.60) through (1.62) each iterated monoidal product is left normalized as in Convention 1.24. ◇

Symmetric Monoidal Closed Structure

Recall the diagram category $\mathcal{D}$-C (Definition 1.15). The following theorem is due to Day [Day70]. A detailed proof can be found in [JY∞, 3.7.22], which covers the more general situation when $\mathcal{D}$ is a small symmetric monoidal V-category. The simpler situation as stated below is sufficient for this monograph.

Theorem 1.63. *Suppose* V *is a symmetric monoidal closed category that is complete and cocomplete. Suppose* $\mathcal{D}$ *is a small symmetric monoidal category. Then the diagram category* $\mathcal{D}$-V *is a symmetric monoidal closed category with the following data.*

- *The monoidal product is given by the Day convolution* (1.51).
- *The braiding is given by* (1.53).
- *The closed structure is given by the hom diagram* (1.54).
- *The monoidal unit is given by the unit diagram* J (1.55).

Definition 1.64 (Tensored and Cotensored)**.** Suppose C is a V-category with $(\mathsf{V}, [,])$ a symmetric monoidal closed category.

Tensored We say that C is *tensored over* V if C is equipped with, for each $X \in \mathsf{C}$ and $A \in \mathsf{V}$, an object $X \otimes A \in \mathsf{C}$ together with isomorphisms in V

$$\mathsf{C}(X \otimes A, Y) \cong [A, \mathsf{C}(X, Y)]$$

for all $Y \in \mathsf{C}$.

Cotensored We say that C is *cotensored over* V if C is equipped with, for each $X \in \mathsf{C}$ and $A \in \mathsf{V}$, an object $X^A \in \mathsf{C}$ together with isomorphisms in V

$$\mathsf{C}(Y, X^A) \cong [A, \mathsf{C}(Y, X)]$$

for all $Y \in \mathsf{C}$. ◇

The following result is a nontrivial consequence of Day's Theorem 1.63, as presented in [JY∞, 3.9.15].

Theorem 1.65. *Suppose* V *is a symmetric monoidal closed category that is complete and cocomplete. Suppose* $\mathcal{D}$ *is a small symmetric monoidal category. Then the symmetric monoidal closed category* $\mathcal{D}$-V *is enriched, tensored, and cotensored over* V*.*

Proof. The reader is referred to [JY∞, Section 3.9] for the detailed proof. Here we provide a brief outline. Since $\mathcal{D}$-V is a symmetric monoidal closed category, it is enriched, tensored, and cotensored over itself. Changing the enrichment along the symmetric monoidal functor

$$\mathcal{D}\text{-}\mathsf{V} \xrightarrow{\mathsf{ev}_e} \mathsf{V},$$

given by evaluation at the monoidal unit e in $\mathcal{D}$, we conclude that $\mathcal{D}$-V is enriched, tensored, and cotensored over V. □

Permutative-Indexed Categories

Recall that a Cat-category is a locally small 2-category (Proposition 1.35).

Explanation 1.66. Let us explain the structure of $\mathcal{D}$-Cat in Theorems 1.63 and 1.65 with

- $\mathsf{V} = (\mathsf{Cat}, \times, \mathbf{1}, [,])$ in Examples 1.16 and 1.33 and
- $(\mathcal{D}, \otimes, \mathbf{1}, \beta)$ a small permutative category.

The 1-category $\mathcal{D}$ is also regarded as a 2-category with only identity 2-cells. We first describe the 2-category structure on $\mathcal{D}$-Cat.

Objects An object in $\mathcal{D}$-Cat is a 2-functor $\mathcal{D} \longrightarrow \mathsf{Cat}$ (Definition 1.36). This is the same thing as a functor $\mathcal{D} \longrightarrow \mathsf{Cat}$, since $\mathcal{D}$ has only identity 2-cells. In other words, an object in $\mathcal{D}$-Cat is a $\mathcal{D}$-indexed category (Definition 1.15).

1-Cells For 2-functors $F, G : \mathcal{D} \longrightarrow \mathsf{Cat}$, the morphism object

$$(\mathcal{D}\text{-}\mathsf{Cat})(F, G)$$

is a small category. An *object* in $(\mathcal{D}\text{-}\mathsf{Cat})(F,G)$ is a 2-natural transformation $\phi : F \longrightarrow G$ (Definition 1.37). This means that ϕ is equipped with a component functor

$$Fa \xrightarrow{\phi_a} Ga \quad \text{for each} \quad a \in \mathcal{D}.$$

For each morphism $f : a \longrightarrow b$ in $\mathcal{D}$, the following diagram of functors, which is the 1-cell naturality condition (1.38), is required to commute.

$$\begin{array}{ccc} Fa & \xrightarrow{\phi_a} & Ga \\ {\scriptstyle Ff}\downarrow & & \downarrow{\scriptstyle Gf} \\ Fb & \xrightarrow{\phi_b} & Gb \end{array} \tag{1.67}$$

The 2-cell naturality condition (1.39) is automatically satisfied because $\mathcal{D}$ has only identity 2-cells. In other words, $\phi : F \longrightarrow G$ is a natural transformation.

The *identity 1-cell* $1_F : F \longrightarrow F$ is the identity natural transformation. It has component identity functors:

$$(1_F)_a = 1_{Fa}.$$

Horizontal Composition for 1-Cells For another 2-functor $H : \mathcal{D} \longrightarrow \mathsf{Cat}$ and a 2-natural transformation $\phi' : G \longrightarrow H$, the *horizontal composite*

$$F \xrightarrow{\phi'\phi} H$$

is the 2-natural transformation as in Definition 1.40. It has component composite functor

$$Fa \xrightarrow{\phi_a} Ga \xrightarrow{\phi'_a} Ha \quad \text{for each} \quad a \in \mathcal{D}.$$

2-Cells For 2-natural transformations $\phi, \varphi : F \longrightarrow G$, a *morphism*

$$\phi \xrightarrow{\Phi} \varphi \quad \text{in} \quad (\mathcal{D}\text{-}\mathsf{Cat})(F, G)$$

is a modification (Definition 1.41). This means that Φ is equipped with a component natural transformation Φ_a for each object $a \in \mathcal{D}$, as displayed below.

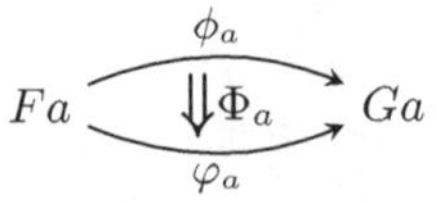

The modification axiom (1.42) in this case requires that, for each morphism $f : a \longrightarrow b$ in $\mathcal{D}$, the following two whiskered natural transformations are equal.

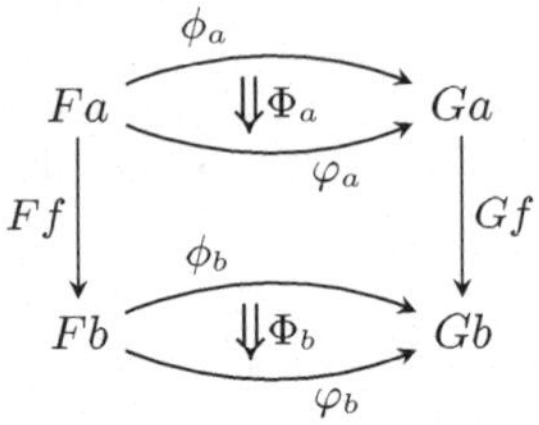

(1.68)

In other words, the modification axiom in this case is the equality

$$1_{Gf} * \Phi_a = \Phi_b * 1_{Ff}$$

of horizontal compositions of natural transformations for each morphism $f : a \longrightarrow b$ in $\mathcal{D}$.

The *identity 2-cell*

$$\phi \xrightarrow{\;1_\phi\;} \phi \quad \text{in} \quad (\mathcal{D}\text{-}\mathsf{Cat})(F, G)$$

is the identity modification. It has component identity natural transformations:

$$(1_\phi)_a = 1_{\phi_a}.$$

Vertical Composition for 2-Cells For another modification

$$\varphi \xrightarrow{\;\Psi\;} \psi \quad \text{in} \quad (\mathcal{D}\text{-}\mathsf{Cat})(F, G)$$

the *vertical composite modification*

$$\phi \xrightarrow{\;\Psi\Phi\;} \psi \quad \text{in} \quad (\mathcal{D}\text{-}\mathsf{Cat})(F, G)$$

is given as in (1.44). For each object $a \in \mathcal{D}$, it has the following component vertical composite natural transformation.

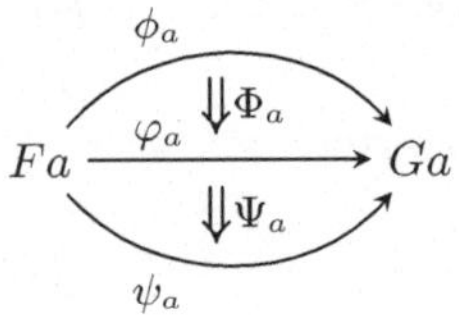

Horizontal Composition for 2-Cells For a modification

$$\phi' \xrightarrow{\Phi'} \varphi' \quad \text{in} \quad (\mathcal{D}\text{-}\mathsf{Cat})(G, H)$$

the *horizontal composite modification*

$$\phi'\phi \xrightarrow{\Phi' * \Phi} \varphi'\varphi \quad \text{in} \quad (\mathcal{D}\text{-}\mathsf{Cat})(F, H)$$

is given as in (1.45). For each object $a \in \mathcal{D}$, it has the following component horizontal composite natural transformation.

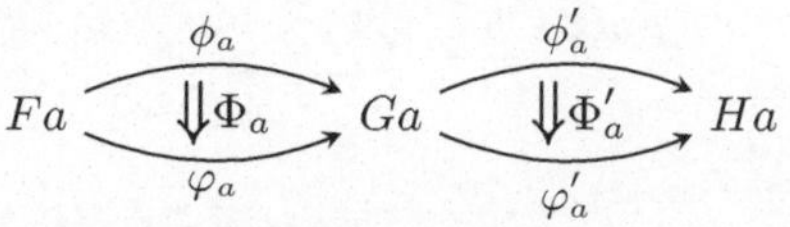

This finishes the description of the 2-category $\mathcal{D}$-Cat.

The (co)tensor structure (Definition 1.64) on $\mathcal{D}$-Cat over Cat is induced by the symmetric monoidal closed structure on Cat, that is, the Cartesian product $\times$ and diagram categories $[-,-]$. More explicitly, suppose $F : \mathcal{D} \longrightarrow \mathsf{Cat}$ is a $\mathcal{D}$-indexed category, and B is a small category.

Tensored The $\mathcal{D}$-indexed category $F \otimes \mathsf{B} : \mathcal{D} \longrightarrow \mathsf{Cat}$ is the composite functor

$$\mathcal{D} \xrightarrow{F} \mathsf{Cat} \xrightarrow{- \times \mathsf{B}} \mathsf{Cat}.$$

Cotensored The $\mathcal{D}$-indexed category $F^{\mathsf{B}} : \mathcal{D} \longrightarrow \mathsf{Cat}$ is the composite functor

$$\mathcal{D} \xrightarrow{F} \mathsf{Cat} \xrightarrow{[\mathsf{B},-]} \mathsf{Cat}.$$

Next we describe the monoidal structure on $\mathcal{D}$-Cat.

Monoidal Unit It is the unit diagram

$$J = \mathcal{D}(1, -) : \mathcal{D} \longrightarrow \mathsf{Cat}. \tag{1.69}$$

For each $a \in \mathcal{D}$ the morphism set $\mathcal{D}(1, a)$ is regarded as a discrete category with only identity morphisms.

Monoidal Product On $\mathcal{D}$-indexed categories and natural transformations, the monoidal product is the Day convolution in (1.51) and (1.52), where $\otimes$ in Cat means the Cartesian product $\times$. Suppose given modifications $\Phi : \phi \longrightarrow \varphi$ and $\Phi' : \phi' \longrightarrow \varphi'$, as displayed below, for $\mathcal{D}$-indexed categories $F, G, F', G' : \mathcal{D} \longrightarrow \mathsf{Cat}$.

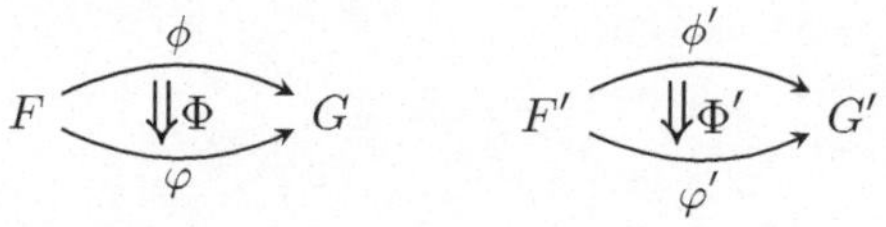

The monoidal product $\Phi \otimes \Phi'$

$$F \otimes F' \underset{\varphi \otimes \varphi'}{\overset{\phi \otimes \phi'}{\Downarrow \Phi \otimes \Phi'}} G \otimes G' \tag{1.70}$$

is the modification whose component natural transformations are induced by the product natural transformations

$$Fa \times F'b \underset{\varphi_a \times \varphi'_b}{\overset{\phi_a \times \phi'_b}{\Downarrow \Phi_a \times \Phi'_b}} Ga \times G'b$$

for $a, b \in \mathcal{D}$.

This completes the description of $\mathcal{D}$-Cat. ◇

2

Symmetric Bimonoidal and Bipermutative Categories

In this chapter we discuss two closely related categorical structures, symmetric bimonoidal categories and bipermutative categories, several detailed examples, and Laplaza's Coherence Theorem for symmetric bimonoidal categories. Below is a summary of some of the definitions and results in this chapter.

- Symmetric bimonoidal categories. Definition 2.1.
- Ring categories. Definition 2.29.
- Bipermutative categories. Definition 2.41.
- Tight bipermutative categories. Theorem 2.45.
- Laplaza's Coherence Theorem. Theorem 2.90.

In later chapters Laplaza's Coherence Theorem is a key part of several important definitions and constructions, including

- additive natural transformations (5.46),
- additive modifications (5.55),
- the Grothendieck construction of an additive natural transformation ((6.54) and (6.56)), and
- opcartesian n-linear functors (7.15).

As references, a detailed discussion of symmetric bimonoidal categories and Laplaza's Coherence Theorem can be found in [Yau∞I]. Bipermutative categories and their relations to symmetric bimonoidal categories are discussed in detail in [Yau∞II].

Organization

In Section 2.1 we review symmetric bimonoidal categories in the sense of Laplaza [Lap72a]. A symmetric bimonoidal category is a categorical analogue of a *commutative rig*, which is a commutative ring without additive inverses.

DOI: 10.1201/9781003451235-2

In Section 2.2 we review bipermutative categories in the sense of Elmendorf-Mandell [EM06]. The main difference between symmetric bimonoidal categories and bipermutative categories is that the former have *distributivity morphisms* (2.3), while the latter have *factorization morphisms* (2.30) that go in the opposite direction as the distributivity morphisms. Since none of these structure morphisms is required to be invertible in general, bipermutative categories do not form a subclass of symmetric bimonoidal categories, or vice versa. On the other hand, a *tight* bipermutative category, which has invertible factorization morphisms, is a *tight* symmetric bimonoidal category, which has invertible distributivity morphisms (Theorem 2.45).

In Section 2.3 we discuss the following tight bipermutative categories.

- $\mathsf{Fin}^{\mathsf{sk}}$. Objects are finite sets. Morphisms are functions. Example 2.48.
- Σ. Objects are finite sets. Morphisms are permutations. Example 2.57.
- $\mathcal{F}$. Objects are pointed finite sets. Morphisms are pointed functions. Example 2.58.
- $\mathcal{A}$. Objects are finite sequences of positive integers. Morphisms are certain maps of unpointed finite sets. Example 2.63.

In particular, Example 2.63 exhibits a bipermutative category structure on Mandell's indexing category $\mathcal{A}$ for inverse K-theory [Man10]. This bipermutative structure on $\mathcal{A}$ is a crucial reason why inverse K-theory is a *pseudo symmetric* Cat-multifunctor, as we will discuss in Chapters 6 and 10.

In Section 2.4 we review Laplaza's Coherence Theorem 2.90 for symmetric bimonoidal categories. Analogous to Mac Lane's Coherence Theorem for monoidal categories [ML98, VII.2], Laplaza's Coherence Theorem states that, in a symmetric bimonoidal category that satisfies a certain monomorphism condition, each formal diagram built from the symmetric bimonoidal structure commutes, provided that the domain is *regular*. The regularity condition excludes generally non-commutative diagrams such as

$$x \otimes x \overset{1}{\underset{\xi^{\otimes}}{\rightrightarrows}} x \otimes x$$

with $\xi^{\otimes}$ the multiplicative braiding. The monomorphism condition in Laplaza's Coherence Theorem is satisfied if the distributivity morphisms are invertible. In particular, by Theorem 2.45 Laplaza's Coherence Theorem applies to tight bipermutative categories, including those in Section 2.3. In Chapter 5 we use Laplaza's Coherence Theorem 2.90 to construct the Cat-enriched multicategory of certain indexed categories when the indexing category is a tight bipermutative category.

2.1 Symmetric Bimonoidal Categories

In this section we review symmetric bimonoidal categories in the sense of Laplaza [Lap72a]. The presentation below is adapted from [Yau∞I, Ch. 2], where much more detailed discussion can be found.

Definition 2.1. A *symmetric bimonoidal category* is a tuple

$$\big(\mathsf{C}, (\oplus, 0, \alpha^{\oplus}, \lambda^{\oplus}, \rho^{\oplus}, \xi^{\oplus}), (\otimes, 1, \alpha^{\otimes}, \lambda^{\otimes}, \rho^{\otimes}, \xi^{\otimes}), (\lambda^{\bullet}, \rho^{\bullet}), (\delta^{l}, \delta^{r})\big)$$

consisting of the following data.

Additive Structure $\mathsf{C}^{\oplus} = (\mathsf{C}, \oplus, 0, \alpha^{\oplus}, \lambda^{\oplus}, \rho^{\oplus}, \xi^{\oplus})$ is a symmetric monoidal category (Definition 1.11).

Multiplicative Structure $\mathsf{C}^{\otimes} = (\mathsf{C}, \otimes, 1, \alpha^{\otimes}, \lambda^{\otimes}, \rho^{\otimes}, \xi^{\otimes})$ is a symmetric monoidal category.

Multiplicative Zeros $\lambda^{\bullet}$ and $\rho^{\bullet}$ are natural isomorphisms

$$0 \otimes A \xrightarrow[\cong]{\lambda^{\bullet}_{A}} 0 \xleftarrow[\cong]{\rho^{\bullet}_{A}} A \otimes 0 \quad \text{for} \quad A \in \mathsf{C}, \tag{2.2}$$

which are called the *left multiplicative zero* and the *right multiplicative zero*, respectively.

Distributivity δ^{l} and δ^{r} are natural monomorphisms

$$\begin{aligned} A \otimes (B \oplus C) &\xrightarrow{\delta^{l}_{A,B,C}} (A \otimes B) \oplus (A \otimes C) \\ (A \oplus B) \otimes C &\xrightarrow{\delta^{r}_{A,B,C}} (A \otimes C) \oplus (B \otimes C) \end{aligned} \tag{2.3}$$

for objects $A, B, C \in \mathsf{C}$, which are called the *left distributivity morphism* and the *right distributivity morphism*, respectively.

We abbreviate $\otimes$ using concatenation and let $\otimes$ take precedence over $\oplus$. For example, the left distributivity morphism is abbreviated to

$$A(B \oplus C) \longrightarrow AB \oplus AC.$$

The following 24 diagrams in C, which are called *Laplaza's Axioms*, are required to commute for all objects A, B, C, and D in C.

Distributivity and Multiplicative Braiding

$$\begin{array}{ccc} (A \oplus B)C & \xrightarrow{\delta^{r}_{A,B,C}} & AC \oplus BC \\ {\scriptstyle \xi^{\otimes}_{A\oplus B,C}}\downarrow & & \downarrow{\scriptstyle \xi^{\otimes}_{A,C} \oplus \xi^{\otimes}_{B,C}} \\ C(A \oplus B) & \xrightarrow{\delta^{l}_{C,A,B}} & CA \oplus CB \end{array} \tag{2.4}$$

Distributivity and Additive Braiding

$$\begin{array}{ccc} A(B \oplus C) & \xrightarrow{\delta^l_{A,B,C}} & AB \oplus AC \\ {\scriptstyle 1_A\xi^{\oplus}_{B,C}}\downarrow & & \downarrow{\scriptstyle \xi^{\oplus}_{AB,AC}} \\ A(C \oplus B) & \xrightarrow{\delta^l_{A,C,B}} & AC \oplus AB \end{array} \tag{2.5}$$

$$\begin{array}{ccc} (A \oplus B)C & \xrightarrow{\delta^r_{A,B,C}} & AC \oplus BC \\ {\scriptstyle \xi^{\oplus}_{A,B}1_C}\downarrow & & \downarrow{\scriptstyle \xi^{\oplus}_{AC,BC}} \\ (B \oplus A)C & \xrightarrow{\delta^r_{B,A,C}} & BC \oplus AC \end{array} \tag{2.6}$$

Distributivity and Additive Associativity

$$\begin{array}{ccccc} [(A \oplus B) \oplus C]D & \xrightarrow{\delta^r_{A\oplus B,C,D}} & (A \oplus B)D \oplus CD & \xrightarrow{\delta^r_{A,B,D}\oplus 1_{CD}} & (AD \oplus BD) \oplus CD \\ {\scriptstyle \alpha^{\oplus}_{A,B,C}1_D}\downarrow & & & & \downarrow{\scriptstyle \alpha^{\oplus}_{AD,BD,CD}} \\ [A \oplus (B \oplus C)]D & \xrightarrow{\delta^r_{A,B\oplus C,D}} & AD \oplus (B \oplus C)D & \xrightarrow{1_{AD}\oplus\delta^r_{B,C,D}} & AD \oplus (BD \oplus CD) \end{array} \tag{2.7}$$

$$\begin{array}{ccccc} A[(B \oplus C) \oplus D] & \xrightarrow{\delta^l_{A,B\oplus C,D}} & A(B \oplus C) \oplus AD & \xrightarrow{\delta^l_{A,B,C}\oplus 1_{AD}} & (AB \oplus AC) \oplus AD \\ {\scriptstyle 1_A\alpha^{\oplus}_{B,C,D}}\downarrow & & & & \downarrow{\scriptstyle \alpha^{\oplus}_{AB,AC,AD}} \\ A[B \oplus (C \oplus D)] & \xrightarrow{\delta^l_{A,B,C\oplus D}} & AB \oplus A(C \oplus D) & \xrightarrow{1_{AB}\oplus\delta^l_{A,C,D}} & AB \oplus (AC \oplus AD) \end{array} \tag{2.8}$$

Distributivity and Multiplicative Associativity

$$\begin{array}{ccccc} (AB)(C \oplus D) & & \xrightarrow{\qquad\qquad\delta^l_{AB,C,D}\qquad\qquad} & & (AB)C \oplus (AB)D \\ {\scriptstyle \alpha^{\otimes}_{A,B,C\oplus D}}\downarrow & & & & \downarrow{\scriptstyle \alpha^{\otimes}_{A,B,C}\oplus\alpha^{\otimes}_{A,B,D}} \\ A[B(C \oplus D)] & \xrightarrow{1_A\delta^l_{B,C,D}} & A(BC \oplus BD) & \xrightarrow{\delta^l_{A,BC,BD}} & A(BC) \oplus A(BD) \end{array} \tag{2.9}$$

$$\begin{array}{ccccc} [(A \oplus B)C]D & \xrightarrow{\delta^r_{A,B,C}1_D} & (AC \oplus BC)D & \xrightarrow{\delta^r_{AC,BC,D}} & (AC)D \oplus (BC)D \\ {\scriptstyle \alpha^{\otimes}_{A\oplus B,C,D}}\downarrow & & & & \downarrow{\scriptstyle \alpha^{\otimes}_{A,C,D}\oplus\alpha^{\otimes}_{B,C,D}} \\ (A \oplus B)(CD) & & \xrightarrow{\qquad\qquad\delta^r_{A,B,CD}\qquad\qquad} & & A(CD) \oplus B(CD) \end{array} \tag{2.10}$$

$$\begin{array}{ccccc} [A(B \oplus C)]D & \xrightarrow{\delta^l_{A,B,C}1_D} & (AB \oplus AC)D & \xrightarrow{\delta^r_{AB,AC,D}} & (AB)D \oplus (AC)D \\ {\scriptstyle \alpha^{\otimes}_{A,B\oplus C,D}}\downarrow & & & & \downarrow{\scriptstyle \alpha^{\otimes}_{A,B,D}\oplus\alpha^{\otimes}_{A,C,D}} \\ A[(B \oplus C)D] & \xrightarrow{1_A\delta^r_{B,C,D}} & A(BD \oplus CD) & \xrightarrow{\delta^l_{A,BD,CD}} & A(BD) \oplus A(CD) \end{array} \tag{2.11}$$

2-By-2 Distributivity

$$\begin{array}{ccc}
(A \oplus B)(C \oplus D) & \xrightarrow{\delta^r_{A,B,C\oplus D}} & A(C \oplus D) \oplus B(C \oplus D) \\
\downarrow{\scriptstyle \delta^l_{A\oplus B,C,D}} & & \downarrow{\scriptstyle \delta^l_{A,C,D} \oplus \delta^l_{B,C,D}} \\
(A \oplus B)C \oplus (A \oplus B)D & & (AC \oplus AD) \oplus (BC \oplus BD) \\
\downarrow{\scriptstyle \delta^r_{A,B,C} \oplus \delta^r_{A,B,D}} & & \downarrow{\scriptstyle \alpha^{\oplus}_{AC,AD,BC\oplus BD}} \\
(AC \oplus BC) \oplus (AD \oplus BD) & & AC \oplus [AD \oplus (BC \oplus BD)] \\
\downarrow{\scriptstyle \alpha^{\oplus}_{AC,BC,AD\oplus BD}} & & \downarrow{\scriptstyle 1_{AC} \oplus (\alpha^{\oplus})^{-1}} \\
AC \oplus [BC \oplus (AD \oplus BD)] & & AC \oplus [(AD \oplus BC) \oplus BD] \\
\downarrow{\scriptstyle 1_{AC} \oplus (\alpha^{\oplus})^{-1}} & & \downarrow{\scriptstyle 1_{AC} \oplus (\xi^{\oplus}_{AD,BC} \oplus 1_{BD})} \\
AC \oplus [(BC \oplus AD) \oplus BD] & = & AC \oplus [(BC \oplus AD) \oplus BD]
\end{array} \tag{2.12}$$

Multiplicative Zero of 0

$$0 \otimes 0 \underset{\rho^{\bullet}_0}{\overset{\lambda^{\bullet}_0}{\rightrightarrows}} 0 \tag{2.13}$$

Multiplicative Zero of a Sum

$$\begin{array}{ccc}
0(A \oplus B) & \xrightarrow{\delta^l_{0,A,B}} & 0A \oplus 0B \\
\downarrow{\scriptstyle \lambda^{\bullet}_{A\oplus B}} & & \downarrow{\scriptstyle \lambda^{\bullet}_A \oplus \lambda^{\bullet}_B} \\
0 & \xleftarrow{\lambda^{\oplus}_0} & 0 \oplus 0
\end{array} \tag{2.14}$$

$$\begin{array}{ccc}
(A \oplus B)0 & \xrightarrow{\delta^r_{A,B,0}} & A0 \oplus B0 \\
\downarrow{\scriptstyle \rho^{\bullet}_{A\oplus B}} & & \downarrow{\scriptstyle \rho^{\bullet}_A \oplus \rho^{\bullet}_B} \\
0 & \xleftarrow{\lambda^{\oplus}_0} & 0 \oplus 0
\end{array} \tag{2.15}$$

Multiplicative Zero and Multiplicative Unit

$$0 \otimes 1 \underset{\rho^{\otimes}_0}{\overset{\lambda^{\bullet}_1}{\rightrightarrows}} 0 \tag{2.16}$$

$$1 \otimes 0 \underset{\lambda^{\otimes}_0}{\overset{\rho^{\bullet}_1}{\rightrightarrows}} 0 \tag{2.17}$$

Braiding of Multiplicative Zero

$$\begin{array}{ccc}
A \otimes 0 & \xrightarrow{\xi^{\otimes}_{A,0}} & 0 \otimes A \\
{\scriptstyle \rho^{\bullet}_A} \searrow & & \swarrow {\scriptstyle \lambda^{\bullet}_A} \\
& 0 &
\end{array} \tag{2.18}$$

Multiplicative Zero and Multiplicative Associativity

$$\begin{array}{ccc} (AB)0 & \xrightarrow{\alpha^{\otimes}_{A,B,0}} & A(B0) \\ {\scriptstyle \rho^{\bullet}_{AB}}\downarrow & & \downarrow{\scriptstyle 1_A \rho^{\bullet}_{B}} \\ 0 & \xleftarrow{\rho^{\bullet}_{A}} & A0 \end{array} \tag{2.19}$$

$$\begin{array}{ccccc} (A0)B & & \xrightarrow{\alpha^{\otimes}_{A,0,B}} & & A(0B) \\ {\scriptstyle \rho^{\bullet}_{A} 1_B}\downarrow & & & & \downarrow{\scriptstyle 1_A \lambda^{\bullet}_{B}} \\ 0B & \xrightarrow{\lambda^{\bullet}_{B}} & 0 & \xleftarrow{\rho^{\bullet}_{A}} & A0 \end{array} \tag{2.20}$$

$$\begin{array}{ccc} (0A)B & \xrightarrow{\lambda^{\bullet}_{A} 1_B} & 0B \\ {\scriptstyle \alpha^{\otimes}_{0,A,B}}\downarrow & & \downarrow{\scriptstyle \lambda^{\bullet}_{B}} \\ 0(AB) & \xrightarrow{\lambda^{\bullet}_{AB}} & 0 \end{array} \tag{2.21}$$

Additive and Multiplicative Zero

$$\begin{array}{ccc} A(0 \oplus B) & \xrightarrow{\delta^{l}_{A,0,B}} & A0 \oplus AB \\ {\scriptstyle 1_A \lambda^{\oplus}_{B}}\downarrow & & \downarrow{\scriptstyle \rho^{\bullet}_{A} \oplus 1_{AB}} \\ AB & \xleftarrow{\lambda^{\oplus}_{AB}} & 0 \oplus AB \end{array} \tag{2.22}$$

$$\begin{array}{ccc} (0 \oplus B)A & \xrightarrow{\delta^{r}_{0,B,A}} & 0A \oplus BA \\ {\scriptstyle \lambda^{\oplus}_{B} 1_A}\downarrow & & \downarrow{\scriptstyle \lambda^{\bullet}_{A} \oplus 1_{BA}} \\ BA & \xleftarrow{\lambda^{\oplus}_{BA}} & 0 \oplus BA \end{array} \tag{2.23}$$

$$\begin{array}{ccc} A(B \oplus 0) & \xrightarrow{\delta^{l}_{A,B,0}} & AB \oplus A0 \\ {\scriptstyle 1_A \rho^{\oplus}_{B}}\downarrow & & \downarrow{\scriptstyle 1_{AB} \oplus \rho^{\bullet}_{A}} \\ AB & \xleftarrow{\rho^{\oplus}_{AB}} & AB \oplus 0 \end{array} \tag{2.24}$$

$$\begin{array}{ccc} (B \oplus 0)A & \xrightarrow{\delta^{r}_{B,0,A}} & BA \oplus 0A \\ {\scriptstyle \rho^{\oplus}_{B} 1_A}\downarrow & & \downarrow{\scriptstyle 1_{BA} \oplus \lambda^{\bullet}_{A}} \\ BA & \xleftarrow{\rho^{\oplus}_{BA}} & BA \oplus 0 \end{array} \tag{2.25}$$

Distributivity and Multiplicative Unit

$$\begin{array}{ccc} 1(A \oplus B) & \xrightarrow{\delta^l_{1,A,B}} & 1A \oplus 1B \\ & {\scriptstyle \lambda^{\otimes}_{A \oplus B}} \searrow \quad \swarrow {\scriptstyle \lambda^{\otimes}_A \oplus \lambda^{\otimes}_B} & \\ & A \oplus B & \end{array} \tag{2.26}$$

$$\begin{array}{ccc} (A \oplus B)1 & \xrightarrow{\delta^r_{A,B,1}} & A1 \oplus B1 \\ & {\scriptstyle \rho^{\otimes}_{A \oplus B}} \searrow \quad \swarrow {\scriptstyle \rho^{\otimes}_A \oplus \rho^{\otimes}_B} & \\ & A \oplus B & \end{array} \tag{2.27}$$

This finishes the definition of a symmetric bimonoidal category.

Moreover, we define the following.

- A symmetric bimonoidal category is *small* if its class of objects is a set.
- A symmetric bimonoidal category is *tight* if both δ^l and δ^r are natural isomorphisms.
- The objects $\mathsf{0}$ and $\mathbb{1}$ are called the *additive zero* and the *multiplicative unit*, respectively.
- $\oplus$ and $\otimes$ are called the *sum* and the *product*, respectively.
- $\alpha^{\oplus}$, $\lambda^{\oplus}$, $\rho^{\oplus}$, and $\xi^{\oplus}$ are called the *additive associativity isomorphism*, the *left additive zero*, the *right additive zero*, and the *additive braiding*, respectively.
- $\alpha^{\otimes}$, $\lambda^{\otimes}$, $\rho^{\otimes}$, and $\xi^{\otimes}$ are called the *multiplicative associativity isomorphism*, the *left multiplicative unit*, the *right multiplicative unit*, and the *multiplicative braiding*, respectively. ◇

Due to the presence of the axioms

- (2.4) relating δ^l and δ^r and
- (2.18) relating $\lambda^{\bullet}$ and $\rho^{\bullet}$,

12 of the 24 axioms in Definition 2.1 follow from the other 12 axioms. The following observation is [Yau∞I, 2.2.13].

Theorem 2.28. *In Definition 2.1 of a symmetric bimonoidal category, it is sufficient to assume the first axiom in each of the twelve groups of axioms, namely,* (2.4), (2.5), (2.7), (2.9), (2.12), (2.13), (2.14), (2.16), (2.18), (2.19), (2.22), *and* (2.26).

2.2 Bipermutative Categories

In this section we review bipermutative categories in the sense of Elmendorf-Mandell [EM06]. The presentation below is adapted from [Yau∞II, Ch. 9]. A bipermutative category is a *ring category*, to be defined below, with further structure and axioms. The following concept of a ring category is due to [EM06, Def. 3.3], as presented in [Yau∞II, Section 9.1]. The main difference between these concepts and symmetric bimonoidal categories (Definition 2.1) is that ring and bipermutative categories have *factorization morphisms* that go in the opposite direction as the distributivity morphisms (2.3).

Definition 2.29. A *ring category* is a tuple

$$\big(\mathsf{C}, (\oplus, 0, \xi^{\oplus}), (\otimes, 1), (\partial^l, \partial^r)\big)$$

consisting of the following data.

Additive Structure $\mathsf{C}^{\oplus} = (\mathsf{C}, \oplus, 0, \xi^{\oplus})$ is a permutative category (Definition 1.11), with $\oplus$, 0, and $\xi^{\oplus}$ called, respectively, the *sum*, the *additive zero*, and the *additive braiding*.

Multiplicative Structure $\mathsf{C}^{\otimes} = (\mathsf{C}, \otimes, 1)$ is a strict monoidal category (Definition 1.3), with $\otimes$ and 1 called, respectively, the *product* and the *multiplicative unit*.

Factorization Morphisms ∂^l and ∂^r are natural transformations

$$\begin{array}{ccc} (A \otimes C) \oplus (B \otimes C) & \xrightarrow{\partial^l_{A,B,C}} & (A \oplus B) \otimes C \\ (A \otimes B) \oplus (A \otimes C) & \xrightarrow{\partial^r_{A,B,C}} & A \otimes (B \oplus C) \end{array} \tag{2.30}$$

for objects $A, B, C \in \mathsf{C}$, which are called the *left factorization morphism* and the *right factorization morphism*, respectively.

We abbreviate $\otimes$ to concatenation, with $\otimes$ taking precedence over $\oplus$ in the absence of clarifying parentheses. For example, the left factorization morphism is abbreviated to

$$AC \oplus BC \longrightarrow (A \oplus B)C.$$

The subscripts in $\xi^{\oplus}$, ∂^l, and ∂^r are sometimes omitted.

The data above are required to satisfy the following seven axioms for all objects A, A', A'', B, B', B'', C, and C' in C, with $\mathbf{1}$ denoting the terminal category with one object $*$ and its identity morphism.

The Multiplicative Zero Axiom The following diagram of functors commutes.

$$\begin{array}{ccccc} \mathbf{1} \times \mathsf{C} & \xrightarrow{\cong} & \mathsf{C} & \xleftarrow{\cong} & \mathsf{C} \times \mathbf{1} \\ {\scriptstyle 0 \times 1_{\mathsf{C}}}\downarrow & & \downarrow{\scriptstyle 0} & & \downarrow{\scriptstyle 1_{\mathsf{C}} \times 0} \\ \mathsf{C} \times \mathsf{C} & \xrightarrow{\otimes} & \mathsf{C} & \xleftarrow{\otimes} & \mathsf{C} \times \mathsf{C} \end{array} \tag{2.31}$$

In (2.31) the top horizontal isomorphisms drop the **1** argument. Each $\mathsf{0}$ denotes the constant functor at $\mathsf{0} \in \mathsf{C}$ and 1_{0}.

The Zero Factorization Axiom

$$\begin{array}{ll} \partial^l_{\mathsf{0},B,C} = 1_{B\otimes C} & \partial^r_{\mathsf{0},B,C} = 1_{\mathsf{0}} \\ \partial^l_{A,\mathsf{0},C} = 1_{A\otimes C} & \partial^r_{A,\mathsf{0},C} = 1_{A\otimes C} \\ \partial^l_{A,B,\mathsf{0}} = 1_{\mathsf{0}} & \partial^r_{A,B,\mathsf{0}} = 1_{A\otimes B} \end{array} \tag{2.32}$$

The three equalities for ∂^l are called the *left zero factorization axioms.* The three equalities for ∂^r are called the *right zero factorization axioms.*

The Unit Factorization Axiom

$$\begin{array}{l} \partial^l_{A,B,\mathbf{1}} = 1_{A\oplus B} \\ \partial^r_{\mathbf{1},B,C} = 1_{B\oplus C} \end{array} \tag{2.33}$$

These are called, respectively, the *left* and the *right* unit factorization axioms.

The Braiding Factorization Axiom The following two diagrams in C commute.

$$\begin{array}{ccc} AC \oplus BC & \xrightarrow{\partial^l} & (A\oplus B)C \\ {\scriptstyle \xi^{\oplus}}\downarrow & & \downarrow{\scriptstyle \xi^{\oplus}1_C} \\ BC \oplus AC & \xrightarrow{\partial^l} & (B\oplus A)C \end{array} \qquad \begin{array}{ccc} AB \oplus AC & \xrightarrow{\partial^r} & A(B\oplus C) \\ {\scriptstyle \xi^{\oplus}}\downarrow & & \downarrow{\scriptstyle 1_A\xi^{\oplus}} \\ AC \oplus AB & \xrightarrow{\partial^r} & A(C\oplus B) \end{array} \tag{2.34}$$

These are called, respectively, the *left* and the *right* symmetry factorization axioms.

The Internal Factorization Axiom The following two diagrams in C commute.

$$\begin{array}{ccc} AB \oplus A'B \oplus A''B & \xrightarrow{\partial^l \oplus 1} & (A\oplus A')B \oplus A''B \\ {\scriptstyle 1\oplus\partial^l}\downarrow & & \downarrow{\scriptstyle \partial^l} \\ AB \oplus (A'\oplus A'')B & \xrightarrow{\partial^l} & (A\oplus A'\oplus A'')B \end{array} \tag{2.35}$$

$$\begin{array}{ccc} AB \oplus AB' \oplus AB'' & \xrightarrow{\partial^r \oplus 1} & A(B\oplus B') \oplus AB'' \\ {\scriptstyle 1\oplus\partial^r}\downarrow & & \downarrow{\scriptstyle \partial^r} \\ AB \oplus A(B'\oplus B'') & \xrightarrow{\partial^r} & A(B\oplus B'\oplus B\text{''}) \end{array} \tag{2.36}$$

These are called, respectively, the *left* and the *right* internal factorization axioms.

The External Factorization Axiom The three diagrams in C below commute.

$$\begin{array}{ccc} ABC \oplus A'BC & \xrightarrow{\partial^l_{A,A',BC}} & (A \oplus A')BC \\ {\scriptstyle \partial^l_{AB,A'B,C}}\big\downarrow & & \big\| \\ (AB \oplus A'B)C & \xrightarrow{\partial^l_{A,A',B}1_C} & (A \oplus A')BC \end{array} \tag{2.37}$$

$$\begin{array}{ccc} ABC \oplus AB'C & \xrightarrow{\partial^l_{AB,AB',C}} & (AB \oplus AB')C \\ {\scriptstyle \partial^r_{A,BC,B'C}}\big\downarrow & & \big\downarrow{\scriptstyle \partial^r 1_C} \\ A(BC \oplus B'C) & \xrightarrow{1_A\partial^l_{B,B',C}} & A(B \oplus B')C \end{array} \tag{2.38}$$

$$\begin{array}{ccc} ABC \oplus ABC' & \xrightarrow{\partial^r_{AB,C,C'}} & AB(C \oplus C') \\ {\scriptstyle \partial^r_{A,BC,BC'}}\big\downarrow & & \big\| \\ A(BC \oplus BC') & \xrightarrow{1_A\partial^r_{B,C,C'}} & AB(C \oplus C') \end{array} \tag{2.39}$$

These are called, respectively, the *left*, the *middle*, and the *right* external factorization axioms.

The 2-By-2 Factorization Axiom The following diagram in C commutes.

$$\begin{array}{ccc} & & A(B \oplus B') \oplus A'(B \oplus B') \\ & \nearrow{\scriptstyle \partial^r \oplus \partial^r} & \big\downarrow{\scriptstyle \partial^l} \\ AB \oplus AB' \oplus A'B \oplus A'B' & & \\ {\scriptstyle 1 \oplus \xi^{\oplus} \oplus 1}\big\downarrow & & (A \oplus A')(B \oplus B') \\ AB \oplus A'B \oplus AB' \oplus A'B' & & \big\uparrow{\scriptstyle \partial^r} \\ & \searrow{\scriptstyle \partial^l \oplus \partial^l} & \\ & & (A \oplus A')B \oplus (A \oplus A')B' \end{array} \tag{2.40}$$

This finishes the definition of a ring category.

Moreover, a ring category as above is said to be

- *small* if it has a set of objects and
- *tight* if ∂^l and ∂^r in (2.30) are natural isomorphisms. ◇

The following concept of a bipermutative category is due to [EM06, Def. 3.6], as presented in [Yau∞II, Section 9.3].

Definition 2.41. A *bipermutative category* is a tuple

$$\big(\mathsf{C}, (\oplus, 0, \xi^{\oplus}), (\otimes, 1, \xi^{\otimes}), (\partial^l, \partial^r)\big)$$

consisting of the following data.

- The tuple
$$\big(\mathsf{C}, (\oplus, 0, \xi^{\oplus}), (\otimes, 1), (\partial^l, \partial^r)\big)$$
is a ring category (Definition 2.29).
- The tuple $\mathsf{C}^{\otimes} = (\mathsf{C}, \otimes, 1, \xi^{\otimes})$ is a permutative category (Definition 1.11), with $\xi^{\otimes}$ called the *multiplicative braiding.*

The data above are required to satisfy the following two axioms for objects $A, B, C \in \mathsf{C}$.

The Zero Braiding Axiom There is an equality of morphisms

$$\xi^{\otimes}_{A,0} = 1_0 : A \otimes 0 = 0 \longrightarrow 0 = 0 \otimes A. \tag{2.42}$$

The Multiplicative Braiding Factorization Axiom The following diagram commutes.

$$\begin{array}{ccc} (A \otimes C) \oplus (B \otimes C) & \xrightarrow{\partial^l_{A,B,C}} & (A \oplus B) \otimes C \\ {\scriptstyle \xi^{\otimes}_{A,C} \oplus \xi^{\otimes}_{B,C}} \downarrow & & \downarrow {\scriptstyle \xi^{\otimes}_{A \oplus B, C}} \\ (C \otimes A) \oplus (C \otimes B) & \xrightarrow{\partial^r_{C,A,B}} & C \otimes (A \oplus B) \end{array} \tag{2.43}$$

This finishes the definition of a bipermutative category. A bipermutative category is *small*, respectively *tight*, if the underlying ring category is so. ◇

Explanation 2.44. Consider Definition 2.41 of a bipermutative category.

1. A bipermutative category C has
   - a permutative category $(\mathsf{C}, \oplus, 0, \xi^{\oplus})$ as its additive structure and
   - a permutative category $(\mathsf{C}, \otimes, 1, \xi^{\otimes})$ as its multiplicative structure.

   This justifies the name *bipermutative* category.

2. By the symmetry axiom (1.12) in the permutative category $(\mathsf{C}, \xi^{\otimes})$, the zero braiding axiom (2.42) is equivalent to the equality
$$\xi^{\otimes}_{0,A} = 1_0 : 0 \otimes A = 0 \longrightarrow 0 = A \otimes 0$$
of morphisms. ◇

Bipermutative categories do *not* form a subclass of symmetric bimonoidal categories, or vice versa, because

- the distributivity morphisms δ^l and δ^r in (2.3) and
- the factorization morphisms ∂^l and ∂^r in (2.30)

are not invertible in general. On the other hand, *tight* bipermutative categories—that is, those with ∂^l and ∂^r invertible—form a subclass of tight symmetric bimonoidal categories. The following result is [Yau∞II, 9.3.7].

Theorem 2.45. *There is a canonical bijective correspondence between*

1. *the class of tight bipermutative categories in Definition 2.41 and*
2. *the class of tight symmetric bimonoidal categories in Definition 2.1 with*
   - *a permutative category as the additive structure,*
   - *a permutative category as the multiplicative structure, and*
   - $\lambda^\bullet = 1$ *and* $\rho^\bullet = 1$.

The correspondence between

- *the factorization morphisms* ∂^l *and* ∂^r *in* (2.30) *and*
- *the distributivity morphisms* δ^l *and* δ^r *in* (2.3)

is given by the equalities

$$\delta^l = (\partial^r)^{-1} \quad and \quad \delta^r = (\partial^l)^{-1}. \tag{2.46}$$

In the bijective correspondence in Theorem 2.45, the additive structure $(\mathsf{C}, \oplus, 0, \xi^{\oplus})$ and the multiplicative structure $(\mathsf{C}, \otimes, 1, \xi^{\otimes})$ remain unchanged. Using Theorem 2.45 we identify tight bipermutative categories with the corresponding subclass of tight symmetric bimonoidal categories with distributivity morphisms given by (2.46).

2.3 Examples of Bipermutative Categories

In this section we discuss several small tight symmetric bimonoidal categories that are also small tight bipermutative categories by Theorem 2.45.

1. In Example 2.48 the underlying category is $\mathsf{Fin}^{\mathsf{sk}}$, which consists of the finite sets $\{1, \ldots, n\}$ and functions. It is the most basic example in this section. Each subsequent example reuses its symmetric bimonoidal structure.
2. In Example 2.57 the underlying category is the subcategory Σ of $\mathsf{Fin}^{\mathsf{sk}}$ consisting of permutations.
3. $\mathcal{F}$ in Example 2.58 is the pointed variant of $\mathsf{Fin}^{\mathsf{sk}}$ with pointed finite sets and pointed functions. This is the indexing category for Γ-categories (Definition 9.11) and is a fundamental part of Segal K-theory; see [JY∞, Ch. 8].

4. Example 2.63 describes Mandell's category $\mathcal{A}$ from [Man10] as a symmetric bimonoidal category. This structure on $\mathcal{A}$ is the ultimate reason why inverse K-theory is a pseudo symmetric Cat-multifunctor, as we discuss in Chapter 10.

Moreover, in each of these small tight symmetric bimonoidal categories the following statements hold:

- Both the additive structure and the multiplicative structure are permutative categories.
- The left multiplicative zero, the right multiplicative zero, and the left distributivity morphism are identities.
- The right distributivity morphism is a natural isomorphism that is not the identity.

By Theorem 2.45 each of these small tight symmetric bimonoidal categories is also a small tight bipermutative category with factorization morphisms given by

$$\partial^l = (\delta^r)^{-1} \quad \text{and} \quad \partial^r = (\delta^l)^{-1}. \tag{2.47}$$

The symmetric group on n letters is denoted Σ_n.

Example 2.48 (Finite Sets)**.** There is a small tight symmetric bimonoidal category $\mathsf{Fin}^{\mathsf{sk}}$ given by the following data. The superscript sk indicates that the category $\mathsf{Fin}^{\mathsf{sk}}$ is a skeleton of the category of all finite sets and functions.

Category Its set of objects is $\mathbb{N}$, the set of natural numbers. For each $n \in \mathbb{N}$, denote by

$$\overline{n} = \begin{cases} \{1, \ldots, n\} & \text{if } n \geq 1, \\ \varnothing & \text{if } n = 0 \end{cases} \tag{2.49}$$

the unpointed finite set with n elements. To simplify the presentation, we identify the object $n \in \mathsf{Fin}^{\mathsf{sk}}$ with $\overline{n}$. The morphisms in $\mathsf{Fin}^{\mathsf{sk}}$ are functions of unpointed finite sets:

$$\mathsf{Fin}^{\mathsf{sk}}(\overline{m}, \overline{n}) = \mathsf{Set}(\overline{m}, \overline{n}).$$

Composition of morphisms is given by composition of functions.

Additive Structure It is given on objects by taking coproducts:

$$\overline{m} \oplus \overline{n} = \overline{m+n} = \overline{m} \coprod \overline{n}.$$

- It is extended to morphisms using the coproduct.
- The sum $\oplus$ is strictly associative with strict additive zero given by the empty set $\overline{0} = \varnothing$.

The additive braiding

$$\overline{m} \oplus \overline{n} \xrightarrow{\xi^{\oplus}_{m,n}} \overline{n} \oplus \overline{m} \tag{2.50}$$

is the *block swapping* in Σ_{m+n} given by

$$\xi^{\oplus}_{m,n}(k) = \begin{cases} n+k & \text{if } 1 \le k \le m, \\ k-m & \text{if } m+1 \le k \le m+n. \end{cases}$$

It swaps the first m elements with the last n elements.

Multiplicative Structure It is given on objects by taking products with the reverse lexicographic ordering that first compares the second entry:

$$\overline{m} \otimes \overline{n} = \overline{mn} = \langle \langle i + (j-1)m \rangle_{i=1}^{m} \rangle_{j=1}^{n}. \tag{2.51}$$

- It is extended to morphisms using the correspondence

$$\overline{m} \times \overline{n} \ni (i,j) \xleftrightarrow{\cong} i + (j-1)m \in \overline{mn}.$$

- The product $\otimes$ is strictly associative with strict multiplicative unit $\overline{1}$.

The multiplicative braiding

$$\overline{m} \otimes \overline{n} \xrightarrow{\xi^{\otimes}_{m,n}} \overline{n} \otimes \overline{m} \tag{2.52}$$

is the *transpose permutation* in Σ_{mn} given by

$$\xi^{\otimes}_{m,n}\big(i + (j-1)m\big) = j + (i-1)n$$

for $1 \le i \le m$ and $1 \le j \le n$. Regarding $\overline{m} \otimes \overline{n}$ as an $n \times m$ matrix with (j,i)-entry $i + (j-1)m$, the multiplicative braiding $\xi^{\otimes}_{m,n}$ in (2.52) corresponds to taking the transpose.

Multiplicative Zeros Both $\lambda^{\bullet}$ and $\rho^{\bullet}$ are identity natural transformations.

Distributivity The left distributivity morphism is the identity function

$$\overline{m} \otimes (\overline{n} \oplus \overline{p}) \xrightarrow[=]{\delta^{l}_{m,n,p}} (\overline{m} \otimes \overline{n}) \oplus (\overline{m} \otimes \overline{p}). \tag{2.53}$$

The right distributivity morphism

$$(\overline{m} \oplus \overline{n}) \otimes \overline{p} \xrightarrow[\cong]{\delta^{r}_{m,n,p}} (\overline{m} \otimes \overline{p}) \oplus (\overline{n} \otimes \overline{p}) \tag{2.54}$$

is the *tetris permutation* in $\Sigma_{(m+n)p}$ given by

$$\begin{cases} \delta^{r}_{m,n,p}\big(i + (k-1)(m+n)\big) = i + (k-1)m \\ \delta^{r}_{m,n,p}\big(j + m + (k-1)(m+n)\big) = j + (k-1)n + pm \end{cases} \tag{2.55}$$

for $1 \le i \le m$, $1 \le j \le n$, and $1 \le k \le p$. Regarding $(\overline{m} \oplus \overline{n}) \otimes \overline{p}$ as a $p \times (m+n)$ matrix of the form $[A \,|\, B]$ with

- A the $p \times m$ matrix consisting of the first m columns and
- B the $p \times n$ matrix consisting of the last n columns,

$\delta^r_{m,n,p}$ rearranges the matrix $[A\,|\,B]$ to the array $\left[\frac{A}{B}\right]$, leaving the order of the entries in each of A and B unchanged. The two rows in (2.55) correspond to the action of $\delta^r_{m,n,p}$ on, respectively, the (k,i)-entry in A and the (k,j)-entry in B.

This finishes the construction of $\mathsf{Fin}^{\mathsf{sk}}$.

All the symmetric bimonoidal category axioms in Definition 2.1 for $\mathsf{Fin}^{\mathsf{sk}}$ follow from

- the fact that $\lambda^{\bullet}$, $\rho^{\bullet}$, and δ^l are identity natural transformations,
- the fact that $\xi^{\oplus}$ is the block swapping, and
- the matrix description of each of $\xi^{\otimes}$ and δ^r.

For example, the axiom (2.4) for $\mathsf{Fin}^{\mathsf{sk}}$ states the commutativity of the following diagram for p, q, r in $\mathbb{N}$.

$$\begin{array}{ccc}
(\overline{p} \oplus \overline{r}) \otimes \overline{q} & \xrightarrow{\delta^r_{p,r,q}} & (\overline{p} \otimes \overline{q}) \oplus (\overline{r} \otimes \overline{q}) \\
{\scriptstyle \xi^{\otimes}_{p+r,q}} \downarrow & & \downarrow {\scriptstyle \xi^{\otimes}_{p,q} \oplus \xi^{\otimes}_{r,q}} \\
\overline{q} \otimes (\overline{p} \oplus \overline{r}) & \xrightarrow[=]{\delta^l_{q,p,r}} & (\overline{q} \otimes \overline{p}) \oplus (\overline{q} \otimes \overline{r})
\end{array} \tag{2.56}$$

As above we regard $(\overline{p} \oplus \overline{r}) \otimes \overline{q}$ as a $q \times (p+r)$ matrix of the form $[A\,|\,B]$ with

- A the $q \times p$ matrix consisting of the first p columns and
- B the $q \times r$ matrix consisting of the last r columns.

Then $\xi^{\otimes}_{p+r,q}$ in (2.56) transposes the matrix $[A\,|\,B]$ to the matrix $\left[\frac{A^T}{B^T}\right]$, with A^T and B^T denoting the transposes of, respectively, A and B. On the other hand, the top-right composite in (2.56) rearranges the matrix $[A\,|\,B]$

- first to the array $\left[\frac{A}{B}\right]$ and
- then to the matrix $\left[\frac{A^T}{B^T}\right]$ by taking the transpose of each of A and B.

This proves that (2.56) is commutative. The axioms (2.5) through (2.7) and (2.10) through (2.12) are checked in a similar manner. In every other axiom in Definition 2.1 for $\mathsf{Fin}^{\mathsf{sk}}$, each morphism is the identity morphism.

Both the additive structure and the multiplicative structure of $\mathsf{Fin}^{\mathsf{sk}}$ are permutative categories, with both distributivity morphisms invertible and strict multiplicative zeros. By Theorem 2.45 $\mathsf{Fin}^{\mathsf{sk}}$ is a small tight bipermutative category with factorization morphisms as in (2.47). ◇

Restricting the morphisms in $\mathsf{Fin}^{\mathsf{sk}}$ to just the permutations, we obtain the following example.

Example 2.57 (Symmetric Groups)**.** There is a small tight symmetric bimonoidal category Σ given by the following data.

Category Its set of objects is $\mathbb{N}$, the set of natural numbers. For $m, n \in \mathbb{N}$, the morphism set is

$$\Sigma(m,n) = \begin{cases} \Sigma_n & \text{if } m = n, \\ \varnothing & \text{if } m \neq n. \end{cases}$$

Symmetric Bimonoidal Structure It is the restriction of the symmetric bimonoidal structure of $\mathsf{Fin}^{\mathsf{sk}}$ in Example 2.48 to permutations. This is well defined because

- $\xi^{\oplus}$ (2.50), $\xi^{\otimes}$ (2.52), and δ^r (2.54) are permutations, and
- $\lambda^{\bullet}$, $\rho^{\bullet}$, and δ^l (2.53) are identity functions.

This example is discussed in detail in [Yau∞I, Section 2.4]. By Theorem 2.45 Σ is a small tight bipermutative category with factorization morphisms as in (2.47). ◇

The next example is a pointed variant of $\mathsf{Fin}^{\mathsf{sk}}$. It is used to define Γ-categories (Definition 9.11) in the discussion of inverse K-theory in Chapters 9 and 10.

Example 2.58 (Pointed Finite Sets)**.** There is a small tight symmetric bimonoidal category $\mathcal{F}$ given by the following data.

Category Its objects are the *pointed* finite sets

$$\underline{n} = \{0, \ldots, n\} \quad \text{for} \quad n \geq 0 \tag{2.59}$$

with basepoint 0. Its morphisms are *pointed* functions, that is, functions preserving the basepoint 0.

Additive Structure It is given on objects by the wedge

$$\underline{m} \vee \underline{n} = \underline{m+n}.$$

- It is the quotient of the disjoint union with the basepoints in $\underline{m}$ and $\underline{n}$ identified.
- It is extended to morphisms using the wedge.
- The wedge is strictly associative with strict additive zero given by $\underline{0}$.

The additive braiding

$$\underline{m} \vee \underline{n} = \underline{m+n} \xrightarrow{\xi^{\vee}_{\underline{m},\underline{n}}} \underline{n+m} = \underline{n} \vee \underline{m}$$

sends 0 to 0 and is given by the block swapping $\xi^{\oplus}_{m,n}$ in $\mathsf{Fin}^{\mathsf{sk}}$ (2.50) for nonzero elements in $\underline{m+n}$.

Multiplicative Structure It is given on objects by the smash product

$$\underline{m} \wedge \underline{n} = \underline{mn}.$$

- It is the quotient of the product $\underline{m} \times \underline{n}$ with $(x, 0)$ and $(0, y)$ identified with $0 \in \underline{mn}$ for $x \in \underline{m}$ and $y \in \underline{n}$.
- For $1 \le i \le m$ and $1 \le j \le n$, we use the identification

$$\underline{m} \wedge \underline{n} \ni (i, j) = i + (j-1)m \in \underline{mn}$$

  as in (2.51).
- It is extended to pointed functions using the basepoint-preserving condition and the expression $i + (j-1)m$ for non-basepoint elements.
- The smash product is strictly associative with strict multiplicative unit given by $\underline{1}$.

The multiplicative braiding

$$\underline{m} \wedge \underline{n} = \underline{mn} \xrightarrow{\xi^{\wedge}_{\underline{m},\underline{n}}} \underline{nm} = \underline{n} \wedge \underline{m} \tag{2.60}$$

sends 0 to 0 and is given by the transpose permutation $\xi^{\otimes}_{m,n}$ in $\mathsf{Fin}^{\mathsf{sk}}$ (2.52) for nonzero elements in $\underline{mn}$.

Multiplicative Zeros Both $\lambda^{\bullet}$ and $\rho^{\bullet}$ are identity natural transformations.

Distributivity The left distributivity morphism is the identity function

$$\underline{m} \wedge (\underline{n} \vee \underline{p}) \xrightarrow[=]{\delta^{l}_{\underline{m},\underline{n},\underline{p}}} (\underline{m} \wedge \underline{n}) \vee (\underline{m} \wedge \underline{p}). \tag{2.61}$$

The right distributivity morphism

$$(\underline{m} \vee \underline{n}) \wedge \underline{p} \xrightarrow[\cong]{\delta^{r}_{\underline{m},\underline{n},\underline{p}}} (\underline{m} \wedge \underline{p}) \vee (\underline{n} \wedge \underline{p}) \tag{2.62}$$

sends 0 to 0 and is given by $\delta^{r}_{m,n,p}$ in $\mathsf{Fin}^{\mathsf{sk}}$ (2.54) for nonzero elements in $\underline{(m+n)p}$.

The symmetric bimonoidal category axioms of $\mathcal{F}$ follow from those of $\mathsf{Fin}^{\mathsf{sk}}$ in Example 2.48. By Theorem 2.45 $\mathcal{F}$ is a small tight bipermutative category with factorization morphisms as in (2.47). In [JY∞, Ch. 8] the multiplicative structure $(\mathcal{F}, \wedge, \underline{1})$ is used to define Segal K-theory. However, in [JY∞, 8.1.5] the smash product $\underline{m} \wedge \underline{n}$ uses the lexicographic order that compares the first entry first. Here, $\underline{m} \wedge \underline{n}$ is defined with the reverse lexicographic order that compares the second entry first. ◇

The next example is crucial for our discussion of inverse K-theory in Chapters 9 and 10.

Example 2.63 (Mandell's Category for Inverse K-Theory)**.** There is a small tight symmetric bimonoidal category $\mathcal{A}$ given by the following data.

Category An object in $\mathcal{A}$ is a finite sequence of positive integers. The empty sequence is denoted by either $\langle\rangle$ or $\varnothing$. With the notation in (2.49) for unpointed finite sets, a morphism

$$m = (m_1, \ldots, m_p) \xrightarrow{\phi} (n_1, \ldots, n_q) = n \quad \text{in} \quad \mathcal{A} \tag{2.64}$$

is a function of unpointed finite sets

$$\coprod_{i=1}^{p} \overline{m_i} \xrightarrow{\phi} \coprod_{j=1}^{q} \overline{n_j} \tag{2.65}$$

such that the preimage of each $\overline{n_j}$ is either empty or contained in a single $\overline{m_i}$, with the index i determined by the index j. Note that, however, each $\overline{m_i}$ may hit several different $\overline{n_j}$. The category $\mathcal{A}$ is Mandell's indexing category for inverse K-theory [Man10]; see also [JY22b] and Chapters 9 and 10.

Additive Structure With the objects m and n as in (2.64), the sum is given by the concatenation of finite sequences:

$$m \oplus n = (m_1, \ldots, m_p, n_1, \ldots, n_q).$$

- It is extended to morphisms by taking coproducts of functions.
- The sum $\oplus$ is strictly associative with strict additive zero given by the empty sequence $\langle\rangle$.

The additive braiding

$$m \oplus n \xrightarrow{\beta^{\oplus}_{m,n}} n \oplus m \tag{2.66}$$

is the block swapping in $\Sigma_{\mathtt{m+n}}$ in (2.50)

$$\overline{\mathtt{m+n}} = \Big(\coprod_{i=1}^{p} \overline{m_i}\Big) \amalg \Big(\coprod_{j=1}^{q} \overline{n_j}\Big) \xrightarrow{\xi^{\oplus}_{\mathtt{m},\mathtt{n}}} \Big(\coprod_{j=1}^{q} \overline{n_j}\Big) \amalg \Big(\coprod_{i=1}^{p} \overline{m_i}\Big) = \overline{\mathtt{n+m}}$$

where

$$\mathtt{m} = m_1 + \cdots + m_p \quad \text{and} \quad \mathtt{n} = n_1 + \cdots + n_q. \tag{2.67}$$

Equivalently,

$$\beta^{\oplus}_{m,n} = \overline{\xi^{\oplus}_{p,q}}$$

is the block permutation induced by the block swapping $\xi^{\oplus}_{p,q} \in \Sigma_{p+q}$ in (2.50) that permutes the $p+q$ blocks of lengths $\{m_1, \ldots, m_p, n_1, \ldots, n_q\}$.

Multiplicative Structure With the objects m and n as in (2.64), the product is given by

$$m \otimes n = \langle \langle m_i n_j \rangle_{i=1}^{p} \rangle_{j=1}^{q}. \tag{2.68}$$

We visualize $m \otimes n$ as the $q \times p$ matrix with (j,i)-entry given by $m_i n_j$, which is also regarded as the unpointed finite set $\overline{m_i} \times \overline{n_j} = \overline{m_i n_j}$.

To define $\otimes$ on morphisms, suppose given morphisms in $\mathcal{A}$ as follows.

$$\begin{aligned} m = (m_1, \ldots, m_p) &\xrightarrow{\ \phi\ } (m_1', \ldots, m_r') = m' \\ n = (n_1, \ldots, n_q) &\xrightarrow{\ \psi\ } (n_1', \ldots, n_s') = n' \end{aligned} \tag{2.69}$$

The morphism

$$\langle \langle m_i n_j \rangle_i \rangle_j = m \otimes n \xrightarrow{\ \phi \otimes \psi\ } m' \otimes n' = \langle \langle m_k' n_\ell' \rangle_k \rangle_\ell \quad \text{in} \quad \mathcal{A}$$

is the function

$$\coprod_{j=1}^{q} \coprod_{i=1}^{p} \overline{m_i n_j} \xrightarrow{\ (\phi,\psi)\ } \coprod_{\ell=1}^{s} \coprod_{k=1}^{r} \overline{m_k' n_\ell'}$$

given by

$$(\phi,\psi)(x,y) = (\phi x, \psi y) \quad \text{for} \quad (x,y) \in \overline{m_i n_j}.$$

In this definition, each of $\overline{m_i n_j}$ and $\overline{m_k' n_\ell'}$ is given the reverse lexicographic ordering in (2.51) that compares the second entry first.

The product $\otimes$ is strictly associative with strict multiplicative unit given by the length-1 sequence (1). The multiplicative braiding

$$\langle \langle m_i n_j \rangle_i \rangle_j = m \otimes n \xrightarrow{\ \beta^{\otimes}_{m,n}\ } n \otimes m = \langle \langle n_j m_i \rangle_j \rangle_i \tag{2.70}$$

is the composite bijection

$$\begin{array}{ccc} \coprod_{j=1}^{q} \coprod_{i=1}^{p} \overline{m_i n_j} & \xrightarrow{\quad \beta^{\otimes}_{m,n} \quad} & \coprod_{i=1}^{p} \coprod_{j=1}^{q} \overline{n_j m_i} \\ & {\scriptstyle \overline{\xi^{\otimes}_{p,q}}} \searrow \qquad \nearrow {\scriptstyle \coprod_{i,j} \xi^{\otimes}_{m_i,n_j}} & \\ & \coprod_{i=1}^{p} \coprod_{j=1}^{q} \overline{m_i n_j} & \end{array}$$

given by

$$\beta^{\otimes}_{m,n}(x,y) = (y,x) \in \overline{n_j m_i} \quad \text{for} \quad (x,y) \in \overline{m_i n_j}.$$

- With $\mathtt{m} = m_1 + \cdots + m_p$ and $\mathtt{n} = n_1 + \cdots + n_q$ as in (2.67), $\overline{\xi^{\otimes}_{p,q}} \in \Sigma_{\mathtt{mn}}$ is the block permutation induced by the transpose permutation $\xi^{\otimes}_{p,q} \in \Sigma_{pq}$ in (2.52) that permutes the pq blocks of lengths $\{\{m_i n_j\}_{i=1}^{p}\}_{j=1}^{q}$.
- For each $1 \le i \le p$ and each $1 \le j \le q$, $\xi^{\otimes}_{m_i,n_j} \in \Sigma_{m_i n_j}$ is the transpose permutation in (2.52).

Multiplicative Zeros Both $\lambda^\bullet$ and $\rho^\bullet$ are identity natural transformations. This is well defined because there are object equalities

$$m \otimes \langle\rangle = \langle\rangle = \langle\rangle \otimes m \quad \text{for} \quad m \in \mathcal{A}.$$

Distributivity Suppose m, m', n, and n' are objects in $\mathcal{A}$ as in (2.69). The left distributivity morphism is the identity morphism

$$m \otimes (n \oplus n') \xrightarrow[=]{\delta^l_{m,n,n'}} (m \otimes n) \oplus (m \otimes n'). \tag{2.71}$$

With $\mathtt{m}$ and $\mathtt{n}$ as in (2.67) and $\mathtt{m}' = m'_1 + \cdots + m'_r$, the right distributivity morphism

$$(m \oplus m') \otimes n \xrightarrow[\cong]{\delta^r_{m,m',n}} (m \otimes n) \oplus (m' \otimes n) \tag{2.72}$$

is the block permutation in $\Sigma_{(\mathtt{m}+\mathtt{m}')\mathtt{n}}$

$$\coprod_{j=1}^{q} \left[\Big(\coprod_{i=1}^{p} \overline{m_i n_j} \Big) \amalg \Big(\coprod_{k=1}^{r} \overline{m'_k n_j} \Big) \right] \xrightarrow{\overline{\delta^r_{p,r,q}}} \Big(\coprod_{j=1}^{q} \coprod_{i=1}^{p} \overline{m_i n_j} \Big) \amalg \Big(\coprod_{j=1}^{q} \coprod_{k=1}^{r} \overline{m'_k n_j} \Big)$$

induced by the tetris permutation (2.54)

$$(\overline{p} \oplus \overline{r}) \otimes \overline{q} \xrightarrow[\cong]{\delta^r_{p,r,q}} (\overline{p} \otimes \overline{q}) \oplus (\overline{r} \otimes \overline{q})$$

that permutes the $(p+r)q$ blocks of lengths

$$\Big\{ \{m_i n_j\}_{i=1}^{p} \, , \, \{m'_k n_j\}_{k=1}^{r} \Big\}_{j=1}^{q}.$$

This finishes the construction of $\mathcal{A}$.

The symmetric bimonoidal category axioms in Definition 2.1 for $\mathcal{A}$ follow from those for $\mathsf{Fin}^{\mathsf{sk}}$ in Example 2.48, applied to induced block permutations. More explicitly, the axiom (2.4) for $\mathcal{A}$ states that the following diagram is commutative for objects $m, m', n \in \mathcal{A}$ as in (2.69).

$$\begin{array}{ccc}
(m \oplus m') \otimes n & \xrightarrow{\delta^r_{m,m',n}} & (m \otimes n) \oplus (m' \otimes n) \\
{\scriptstyle \beta^{\otimes}_{m \oplus m', n}} \big\downarrow & & \big\downarrow {\scriptstyle \beta^{\otimes}_{m,n} \oplus \beta^{\otimes}_{m',n}} \\
n \otimes (m \oplus m') & \xrightarrow[=]{\delta^l_{n,m,m'}} & (n \otimes m) \oplus (n \otimes m')
\end{array}$$

The two composites in the previous diagram are the following two composite functions, where the coproduct indices run through $i \in \{1, \ldots, p\}$, $j \in \{1, \ldots, q\}$, and $k \in \{1, \ldots, r\}$.

$$\begin{array}{ccc}
\amalg_j \left[(\amalg_i \overline{m_i n_j}) \amalg (\amalg_k \overline{m'_k n_j}) \right] & \xrightarrow{\overline{\delta^r_{p,r,q}}} & (\amalg_j \amalg_i \overline{m_i n_j}) \amalg (\amalg_j \amalg_k \overline{m'_k n_j}) \\
{\scriptstyle \overline{\xi^{\otimes}_{p+r,q}}} \searrow & & \swarrow {\scriptstyle \overline{\xi^{\otimes}_{p,q}} \amalg \overline{\xi^{\otimes}_{r,q}}} \\
& (\amalg_i \amalg_j \overline{m_i n_j}) \amalg (\amalg_k \amalg_j \overline{m'_k n_j}) & \\
& {\scriptstyle (\amalg_{i,j} \xi^{\otimes}_{m_i, n_j}) \amalg (\amalg_{k,j} \xi^{\otimes}_{m'_k, n_j})} \big\downarrow & \\
& (\amalg_i \amalg_j \overline{n_j m_i}) \amalg (\amalg_k \amalg_j \overline{n_j m'_k}) &
\end{array}$$

In the previous diagram, the top triangle commutes by (2.56), which is the axiom (2.4) for $\mathsf{Fin}^{\mathsf{sk}}$, applied to induced block permutations. This proves the axiom (2.4) for $\mathcal{A}$. The only other axiom that involves the multiplicative braiding is (2.18), which holds in $\mathcal{A}$ because $\lambda^{\bullet}$ and $\rho^{\bullet}$ are both the identity natural transformations. Every other axiom in Definition 2.1 for $\mathcal{A}$ holds as they do for $\mathsf{Fin}^{\mathsf{sk}}$, applied to induced block permutations. By Theorem 2.45 $\mathcal{A}$ is a small tight bipermutative category with factorization morphisms as in (2.47). ◇

2.4 Laplaza's Coherence Theorem

In this section we recall the statement of Laplaza's Coherence Theorem 2.90 for symmetric bimonoidal categories [Lap72a]. A detailed proof, along with corrections of the original argument in [Lap72a], can be found in [Yau∞I, Ch. 3]. The statement of this coherence theorem requires some preliminary definitions about graphs, which we first discuss.

Elementary Graph

Definition 2.73. Suppose S is a set. The *free* $\{\oplus, \otimes\}$*-algebra* of S is the set S^{fr} defined inductively by the following two conditions.

- $S \subset S^{\mathsf{fr}}$.
- If $a, b \in S^{\mathsf{fr}}$, then the symbols

$$a \oplus b \quad \text{and} \quad a \otimes b$$

  also belong to S^{fr}. They are called, respectively, the *sum* and the *product* of a and b.

To simplify the presentation, we sometimes abbreviate $a \otimes b$ to ab. In the absence of clarifying parentheses, $\otimes$ takes precedence over $\oplus$. ◇

Definition 2.74. A *graph* $G = (V, E)$ is a pair consisting of the following data.

- V is a class. An element in V is called a *vertex* in G.
- E assigns to each ordered pair (u, v) with $u, v \in V$ a set $E(u, v)$, an element of which is called an *edge* with *domain* u and *codomain* v. We also denote such an edge by

$$u \longrightarrow v, \quad e : u \longrightarrow v, \quad \text{or} \quad u \xrightarrow{\;e\;} v$$

  if e is the name of the edge.

A *path* in such a graph is a nonempty finite sequence of edges $(e_n, \ldots, e_1)$ as in

$$v_0 \xrightarrow{e_1} v_1 \xrightarrow{e_2} \cdots \xrightarrow{e_n} v_n.$$

Such a path is denoted by $v_0 \twoheadrightarrow v_n$ and is said to have *length* n, domain v_0, and codomain v_n. ◇

Example 2.75. Each category C has an associated graph (V, E), with V the class of objects in C, and $E(u,v) = \mathsf{C}(u,v)$ for objects $u, v \in \mathsf{C}$. A nonempty finite sequence of composable morphisms in C yields a path in the associated graph. ◇

Definition 2.76. Suppose X is a set with two distinguished elements 0^X and 1^X, called the *additive zero* and the *multiplicative unit*, respectively. The *elementary graph of* X, denoted by $\mathsf{Gr}^{\mathsf{el}}(X)$, is the graph defined as follows.

Vertices The set of vertices in $\mathsf{Gr}^{\mathsf{el}}(X)$ is the free $\{\oplus, \otimes\}$-algebra X^{fr} of X.

Edges Edges in $\mathsf{Gr}^{\mathsf{el}}(X)$ are of the following types for all $x, y, z \in X^{\mathsf{fr}}$.

The Additive Structure

$$(x \oplus y) \oplus z \underset{\alpha^{-\oplus}_{x,y,z}}{\overset{\alpha^{\oplus}_{x,y,z}}{\rightleftarrows}} x \oplus (y \oplus z) \qquad x \oplus y \underset{\xi^{-\oplus}_{x,y}}{\overset{\xi^{\oplus}_{x,y}}{\rightleftarrows}} y \oplus x$$

$$0^X \oplus x \underset{\lambda^{-\oplus}_{x}}{\overset{\lambda^{\oplus}_{x}}{\rightleftarrows}} x \underset{\rho^{-\oplus}_{x}}{\overset{\rho^{\oplus}_{x}}{\leftrightarrows}} x \oplus 0^X$$

The Multiplicative Structure

$$(x \otimes y) \otimes z \underset{\alpha^{-\otimes}_{x,y,z}}{\overset{\alpha^{\otimes}_{x,y,z}}{\rightleftarrows}} x \otimes (y \otimes z) \qquad x \otimes y \underset{\xi^{-\otimes}_{x,y}}{\overset{\xi^{\otimes}_{x,y}}{\rightleftarrows}} y \otimes x$$

$$1^X \otimes x \underset{\lambda^{-\otimes}_{x}}{\overset{\lambda^{\otimes}_{x}}{\rightleftarrows}} x \underset{\rho^{-\otimes}_{x}}{\overset{\rho^{\otimes}_{x}}{\leftrightarrows}} x \otimes 1^X$$

The Multiplicative Zeros

$$0^X \otimes x \underset{\lambda^{-\bullet}_{x}}{\overset{\lambda^{\bullet}_{x}}{\rightleftarrows}} 0^X \underset{\rho^{-\bullet}_{x}}{\overset{\rho^{\bullet}_{x}}{\leftrightarrows}} x \otimes 0^X$$

Distributivity

$$x \otimes (y \oplus z) \xrightarrow{\delta^l_{x,y,z}} (x \otimes y) \oplus (x \otimes z)$$

$$(x \oplus y) \otimes z \xrightarrow{\delta^r_{x,y,z}} (x \otimes z) \oplus (y \otimes z)$$

Identity

$$x \xrightarrow{1_x} x$$

This finishes the definition of $\mathsf{Gr}^{\mathsf{el}}(X)$.

Moreover, we define the following.

- The set of edges in $\mathsf{Gr}^{\mathsf{el}}(X)$ is denoted by $\mathsf{E}_{\mathsf{el}}(X)$, the elements of which are called *elementary edges.*
- $\alpha^{\oplus}$ and $\alpha^{-\oplus}$ are *formal inverses* of each other, and similarly for the other 9 pairs of elementary edges in the first three groups above.
- 1_x is called the *identity* of x.
- The names in Definition 2.1 are reused for elementary edges. For example, $\lambda^{\bullet}$ is called the left multiplicative zero, and δ^r is called the right distributivity. ◇

Graph

The next two definitions will be combined to yield the graph of X in Definition 2.79.

Definition 2.77. With $(X, 0^X, 1^X)$ as in Definition 2.76, consider the free $\{\oplus, \otimes\}$-algebra $\mathsf{E}^{\mathsf{fr}}_{\mathsf{el}}(X)$ of the set $\mathsf{E}_{\mathsf{el}}(X)$ of elementary edges. The *domain* and *codomain* of an element $f \in \mathsf{E}^{\mathsf{fr}}_{\mathsf{el}}(X)$ are elements in X^{fr} defined inductively as follows.

- For an elementary edge $f \in \mathsf{E}_{\mathsf{el}}(X)$, its (co)domain are those of f in the elementary graph $\mathsf{Gr}^{\mathsf{el}}(X)$.
- Suppose $f_1, f_2 \in \mathsf{E}^{\mathsf{fr}}_{\mathsf{el}}(X)$ with
  - $u_i \in X^{\mathsf{fr}}$ the domain of f_i and
  - $v_i \in X^{\mathsf{fr}}$ the codomain of f_i

  already defined for $i = 1, 2$. Then
  - $f_1 \oplus f_2$ has *domain* $u_1 \oplus u_2$ and *codomain* $v_1 \oplus v_2$, and
  - $f_1 \otimes f_2$ has *domain* $u_1 \otimes u_2$ and *codomain* $v_1 \otimes v_2$. ◇

Definition 2.78. Continuing Definition 2.77, *identity prime edges* and *nonidentity prime edges* are elements in $\mathsf{E}^{\mathsf{fr}}_{\mathsf{el}}(X)$ defined inductively by the following four conditions.

- Elementary edges of the type 1_x for $x \in X^{\mathsf{fr}}$ are identity prime edges.
- Elementary edges not of the type 1_x for $x \in X^{\mathsf{fr}}$ are nonidentity prime edges.

- If $e_1, e_2 \in \mathsf{E}^{\mathsf{fr}}_{\mathsf{el}}(X)$ are identity prime edges, then so are $e_1 \oplus e_2$ and $e_1 \otimes e_2$.
- If f is a nonidentity prime edge and if e is an identity prime edge, then

$$f \oplus e, \quad e \oplus f, \quad f \otimes e, \quad \text{and} \quad e \otimes f$$

  are nonidentity prime edges.

Moreover, we define the following.

- A *prime edge* means either an identity prime edge or a nonidentity prime edge. The set of prime edges is denoted by $\mathsf{E}^{\mathsf{pr}}(X)$.
- A *δ-prime edge* is a prime edge that involves either δ^l or δ^r. ◇

Definition 2.79. With $(X, 0^X, 1^X)$ as in Definition 2.76, the *graph of* X, which is denoted by $\mathsf{Gr}(X)$, is the graph defined as follows.

Vertices The set of vertices in $\mathsf{Gr}(X)$ is the free $\{\oplus, \otimes\}$-algebra X^{fr} of X.

Edges The set of edges in $\mathsf{Gr}(X)$ is the set $\mathsf{E}^{\mathsf{pr}}(X)$ of prime edges as in Definition 2.78, with (co)domain as in Definition 2.77. ◇

The Value of a Path

The graph of X is interpreted in a symmetric bimonoidal category via the following concept.

Definition 2.80. Suppose $G_1 = (V_1, E_1)$ and $G_2 = (V_2, E_2)$ are graphs (Definition 2.74). A *graph morphism* $f : G_1 \longrightarrow G_2$ consists of functions

- $f_V : V_1 \longrightarrow V_2$ and
- $f_E : E_1(u, v) \longrightarrow E_2\big(f_V u, f_V v\big)$ for $u, v \in V_1$.

To simplify the notation, both f_V and f_E are denoted by f below. ◇

Definition 2.81. Suppose given the data (X, C, φ) as follows.

- X is a set with two distinguished elements 0^X and 1^X as in Definition 2.76.
- C is a symmetric bimonoidal category as in Definition 2.1, equipped with the graph structure in Example 2.75.
- $\varphi : X \longrightarrow \mathsf{Ob}(\mathsf{C})$ is any function such that

$$\varphi(0^X) = \mathsf{0} \quad \text{and} \quad \varphi(1^X) = \mathsf{1}. \tag{2.82}$$

Extend φ to a graph morphism

$$\mathsf{Gr}(X) \xrightarrow{\ \varphi\ } \mathsf{C}$$

as follows.

Vertices For $x, y \in X^{\mathsf{fr}}$ such that $\varphi x, \varphi y \in \mathsf{Ob}(\mathsf{C})$ are already defined, we define

$$\begin{aligned} \varphi(x \oplus y) &= \varphi x \oplus \varphi y \quad \text{and} \\ \varphi(x \otimes y) &= \varphi x \otimes \varphi y. \end{aligned} \tag{2.83}$$

Elementary Edges φ sends each elementary edge to the structure morphism in C with the same name and with the subscripts replaced by their images under φ.

Prime Edges If $f_1, f_2 \in \mathsf{E}^{\mathsf{pr}}(X)$ are prime edges with at most one of them nonidentity and with $\varphi(f_1)$ and $\varphi(f_2)$ already defined, then we define

$$\begin{aligned} \varphi(f_1 \oplus f_2) &= \varphi(f_1) \oplus \varphi(f_2) \quad \text{and} \\ \varphi(f_1 \otimes f_2) &= \varphi(f_1) \otimes \varphi(f_2). \end{aligned} \tag{2.84}$$

This finishes the definition of the graph morphism φ.

Moreover, for a path $P = (f_n, \ldots, f_1)$ in $\mathsf{Gr}(X)$ with domain u and codomain v, its *value* in C is the composite

$$\varphi P = \varphi(f_n) \circ \cdots \circ \varphi(f_1) \tag{2.85}$$

in $\mathsf{C}(\varphi u, \varphi v)$. ◇

Regularity

Next we introduce a restriction, called regularity, on the elements in X^{fr} for the coherence theorem of symmetric bimonoidal categories.

Definition 2.86. Suppose $(X, 0^X, 1^X)$ is as in Definition 2.76. Define its *strict* $\{\oplus, \otimes\}$*-algebra* X^{st} as the quotient set of X^{fr} in Definition 2.73 by the smallest relation that contains the following identifications for all elements $x, y, z \in X^{\mathsf{fr}}$.

Additive Structure

$$(x \oplus y) \oplus z = x \oplus (y \oplus z)$$
$$0^X \oplus x = x = x \oplus 0^X$$
$$x \oplus y = y \oplus x$$

Multiplicative Structure

$$(x \otimes y) \otimes z = x \otimes (y \otimes z)$$
$$1^X \otimes x = x = x \otimes 1^X$$
$$x \otimes y = y \otimes x$$

Multiplicative Zeros

$$0^X \otimes x = 0^X = x \otimes 0^X$$

Distributivity

$$x \otimes (y \oplus z) = (x \otimes y) \oplus (x \otimes z)$$
$$(x \oplus y) \otimes z = (x \otimes z) \oplus (y \otimes z)$$

This finishes the definition of X^{st}. Moreover, denote by

$$X^{\mathsf{fr}} \xrightarrow{\mathsf{supp}} X^{\mathsf{st}} \tag{2.87}$$

the quotient map, called the *support.* ◇

Definition 2.88. An element $x \in X^{\mathsf{fr}}$ is *regular* if there exist elements $x^i_j \in X$ for $1 \leq i \leq m$ and $1 \leq j \leq k_i$ for each i with $m, k_1, \ldots, k_m > 0$, such that the following three conditions hold.

1. The equality
$$\mathsf{supp}(x) = \mathsf{supp}\Bigg(\bigoplus_{i=1}^{m} \big(x^i_1 \otimes \cdots \otimes x^i_{k_i} \big) \Bigg) \tag{2.89}$$
holds in X^{st}. The iterated sum $\bigoplus_{i=1}^{m}$ and each of the m iterated products $x^i_1 \otimes \cdots \otimes x^i_{k_i}$ have some bracketings. By the definition of X^{st}, different bracketings yield the same support.

2. For each $1 \leq i \leq m$, the elements $x^i_1, \ldots, x^i_{k_i} \in X$ are distinct.

3. The m elements
$$\mathsf{supp}\big(x^i_1 \otimes \cdots \otimes x^i_{k_i}\big) \in X^{\mathsf{st}}$$
for $1 \leq i \leq m$ are distinct. ◇

We are now ready to state the following coherence theorem for symmetric bimonoidal categories. See [Yau∞I, 3.9.1] for a detailed proof.

Theorem 2.90 (Laplaza's Coherence)**.** *In the context of Definition 2.81, suppose* C *is a symmetric bimonoidal category in which the value of each* δ*-prime edge is a monomorphism. Suppose*

$$a \overset{P_1}{\underset{P_2}{\rightrightarrows}} b$$

are two paths in $\mathsf{Gr}(X)$ *with* $a \in X^{\mathsf{fr}}$ *regular. Then the values of* P_1 *and* P_2 *in* C *are equal.*

Example 2.91. If C is a *tight* symmetric bimonoidal category—that is, both distributivity morphisms δ^l and δ^r in C are invertible—then the value of each δ-prime edge is an isomorphism, which is, in particular, a monomorphism. In this case Theorem 2.90 applies to C. The value in C of each path in Theorem 2.90 is called a *Laplaza coherence isomorphism.*

- Via Theorem 2.45 each tight bipermutative category is also a tight symmetric bimonoidal category, so Laplaza's Coherence Theorem 2.90 applies to tight bipermutative categories.

- In particular, Theorem 2.90 applies to the tight bipermutative categories $\mathsf{Fin}^{\mathsf{sk}}$, Σ, $\mathcal{F}$, and $\mathcal{A}$ in, respectively, Examples 2.48, 2.57, 2.58, and 2.63. $\diamond$

There is another coherence theorem for symmetric bimonoidal categories that is also due to Laplaza [Lap72b]. Since we will not use that coherence theorem in this monograph, the interested reader is referred to [Yau∞I, Ch. 4] for a detailed discussion.

3

Enriched Multicategories and Multiequivalences

This chapter has two main purposes:

1. We review enriched multicategories and related concepts that are used in subsequent chapters.
2. We characterize categorically-enriched multiequivalences in terms of essential surjectivity on objects and invertibility on multimorphism categories (Theorem 3.69). This characterization holds in both the symmetric and the non-symmetric contexts. Theorem 3.69 has a pseudo symmetric variant in Theorem 4.64.

Below is a summary of some of the definitions and results in this chapter.

- Enriched multicategories, multifunctors, and multinatural transformations. Definitions 3.2, 3.19, and 3.23
- 2-categories of enriched multicategories. Theorem 3.28.
- Cat-multicategories and multinatural transformations. Definition 3.31 and Explanation 3.33.
- Cat-multiequivalences. Definition 3.42.
- Characterizations of Cat-multiequivalences and 2-equivalences. Theorems 3.69 and 3.70.

In Theorem 8.76 we use Theorem 3.69 to show that the Grothendieck construction on a small tight bipermutative category is a non-symmetric Cat-multiequivalence. An important consequence of this fact is that inverse K-theory factors through a non-symmetric Cat-multiequivalence (10.14).

Organization

In Section 3.1 we review multicategories enriched in a symmetric monoidal category V. In Section 3.2 we review V-enriched multifunctors, V-enriched multinatural transformations, and the 2-category V-Multicat that they form

DOI: 10.1201/9781003451235-3

with small V-multicategories as objects. Most of the material in Sections 3.1 and 3.2 is presented with much more detail in [JY∞, Ch. 5 and 6] and [Yau16].

In Section 3.3 we show that Cat-multiequivalences have the properties of being essentially surjective on objects and invertible on multimorphism categories. Theorem 3.69 in Section 3.4 shows that these properties precisely characterize Cat-multiequivalences. Moreover, restricting this proof to the 1-ary multimorphism categories recovers an analogous characterization of 2-equivalences between 2-categories (Theorem 3.70).

Throughout this chapter we assume that

$$\mathsf{V} = (\mathsf{V}, \otimes, \mathbb{1}, \alpha, \lambda, \rho, \xi)$$

is a symmetric monoidal category (Definition 1.11). We remind the reader of our left normalized bracketing Convention 1.24.

3.1 Enriched Multicategories

In this section we review the definition of an enriched multicategory and a few basic examples. More elaborate examples appear in later chapters; see Explanation 5.2 and Theorems 5.71, 6.31, 7.51, and 9.23. We begin with some notation.

Definition 3.1. Suppose D is a class.

- Denote by

$$\mathsf{Prof}(D) = \coprod_{n \geq 0} D^{\times n}$$

the class of finite tuples in D. An element in $\mathsf{Prof}(D)$ is called a *D-profile.*

- A D-profile of length $n = \mathrm{len}\langle d \rangle$ is denoted by $\langle d \rangle = (d_1, \ldots, d_n) \in D^n$ or by $\langle d_i \rangle_{i=1}^n$. The empty D-profile is denoted by $\langle \rangle$.

- We let $\oplus$ denote the concatenation of profiles. It is associative with unit given by the empty tuple $\langle \rangle$.

- An element in $\mathsf{Prof}(D) \times D$ is denoted by $(\langle d \rangle \,; d')$ with $d' \in D$ and $\langle d \rangle \in \mathsf{Prof}(D)$. ◇

The symmetric group on n letters is denoted Σ_n.

Definition 3.2. A V-*multicategory* $(\mathsf{M}, \gamma, \mathbb{1})$ consists of the following data.

- M is equipped with a class $\mathsf{Ob}\,\mathsf{M}$ of *objects.* We abbreviate $\mathsf{Prof}(\mathsf{Ob}\,\mathsf{M})$ to $\mathsf{Prof}(\mathsf{M})$.

- For $x' \in \mathrm{Ob}\,\mathsf{M}$ and $\langle x\rangle = \langle x_j\rangle_{j=1}^n \in \mathsf{Prof}(\mathsf{M})$, M is equipped with an object

$$\mathsf{M}\big(\langle x\rangle\,;x'\big) = \mathsf{M}\big(x_1,\ldots,x_n\,;x'\big) \in \mathsf{V},$$

called the *n-ary operation object* or *n-ary multimorphism object* with *input profile* $\langle x\rangle$ and *output* x'.

- For $(\langle x\rangle\,;x') \in \mathsf{Prof}(\mathsf{M}) \times \mathrm{Ob}\,\mathsf{M}$ as above and a permutation $\sigma \in \Sigma_n$, M is equipped with an isomorphism in V

$$\mathsf{M}\big(\langle x\rangle\,;x'\big) \xrightarrow[\cong]{\sigma} \mathsf{M}\big(\langle x\rangle\sigma\,;x'\big), \tag{3.3}$$

called the *right σ-action* or the *symmetric group action*, with

$$\langle x\rangle\sigma = \big(x_{\sigma(1)},\ldots,x_{\sigma(n)}\big) \in \mathsf{Prof}(\mathsf{M})$$

the right permutation of $\langle x\rangle$ by σ.

- Each object x in M is equipped with a morphism

$$\mathbb{1} \xrightarrow{1_x} \mathsf{M}\big(x\,;x\big), \tag{3.4}$$

called the *x-colored unit.*

- For
  - $x'' \in \mathrm{Ob}\,\mathsf{M}$,
  - $\langle x'\rangle = \langle x'_j\rangle_{j=1}^n \in \mathsf{Prof}(\mathsf{M})$, and
  - $\langle x_j\rangle = \langle x_{j,i}\rangle_{i=1}^{k_j} \in \mathsf{Prof}(\mathsf{M})$ for each $j \in \{1,\ldots,n\}$ with $\langle x\rangle = \oplus_{j=1}^n \langle x_j\rangle$,

  M is equipped with a morphism in V

$$\mathsf{M}\big(\langle x'\rangle\,;x''\big) \otimes \textstyle\bigotimes_{j=1}^n \mathsf{M}\big(\langle x_j\rangle\,;x'_j\big) \xrightarrow{\gamma} \mathsf{M}\big(\langle x\rangle\,;x''\big), \tag{3.5}$$

called the *composition* or *multicategorical composition.*

The data above are required to satisfy the following axioms.

Symmetric Group Action For $\big(\langle x\rangle\,;x'\big) \in \mathsf{Prof}(\mathsf{M}) \times \mathrm{Ob}\,\mathsf{M}$ with $n = \mathrm{len}\langle x\rangle$ and $\sigma, \tau \in \Sigma_n$, the following diagram in V commutes.

$$\begin{array}{ccc} \mathsf{M}\big(\langle x\rangle\,;x'\big) & \xrightarrow{\sigma} & \mathsf{M}\big(\langle x\rangle\sigma\,;x'\big) \\ & \searrow^{\sigma\tau} & \downarrow{\scriptstyle\tau} \\ & & \mathsf{M}\big(\langle x\rangle\sigma\tau\,;x'\big) \end{array} \tag{3.6}$$

The identity in Σ_n acts as the identity morphism on $\mathsf{M}\big(\langle x\rangle\,;x'\big)$.

Associativity Suppose given

- $x''' \in \mathsf{Ob}\,\mathsf{M}$,
- $\langle x'' \rangle = \langle x''_j \rangle_{j=1}^n \in \mathsf{Prof}(\mathsf{M})$,
- $\langle x'_j \rangle = \langle x'_{j,i} \rangle_{i=1}^{k_j} \in \mathsf{Prof}(\mathsf{M})$ for each $j \in \{1, \ldots, n\}$ with $\langle x' \rangle = \oplus_{j=1}^n \langle x'_j \rangle$ and $k_j > 0$ for at least one j, and
- $\langle x_{j,i} \rangle = \langle x_{j,i,p} \rangle_{p=1}^{\ell_{j,i}} \in \mathsf{Prof}(\mathsf{M})$ for each $j \in \{1, \ldots, n\}$ and each $i \in \{1, \ldots, k_j\}$ with $\langle x_j \rangle = \oplus_{i=1}^{k_j} \langle x_{j,i} \rangle$ and $\langle x \rangle = \oplus_{j=1}^n \langle x_j \rangle$.

Then the *associativity diagram* below commutes.

$$\begin{array}{ccc}
\mathsf{M}\big(\langle x'' \rangle ; x'''\big) \otimes \Big[\bigotimes_{j=1}^{n} \mathsf{M}\big(\langle x'_j \rangle ; x''_j\big)\Big] \otimes \bigotimes_{j=1}^{n}\Big[\bigotimes_{i=1}^{k_j} \mathsf{M}\big(\langle x_{j,i} \rangle ; x'_{j,i}\big)\Big] & \xrightarrow{(\gamma,1)} & \mathsf{M}\big(\langle x' \rangle ; x'''\big) \otimes \bigotimes_{j=1}^{n}\Big[\bigotimes_{i=1}^{k_j} \mathsf{M}\big(\langle x_{j,i} \rangle ; x'_{j,i}\big)\Big] \\
\Big\downarrow {\scriptstyle \text{permute}\ \cong} & & \Big\downarrow {\scriptstyle \gamma} \\
\mathsf{M}\big(\langle x'' \rangle ; x'''\big) \otimes \bigotimes_{j=1}^{n}\Big[\mathsf{M}\big(\langle x'_j \rangle ; x''_j\big) \otimes \bigotimes_{i=1}^{k_j} \mathsf{M}\big(\langle x_{j,i} \rangle ; x'_{j,i}\big)\Big] & & \mathsf{M}\big(\langle x \rangle ; x'''\big) \\
\searrow {\scriptstyle (1, \otimes_j \gamma)} & & \Big\uparrow {\scriptstyle \gamma} \\
 & & \mathsf{M}\big(\langle x'' \rangle ; x'''\big) \otimes \bigotimes_{j=1}^{n} \mathsf{M}\big(\langle x_j \rangle ; x''_j\big)
\end{array} \tag{3.7}$$

Unity Suppose $x' \in \mathsf{Ob}\,\mathsf{M}$.

1. If $\langle x \rangle = \langle x_j \rangle_{j=1}^n \in \mathsf{Prof}(\mathsf{M})$ has length $n \geq 1$, then the following *right unity diagram* commutes, with 1^n the n-fold monoidal product of 1 with itself.

$$\begin{array}{ccc}
\mathsf{M}\big(\langle x \rangle ; x'\big) \otimes 1^n & \xrightarrow{\ \rho\ } & \mathsf{M}\big(\langle x \rangle ; x'\big) \\
{\scriptstyle 1 \otimes (\otimes_j 1_{x_j})} \Big\downarrow & & \Big\downarrow {\scriptstyle 1} \\
\mathsf{M}\big(\langle x \rangle ; x'\big) \otimes \bigotimes_{j=1}^n \mathsf{M}\big(x_j ; x_j\big) & \xrightarrow{\ \gamma\ } & \mathsf{M}\big(\langle x \rangle ; x'\big)
\end{array} \tag{3.8}$$

2. For any $\langle x \rangle \in \mathsf{Prof}(\mathsf{M})$, the *left unity diagram* below commutes.

$$\begin{array}{ccc}
1 \otimes \mathsf{M}\big(\langle x \rangle ; x'\big) & \xrightarrow{\ \lambda\ } & \mathsf{M}\big(\langle x \rangle ; x'\big) \\
{\scriptstyle 1_{x'} \otimes 1} \Big\downarrow & & \Big\downarrow {\scriptstyle 1} \\
\mathsf{M}\big(x' ; x'\big) \otimes \mathsf{M}\big(\langle x \rangle ; x'\big) & \xrightarrow{\ \gamma\ } & \mathsf{M}\big(\langle x \rangle ; x'\big)
\end{array} \tag{3.9}$$

Equivariance Suppose that in the definition of γ (3.5), $\mathrm{len}\langle x_j \rangle = k_j \geq 0$.

1. For each $\sigma \in \Sigma_n$, the following *top equivariance diagram* commutes.

$$\begin{array}{ccc} \mathsf{M}\big(\langle x'\rangle\,;\,x''\big) \otimes \bigotimes_{j=1}^{n} \mathsf{M}\big(\langle x_j\rangle\,;\,x'_j\big) & \xrightarrow{(\sigma,\sigma^{-1})} & \mathsf{M}\big(\langle x'\rangle\sigma\,;\,x''\big) \otimes \bigotimes_{j=1}^{n} \mathsf{M}\big(\langle x_{\sigma(j)}\rangle\,;\,x'_{\sigma(j)}\big) \\ {\scriptstyle\gamma}\big\downarrow & & \big\downarrow{\scriptstyle\gamma} \\ \mathsf{M}\big(\langle x_1\rangle,\ldots,\langle x_n\rangle\,;\,x''\big) & \xrightarrow{\sigma\langle k_{\sigma(1)},\ldots,k_{\sigma(n)}\rangle} & \mathsf{M}\big(\langle x_{\sigma(1)}\rangle,\ldots,\langle x_{\sigma(n)}\rangle\,;\,x''\big) \end{array} \tag{3.10}$$

Here

$$\sigma\langle k_{\sigma(1)},\ldots,k_{\sigma(n)}\rangle \in \Sigma_{k_1+\cdots+k_n} \tag{3.11}$$

is the block permutation that permutes n consecutive blocks of lengths $k_{\sigma(1)},\ldots,k_{\sigma(n)}$ as σ permutes $\{1,\ldots,n\}$, leaving the relative order within each block unchanged.

2. Given permutations $\tau_j \in \Sigma_{k_j}$ for $1 \leq j \leq n$, the following *bottom equivariance diagram* commutes.

$$\begin{array}{ccc} \mathsf{M}\big(\langle x'\rangle\,;\,x''\big) \otimes \bigotimes_{j=1}^{n} \mathsf{M}\big(\langle x_j\rangle\,;\,x'_j\big) & \xrightarrow{(1,\otimes_j \tau_j)} & \mathsf{M}\big(\langle x'\rangle\,;\,x''\big) \otimes \bigotimes_{j=1}^{n} \mathsf{M}\big(\langle x_j\rangle\tau_j\,;\,x'_j\big) \\ {\scriptstyle\gamma}\big\downarrow & & \big\downarrow{\scriptstyle\gamma} \\ \mathsf{M}\big(\langle x_1\rangle,\ldots,\langle x_n\rangle\,;\,x''\big) & \xrightarrow{\tau_1\times\cdots\times\tau_n} & \mathsf{M}\big(\langle x_1\rangle\tau_1,\ldots,\langle x_n\rangle\tau_n\,;\,x''\big) \end{array} \tag{3.12}$$

Here the block sum

$$\tau_1 \times \cdots \times \tau_n \in \Sigma_{k_1+\cdots+k_n} \tag{3.13}$$

is the image of $(\tau_1,\ldots,\tau_n)$ under the canonical inclusion

$$\Sigma_{k_1} \times \cdots \times \Sigma_{k_n} \longrightarrow \Sigma_{k_1+\cdots+k_n}.$$

This finishes the definition of a V-multicategory.

Moreover, we define the following.

- A V-multicategory is *small* if its class of objects is a set.
- A *non-symmetric* V*-multicategory* is defined in the same way as a V-multicategory but without the symmetric group action and the axioms (3.6), (3.10), and (3.12).
- A *(non-symmetric) multicategory* is a (non-symmetric) Set-multicategory, where $(\mathsf{Set}, \times, *)$ is the symmetric monoidal category of sets and functions with the Cartesian product.
- A V*-operad* is a V-multicategory with one object. If M is a V-operad, then its object of n-ary operations is denoted by $\mathsf{M}_n \in \mathsf{V}$.
- An *operad* is a Set-operad, that is, a multicategory with one object. ◇

Remark 3.14. A non-symmetric multicategory as in Definition 3.2 is the same as a multicategory as originally defined in [Lam69]. ◇

Example 3.15. The *terminal multicategory* T consists of a single object $*$ and a single n-ary operation ι_n for each $n \geq 0$. ◇

Example 3.16 (Endomorphism Operad)**.** For each V-multicategory M and object x of M, the *endomorphism* V*-operad* $\mathsf{End}(x)$ consists of the single object x and n-ary operation object

$$\mathsf{End}(x)_n = \mathsf{M}\big(\langle x\rangle\,;\, x\big) \in \mathsf{V},$$

with $\langle x\rangle$ the constant n-tuple at x. The symmetric group action, unit, and composition of $\mathsf{End}(x)$ are given by those of M. ◇

Example 3.17 (Endomorphism Multicategory)**.** Each permutative category $(\mathsf{C}, \oplus, e, \xi)$ as in Definition 1.11 has an associated *endomorphism multicategory* $\mathsf{End}(\mathsf{C})$ with object set $\mathsf{Ob}\,\mathsf{C}$ and n-ary multimorphism set

$$\mathsf{End}(\mathsf{C})\big(\langle x\rangle\,;\, y\big) = \mathsf{C}(x_1 \oplus \cdots \oplus x_n, y)$$

for $y \in \mathsf{Ob}\,\mathsf{C}$ and $\langle x\rangle = \langle x_j\rangle_{j=1}^n \in (\mathsf{Ob}\,\mathsf{C})^{\times n}$. An empty $\oplus$ is, by definition, the unit object e. ◇

Example 3.18 (Underlying V-Category)**.** A non-symmetric V-multicategory $(\mathsf{M}, \gamma, 1)$ has an underlying V-category (Definition 1.25) defined by the following data.

- It has the same class of objects as M.
- It has hom object $\mathsf{M}(a\,;\, b)$ with domain a and codomain b.
- Its identities are given by the colored units in M.
- Its composition is given by

$$\mathsf{M}\big(b\,;\, c\big) \otimes \mathsf{M}\big(a\,;\, b\big) \xrightarrow{\quad\gamma\quad} \mathsf{M}\big(a\,;\, c\big)$$

  for objects $a, b, c \in \mathsf{M}$.

The associativity diagram (1.26) and the unity diagram (1.27) of a V-category are the 1-ary special cases of, respectively, the associativity diagram (3.7) and the unity diagrams (3.8) and (3.9) of a V-multicategory. ◇

3.2 The 2-Category of Enriched Multicategories

As in Section 3.1 we assume that V is a symmetric monoidal category. In this section we review the 2-category of small V-enriched multicategories, V-enriched multifunctors, and V-enriched multinatural transformations. In Explanation 3.33 we explicitly describe the data and axioms of a Cat-enriched multinatural transformation.

Enriched Multifunctors and Multinatural Transformations

Definition 3.19. A V-*multifunctor* $F : \mathsf{M} \longrightarrow \mathsf{N}$ between V-multicategories M and N consists of the following data and axioms.

- It is equipped with an object assignment $F : \mathsf{Ob}\,\mathsf{M} \longrightarrow \mathsf{Ob}\,\mathsf{N}$.
- For each $(\langle x \rangle\,;x') \in \mathsf{Prof}(\mathsf{M}) \times \mathsf{Ob}\,\mathsf{M}$ with $\langle x \rangle = \langle x_j \rangle_{j=1}^n$, it is equipped with a component morphism in V

$$\mathsf{M}\big(\langle x \rangle\,; x'\big) \xrightarrow{\;F\;} \mathsf{N}\big(F\langle x \rangle\,; Fx'\big)$$

  with $F\langle x \rangle = \langle Fx_j \rangle_{j=1}^n$.

The data above are required to satisfy the following three axioms.

Symmetric Group Action For each $(\langle x \rangle\,; x')$ as above and permutation $\sigma \in \Sigma_n$, the following diagram in V commutes.

$$\begin{array}{ccc} \mathsf{M}\big(\langle x \rangle\,; x'\big) & \xrightarrow{\;F\;} & \mathsf{N}\big(F\langle x \rangle\,; Fx'\big) \\ {\scriptstyle\sigma}\big\downarrow{\scriptstyle\cong} & & {\scriptstyle\sigma}\big\downarrow{\scriptstyle\cong} \\ \mathsf{M}\big(\langle x \rangle\sigma\,; x'\big) & \xrightarrow{\;F\;} & \mathsf{N}\big(F\langle x \rangle\sigma\,; x'\big) \end{array} \tag{3.20}$$

Units For each object x in M, the following diagram in V commutes.

$$\begin{array}{ccc} & & \mathsf{M}\big(x\,; x\big) \\ & \overset{1_x}{\nearrow} & \\ \mathbb{1} & & \big\downarrow{\scriptstyle F} \\ & \underset{1_{Fx}}{\searrow} & \\ & & \mathsf{N}\big(Fx\,; Fx\big) \end{array} \tag{3.21}$$

Composition For x'', $\langle x' \rangle$, and $\langle x \rangle = \oplus_{j=1}^n \langle x_j \rangle$ as in (3.5), the following diagram in V commutes.

$$\begin{array}{ccc} \mathsf{M}\big(\langle x' \rangle\,; x''\big) \otimes \bigotimes_{j=1}^n \mathsf{M}\big(\langle x_j \rangle\,; x_j'\big) & \xrightarrow{(F, \otimes_j F)} & \mathsf{N}\big(F\langle x' \rangle\,; Fx''\big) \otimes \bigotimes_{j=1}^n \mathsf{N}\big(F\langle x_j \rangle\,; Fx_j'\big) \\ {\scriptstyle\gamma}\big\downarrow & & \big\downarrow{\scriptstyle\gamma} \\ \mathsf{M}\big(\langle x \rangle\,; x''\big) & \xrightarrow{\quad F \quad} & \mathsf{N}\big(F\langle x \rangle\,; Fx''\big) \end{array} \tag{3.22}$$

This finishes the definition of a V-multifunctor.

Moreover, we define the following.

- For another V-multifunctor $G : \mathsf{N} \longrightarrow \mathsf{P}$ between V-multicategories, where P has object class $\mathsf{Ob}\,\mathsf{P}$, the *composition* $GF : \mathsf{M} \longrightarrow \mathsf{P}$ is the V-multifunctor defined by composing the object assignments

$$\mathsf{Ob}\,\mathsf{M} \xrightarrow{\;F\;} \mathsf{Ob}\,\mathsf{N} \xrightarrow{\;G\;} \mathsf{Ob}\,\mathsf{P}$$

and the morphisms on n-ary multimorphism objects

$$\mathsf{M}(\langle x\rangle; x') \xrightarrow{F} \mathsf{N}(F\langle x\rangle; Fx') \xrightarrow{G} \mathsf{P}(GF\langle x\rangle; GFx').$$

- The *identity* V-*multifunctor* $1_{\mathsf{M}} : \mathsf{M} \longrightarrow \mathsf{M}$ is defined by the identity assignment on objects and the identity morphisms on n-ary multimorphism objects.
- A V-*operad morphism* is a V-multifunctor between two V-multicategories with one object.
- A *non-symmetric* V-*multifunctor* $F : \mathsf{M} \longrightarrow \mathsf{N}$ between non-symmetric V-multicategories M and N is defined in the same way as a V-multifunctor but without the symmetric group action axiom (3.20). Composition and identities are defined as above.
- A *(non-symmetric) multifunctor* is a (non-symmetric) Set-multifunctor.

This finishes the definition. ◇

Definition 3.23. For V-multifunctors $F, G : \mathsf{M} \longrightarrow \mathsf{N}$, a V-*multinatural transformation* $\theta : F \longrightarrow G$ consists of, for each object x in M, a component morphism in V

$$1 \xrightarrow{\theta_x} \mathsf{N}(Fx; Gx)$$

such that the following V-*naturality diagram* in V commutes for each $(\langle x\rangle; x') \in \mathsf{Prof}(\mathsf{M}) \times \mathsf{Ob}\,\mathsf{M}$ with $\langle x\rangle = \langle x_j\rangle_{j=1}^n$.

$$\begin{array}{ccc}
1 \otimes \mathsf{M}(\langle x\rangle; x') & \xrightarrow{\theta_{x'} \otimes F} & \mathsf{N}(Fx'; Gx') \otimes \mathsf{N}(F\langle x\rangle; Fx') \\
\lambda^{-1} \nearrow & & \searrow \gamma \\
\mathsf{M}(\langle x\rangle; x') & & \mathsf{N}(F\langle x\rangle; Gx') \\
\rho^{-1} \searrow & & \nearrow \gamma \\
\mathsf{M}(\langle x\rangle; x') \otimes \bigotimes_{j=1}^n 1 & \xrightarrow{G \otimes \bigotimes_{j=1}^n \theta_{x_j}} & \mathsf{N}(G\langle x\rangle; Gx') \otimes \bigotimes_{j=1}^n \mathsf{N}(Fx_j; Gx_j)
\end{array} \tag{3.24}$$

This finishes the definition of a V-multinatural transformation.

Moreover, we define the following.

- The *identity* V-*multinatural transformation* $1_F : F \longrightarrow F$ has components

$$(1_F)_x = 1_{Fx} \quad \text{for} \quad x \in \mathsf{Ob}\,\mathsf{M}.$$

- A *multinatural transformation* is a Set-multinatural transformation.
- A *non-symmetric* V-*multinatural transformation* $\theta : F \longrightarrow G$ between non-symmetric V-multifunctors $F, G : \mathsf{M} \longrightarrow \mathsf{N}$ is defined as above. ◇

Definition 3.25. Suppose $\theta : F \longrightarrow G$ is a V-multinatural transformation between V-multifunctors as in Definition 3.23.

- Suppose $\psi : G \longrightarrow H$ is a V-multinatural transformation with $H : \mathsf{M} \longrightarrow \mathsf{N}$ a V-multifunctor. The *vertical composition*

$$F \xrightarrow{\psi\theta} H \tag{3.26}$$

  is the V-multinatural transformation with component at each $x \in \mathsf{Ob}\,\mathsf{M}$ given by the following composite in V.

$$\begin{array}{ccc} 1 & \xrightarrow{(\psi\theta)_x} & \mathsf{N}(Fx\,;Hx) \\ {\scriptstyle \lambda^{-1}}\downarrow & & \uparrow{\scriptstyle \gamma} \\ 1 \otimes 1 & \xrightarrow{\psi_x \otimes \theta_x} & \mathsf{N}(Gx\,;Hx) \otimes \mathsf{N}(Fx\,;Gx) \end{array}$$

- Suppose $F', G' : \mathsf{N} \longrightarrow \mathsf{P}$ are V-multifunctors, and $\theta' : F' \longrightarrow G'$ is a V-multinatural transformation. The *horizontal composition*

$$F'F \xrightarrow{\theta' * \theta} G'G \tag{3.27}$$

  is the V-multinatural transformation with component at each $x \in \mathsf{Ob}\,\mathsf{M}$ given by the following composite in V.

$$\begin{array}{ccc} 1 & \xrightarrow{(\theta' * \theta)_x} & \mathsf{P}(F'Fx\,;G'Gx) \\ & & \uparrow{\scriptstyle \gamma} \\ {\scriptstyle \lambda^{-1}}\Big\downarrow & & \mathsf{P}(F'Gx\,;G'Gx) \otimes \mathsf{P}(F'Fx\,;F'Gx) \\ & & \uparrow{\scriptstyle 1 \otimes F'} \\ 1 \otimes 1 & \xrightarrow{\theta'_{Gx} \otimes \theta_x} & \mathsf{P}(F'Gx\,;G'Gx) \otimes \mathsf{N}(Fx\,;Gx) \end{array}$$

Vertical and horizontal compositions of non-symmetric V-multinatural transformations are defined as above. This finishes the definition. ◇

The 2-Category Structure

Theorem 3.28. *For each symmetric monoidal category* V*, there is a 2-category*

$$\mathsf{V\text{-}Multicat}$$

consisting of the following data.

- *Its objects are small* V*-multicategories.*
- *For small* V*-multicategories* M *and* N*, the hom category*

$$\mathsf{V}\text{-}\mathsf{Multicat}(\mathsf{M},\mathsf{N})$$

  is defined as follows.
  - *Its objects are* V*-multifunctors* $\mathsf{M} \longrightarrow \mathsf{N}$.
  - *Its morphisms are* V*-multinatural transformations.*
  - *Composition is vertical composition of* V*-multinatural transformations.*
  - *Identity morphisms are identity* V*-multinatural transformations.*
- *The identity 1-cell* 1_{M} *is the identity* V*-multifunctor* 1_{M}.
- *Horizontal composition of 1-cells is the composition of* V*-multifunctors.*
- *Horizontal composition of 2-cells is that of* V*-multinatural transformations.*

Moreover, there is a 2-category $\mathsf{V}\text{-}\mathsf{Multicat}^{\mathsf{ns}}$ *consisting of the following data.*

- *Its objects are small non-symmetric* V*-multicategories.*
- *Its 1-cells are non-symmetric* V*-multifunctors.*
- *Its 2-cells are non-symmetric* V*-multinatural transformations.*
- *The rest of the 2-category structure is given as in* $\mathsf{V}\text{-}\mathsf{Multicat}$.

Proof. The special case $\mathsf{V} = \mathsf{Set}$ is [JY21, 2.4.26]. That proof can be adapted to prove the V-enriched version essentially without change. The proof for the non-symmetric variant is obtained by omitting the parts involving the symmetric group action. □

Recall from Definition 1.32 the *underlying 1-category* of a 2-category. Next is the V-enriched analogue of [JY∞, 5.5.14], whose proof naturally extends to the V-enriched setting.

Theorem 3.29. *Suppose* V *is a complete and cocomplete symmetric monoidal closed category. Then the underlying 1-categories of the 2-categories*

$$\mathsf{V}\text{-}\mathsf{Multicat} \quad \textit{and} \quad \mathsf{V}\text{-}\mathsf{Multicat}^{\mathsf{ns}}$$

are complete and cocomplete.

Example 3.30 (Initial and Terminal Objects). The initial V-multicategory has an empty set of objects. If V has a terminal object $\top$, then a terminal V-multicategory T has a single object $*$ and n-ary multimorphism object

$$\mathsf{T}_n = \mathsf{T}(\overbrace{*,\ldots,*}^{n \text{ terms}};*) = \top$$

for each $n \geq 0$. ◇

Categorically-Enriched Multicategories

Recall the symmetric monoidal closed category $(\mathsf{Cat}, \times, \mathbf{1})$ of small categories and functors with the Cartesian product (Example 1.16).

Definition 3.31. Suppose $\langle x\rangle = \langle x_j\rangle_{j=1}^n, y, z$ are objects in a Cat-multicategory $(\mathsf{M}, \gamma, 1)$.

- The category $\mathsf{M}(\langle x\rangle; y)$ is called a *multimorphism category.*
- An object in the multimorphism category $\mathsf{M}(\langle x\rangle; y)$ is called an *n-ary 1-cell* and is denoted $\langle x\rangle \longrightarrow y$.
- A 1-ary 1-cell $f : y \longrightarrow z$ is an *isomorphism* if there exists a 1-ary 1-cell $g : z \longrightarrow y$ such that

  $$\gamma(g; f) = 1_y \quad \text{and} \quad \gamma(f; g) = 1_z.$$

  In this case, we say that f is *invertible* with *inverse* g and denote it by $y \xrightarrow{\cong} z$.
- A Cat-multifunctor $F : \mathsf{M} \longrightarrow \mathsf{N}$ is *essentially surjective* on objects if, for each object $w \in \mathsf{N}$, there exist
  - an object $v \in \mathsf{M}$ and
  - an isomorphism $h : Fv \xrightarrow{\cong} w$ in N.
- A morphism $\theta : f \longrightarrow g$ in the multimorphism category $\mathsf{M}(\langle x\rangle; y)$ is called an *n-ary 2-cell.* It is also denoted by

  $$\langle x\rangle \underset{g}{\overset{f}{\rightrightarrows}} y \quad \Downarrow\theta \qquad (3.32)$$

  using the 2-cell notation in (1.23).

We use the same terminology and notation for non-symmetric Cat-multicategories and Cat-multifunctors. ◇

Explanation 3.33 (Cat-Multinatural Transformations)**.** Suppose M and N are Cat-multicategories, and

$$F, G : \mathsf{M} \longrightarrow \mathsf{N}$$

are Cat-multifunctors. A Cat-*multinatural transformation* $\theta : F \longrightarrow G$ consists of a component 1-ary 1-cell

$$Fc \xrightarrow{\theta_c} Gc \qquad \text{in } \mathsf{N} \text{ for each } c \in \operatorname{Ob}\mathsf{M} \qquad (3.34)$$

such that the following two Cat-*naturality conditions* hold:

Objects For each k-ary 1-cell $p : \langle c\rangle \longrightarrow c'$ in M with $\langle c\rangle = \langle c_j\rangle_{j=1}^k$, denote by

$$\begin{cases} F\langle c\rangle = \langle Fc_j\rangle_{j=1}^k \in (\mathsf{Ob}\,\mathsf{N})^k \\ \theta_{\langle c\rangle} = \langle \theta_{c_j}\rangle_{j=1}^k \in \prod_{j=1}^k \mathsf{N}\big(Fc_j\,;\,Gc_j\big). \end{cases} \tag{3.35}$$

Then the following k-ary 1-cell equality holds, with composition taken in the Cat-multicategory N:

$$\gamma\big(Gp\,;\,\theta_{\langle c\rangle}\big) = \gamma\big(\theta_{c'}\,;\,Fp\big) \quad \text{in} \quad \mathsf{N}\big(F\langle c\rangle\,;\,Gc'\big). \tag{3.36}$$

Morphisms For each k-ary 2-cell $f : p \longrightarrow q$ in $\mathsf{M}\big(\langle c\rangle\,;\,c'\big)$, the following k-ary 2-cell equality holds, with $1_{\theta_{\langle c\rangle}} = \langle 1_{\theta_{c_j}}\rangle_{j=1}^k$:

$$\gamma\big(Gf\,;\,1_{\theta_{\langle c\rangle}}\big) = \gamma\big(1_{\theta_{c'}}\,;\,Ff\big) \quad \text{in} \quad \mathsf{N}\big(F\langle c\rangle\,;\,Gc'\big). \tag{3.37}$$

The conditions (3.36) and (3.37) together constitute the V-naturality condition (3.24) when V is Cat.

The two sides of the object Cat-naturality condition (3.36) use the following two compositions in N, applied to objects.

$$\begin{array}{ccc} \mathsf{N}\big(G\langle c\rangle\,;\,Gc'\big) \times \prod_{j=1}^k \mathsf{N}\big(Fc_j\,;\,Gc_j\big) & & \mathsf{N}\big(Fc'\,;\,Gc'\big) \times \mathsf{N}\big(F\langle c\rangle\,;\,Fc'\big) \\ & \searrow^{\gamma} \quad {}^{\gamma}\swarrow & \\ & \mathsf{N}\big(F\langle c\rangle\,;\,Gc'\big) & \end{array} \tag{3.38}$$

The object Cat-naturality condition (3.36) is the commutative diagram

$$\begin{array}{ccc} F\langle c\rangle & \xrightarrow{\theta_{\langle c\rangle}} & G\langle c\rangle \\ {\scriptstyle Fp}\downarrow & & \downarrow{\scriptstyle Gp} \\ Fc' & \xrightarrow{\theta_{c'}} & Gc' \end{array} \tag{3.39}$$

with the understanding that it is a k-ary 1-cell equality involving the compositions (3.38) in N.

Similarly, the two sides of the morphism Cat-naturality condition (3.37) use the compositions in (3.38), applied to morphisms. The condition (3.37) is the equality of multicategorical pasting diagrams

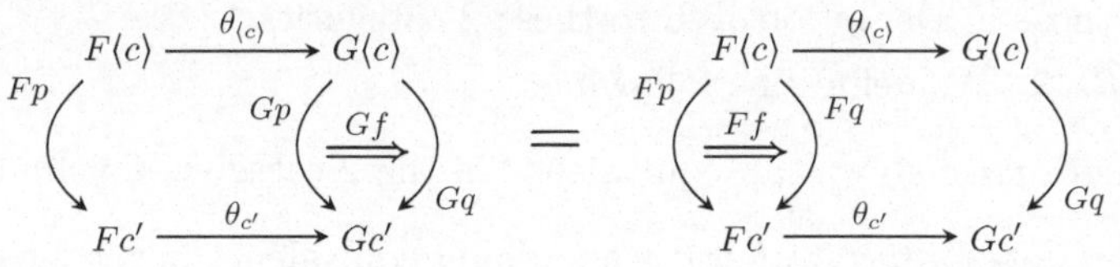

(3.40)

with the understanding that it is a k-ary 2-cell equality involving the composition in N.

By the object Cat-naturality condition (3.36) for $\theta' : F' \longrightarrow G'$, there are two equal ways to write the c-component of the horizontal composite for each object c in M:

$$\begin{aligned}(\theta' * \theta)_c &= \gamma\big(\theta'_{Gc}\,;\, F'\theta_c\big)\\ &= \gamma\big(G'\theta_c\,;\, \theta'_{Fc}\big).\end{aligned} \tag{3.41}$$

A Cat-multinatural transformation $\theta : F \longrightarrow G$ is called a Cat-*multinatural isomorphism* if θ is invertible with respect to the vertical composition. This happens if and only if each component 1-ary 1-cell θ_c (3.34) is invertible with respect to the composition in N. In other words, θ is a Cat-multinatural isomorphism if and only if each component 1-ary 1-cell θ_c is an isomorphism in the underlying 1-ary 1-category of N (Definition 1.32, Proposition 1.35, and Example 3.18).

A non-symmetric Cat-multinatural transformation or isomorphism admits the same description as above. ◇

3.3 Categorically-Enriched Multiequivalences

In this section we show that equivalences in the 2-categories Cat-Multicat and Cat-Multicat^{ns} are

1. essentially surjective on objects and
2. isomorphisms between multimorphism categories.

See Lemma 3.45. In Section 3.4 we show that equivalences are actually characterized by these two properties.

Recall from Theorem 3.28 the 2-category Cat-Multicat with

- small Cat-multicategories as objects,
- Cat-multifunctors as 1-cells, and
- Cat-multinatural transformations as 2-cells.

It has a non-symmetric variant Cat-Multicat^{ns}. The following notions are restrictions of those in Definition 1.46 to these 2-categories.

Definition 3.42. We define the following.

- A Cat-*multiequivalence* is an equivalence in the 2-category Cat-Multicat.
- An *adjoint* Cat-*multiequivalence* is an adjoint equivalence in the 2-category Cat-Multicat.

Moreover, their non-symmetric variants are defined in the same way in the 2-category $\mathsf{Cat}\text{-}\mathsf{Multicat}^{\mathsf{ns}}$. ◇

Explanation 3.43. Adjoint Cat-multiequivalences can be defined directly without mentioning the 2-category Cat-Multicat. In more detail, a Cat-multifunctor $F : \mathsf{M} \longrightarrow \mathsf{N}$ is a Cat-*multiequivalence* if there exist

- a Cat-multifunctor $G : \mathsf{N} \longrightarrow \mathsf{M}$ and
- Cat-multinatural isomorphisms (Explanation 3.33)

$$1_{\mathsf{M}} \xrightarrow[\cong]{\eta} GF \quad \text{and} \quad FG \xrightarrow[\cong]{\varepsilon} 1_{\mathsf{N}}. \tag{3.44}$$

The quadruple $(F, G, \eta, \varepsilon)$ is, furthermore, an *adjoint* Cat-*multiequivalence* if it satisfies the two triangle identities (1.47), which involve vertical and horizontal compositions of Cat-multinatural transformations. The same remarks apply to the non-symmetric variants. In other words, in Definition 3.42 it is *not* necessary to assume that the Cat-multicategories M and N are small. ◇

Recall from Definition 3.31 that n-ary 1-cells and n-ary 2-cells are, respectively, the objects and morphisms in the n-ary multimorphism categories.

Lemma 3.45. *Suppose*

$$\mathsf{N} \xrightarrow{G} \mathsf{M}$$

is a Cat-*multiequivalence. Then G is*

- *essentially surjective on objects and*
- *an isomorphism on each multimorphism category.*

Moreover, the non-symmetric variant is also true.

Proof. Since G is a Cat-multiequivalence, by definition there exist

- a Cat-multifunctor $F : \mathsf{M} \longrightarrow \mathsf{N}$ and
- Cat-multinatural isomorphisms $\eta : 1_{\mathsf{M}} \xrightarrow{\cong} GF$ and $\varepsilon : FG \xrightarrow{\cong} 1_{\mathsf{N}}$ as in (3.44).

We consider essential surjectivity first and then the invertibility of the multimorphism functors of G.

Step 1: Essential Surjectivity on Objects

For each object $x \in \mathsf{M}$, there is an invertible component 1-ary 1-cell

$$x \xrightarrow[\cong]{\eta_x} G(Fx) \quad \text{in} \quad \mathsf{M}(x\,; GFx).$$

This shows that G is essentially surjective on the underlying 1-ary 1-categories.

Step 2: Injectivity on n-Ary 1-Cells

Suppose given an n-tuple of objects $\langle x\rangle = \langle x_j\rangle_{j=1}^n$ with $n \geq 0$ and an object y in N. We want to show that the multimorphism functor

$$\mathsf{N}\big(\langle x\rangle\,;y\big) \xrightarrow{\;G\;} \mathsf{M}\big(G\langle x\rangle\,;Gy\big) \tag{3.46}$$

is injective on objects. For each n-ary 1-cell $f : \langle x\rangle \longrightarrow y$ in N, the following computation shows that f is uniquely determined by Gf, with γ denoting the composition (3.5) in N.

$$\begin{cases} f = \gamma\big(f\,;\langle 1_{x_j}\rangle\big) & \text{by right unity (3.8)}\\ = \gamma\big(f\,;\big\langle\gamma(\varepsilon_{x_j}\,;\varepsilon_{x_j}^{-1})\big\rangle\big) & \text{by invertibility of } \varepsilon\\ = \gamma\big(\gamma(f\,;\varepsilon_{\langle x\rangle})\,;\varepsilon_{\langle x\rangle}^{-1}\big) & \text{by associativity (3.7)}\\ = \gamma\big(\gamma(\varepsilon_y\,;FGf)\,;\varepsilon_{\langle x\rangle}^{-1}\big) & \text{by object } \mathsf{Cat}\text{-naturality (3.36)}\end{cases} \tag{3.47}$$

This shows that the multimorphism functor G in (3.46) is injective on objects.

The computation (3.47) with ε replaced by

$$\eta^{-1} : GF \xrightarrow{\;\cong\;} 1_{\mathsf{M}}$$

also shows that the Cat-multiequivalence $F : \mathsf{M} \longrightarrow \mathsf{N}$ is injective on n-ary 1-cells.

Step 3: Injectivity on n-Ary 2-Cells

Suppose $\theta : f \longrightarrow f'$ is an n-ary 2-cell in $\mathsf{N}\big(\langle x\rangle\,;y\big)$. The computation (3.47)—with f, ε, and (3.36) replaced by, respectively, θ, 1_ε, and morphism Cat-naturality (3.37)—shows that θ is uniquely determined by $G\theta$. This shows that the multimorphism functor G in (3.46) is injective on morphisms. With 1_ε replaced by $1_{\eta^{-1}}$, the same computation also shows that the Cat-multiequivalence $F : \mathsf{M} \longrightarrow \mathsf{N}$ is injective on n-ary 2-cells.

Step 4: Surjectivity on n-Ary 1-Cells

In the context of (3.46), suppose given an n-ary 1-cell

$$h : G\langle x\rangle \longrightarrow Gy \quad \text{in} \quad \mathsf{M}.$$

We must show that there exists an n-ary 1-cell $f : \langle x\rangle \longrightarrow y$ in N such that $Gf = h$. We define the n-ary 1-cell

$$f = \gamma\big(\gamma(\varepsilon_y\,;Fh)\,;\varepsilon_{\langle x\rangle}^{-1}\big) \quad \text{in} \quad \mathsf{N}\big(\langle x\rangle\,;y\big),$$

which we visualize as the long composite below.

$$\begin{array}{ccc} FG\langle x\rangle & \xleftarrow[\cong]{\;\varepsilon_{\langle x\rangle}^{-1}\;} & \langle x\rangle \\ {\scriptstyle Fh}\big\downarrow & & \big\downarrow{\scriptstyle f} \\ FGy & \xrightarrow[\cong]{\;\varepsilon_y\;} & y \end{array}$$

Similar to (3.47), we compute as follows.

$$\begin{cases} Fh = \gamma\big(\varepsilon_y^{-1}\,;\, \gamma(f\,;\, \varepsilon_{\langle x\rangle})\big) & \text{by (3.7) through (3.9)} \\ = \gamma\big(\varepsilon_y^{-1}\,;\, \gamma(\varepsilon_y\,;\, FGf)\big) & \text{by (3.36)} \\ = FGf & \text{by (3.7) and (3.9)} \end{cases} \tag{3.48}$$

The injectivity of F on n-ary 1-cells established above and (3.48) imply the desired equality, $Gf = h$.

Step 5: Surjectivity on n-Ary 2-Cells

In the context of (3.46), suppose given an n-ary 2-cell

$$h \xrightarrow{\;\varphi\;} h' \quad \text{in} \quad \mathsf{M}\big(G\langle x\rangle\,;\, Gy\big).$$

We must show that there exists an n-ary 2-cell $\theta : f \longrightarrow f'$ in $\mathsf{N}(\langle x\rangle\,;\, y)$ such that $G\theta = \varphi$. We define the n-ary 2-cell

$$\theta = \gamma\big(\gamma(1_{\varepsilon_y}\,;\, F\varphi)\,;\, 1_{\varepsilon^{-1}_{\langle x\rangle}}\big) \quad \text{in} \quad \mathsf{N}\big(\langle x\rangle\,;\, y\big),$$

which we visualize as the long composite below.

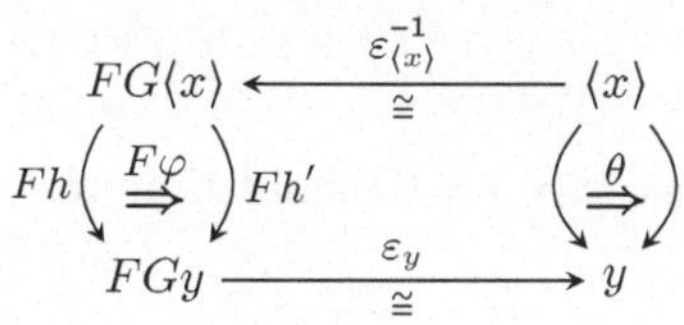

The computation (3.48)—with h, f, ε, and (3.36) replaced by, respectively, φ, θ, 1_ε, and morphism Cat-naturality (3.37)—shows that

$$F\varphi = FG\theta.$$

The injectivity of F on n-ary 2-cells established above now implies the desired equality, $G\theta = \varphi$. This finishes the proof that each multimorphism functor (3.46) of G is an isomorphism.

For the non-symmetric variant, we reuse the proof above by noting that the symmetric group action is not used. □

3.4 Characterization of Cat-Multiequivalences

In Lemma 3.45 we observe that (non-symmetric) Cat-multiequivalences are essentially surjective on objects and isomorphisms on multimorphism categories. In this section we prove that these properties precisely characterize (non-symmetric) Cat-multiequivalences; see Theorem 3.69. In Theorem 8.76 we apply Theorem 3.69 to show that the Grothendieck construction on bipermutative-indexed categories is a non-symmetric Cat-multiequivalence.

Here is an outline of this section.

1. For a Cat-multifunctor G that satisfies the conclusion of Lemma 3.45, we construct the left adjoint F and the unit $\eta : 1 \longrightarrow GF$ in Definition 3.49.
   - Lemmas 3.55 and 3.56 show that F is a Cat-multifunctor, which is non-symmetric if G is.
   - Lemma 3.61 shows that η is a Cat-multinatural isomorphism, which is non-symmetric if G is.
2. The counit $\varepsilon : FG \longrightarrow 1$ is constructed in Definition 3.63.
   - Lemma 3.66 shows that ε is a Cat-multinatural isomorphism, which is non-symmetric if G is.
   - Lemma 3.68 verifies the triangle identities for the quadruple $(F, G, \eta, \varepsilon)$, providing the last part of the proof of Theorem 3.69.
3. Theorem 3.70 is the 1-ary special case that characterizes 2-equivalences between 2-categories.

Left Adjoint and Unit

Definition 3.49. Suppose given a Cat-multifunctor

$$\mathsf{N} \xrightarrow{G} \mathsf{M}$$

between Cat-multicategories that is

- essentially surjective on objects and
- an isomorphism on each multimorphism category.

Define the data of

- a Cat-multifunctor $F : \mathsf{M} \longrightarrow \mathsf{N}$ and
- a Cat-multinatural isomorphism $\eta : 1_{\mathsf{M}} \xrightarrow{\cong} GF$

as follows.

F on objects and components of η Given an object $x \in \mathsf{M}$, by the essential surjectivity of G and the Axiom of Choice, we choose an object $Fx \in \mathsf{N}$ and an isomorphism

$$x \xrightarrow[\cong]{\eta_x} G(Fx) \quad \text{in} \quad \mathsf{M}(x; GFx). \tag{3.50}$$

These choices define the object assignment, $x \longmapsto Fx$, of F and the component 1-ary 1-cells of η.

F on n-ary 1-cells Given an n-ary 1-cell $f : \langle x \rangle \longrightarrow y$ in M with $\langle x \rangle = \langle x_j \rangle_{j=1}^{n}$, consider the n-ary 1-cell

$$\gamma\big(\eta_y\,;\gamma\big(f\,;\eta_{\langle x \rangle}^{-1}\big)\big) \quad \text{in} \quad \mathsf{M}\big(GF\langle x \rangle\,; GFy\big),$$

which we visualize as the long composite below.

$$\begin{array}{ccc} \langle x \rangle & \xleftarrow[\cong]{\eta_{\langle x \rangle}^{-1}} & GF\langle x \rangle \\ {\scriptstyle f}\downarrow & & \downarrow{\scriptstyle G(Ff)} \\ y & \xrightarrow[\cong]{\eta_y} & GFy \end{array}$$

Since the multimorphism functor

$$\mathsf{N}\big(F\langle x \rangle\,; Fy\big) \xrightarrow[\cong]{G} \mathsf{M}\big(GF\langle x \rangle\,; GFy\big) \tag{3.51}$$

is an isomorphism, there exists a unique n-ary 1-cell

$$F\langle x \rangle \xrightarrow{Ff} Fy \quad \text{in} \quad \mathsf{N}\big(F\langle x \rangle\,; Fy\big)$$

such that

$$G(Ff) = \gamma\big(\eta_y\,;\gamma\big(f\,;\eta_{\langle x \rangle}^{-1}\big)\big). \tag{3.52}$$

This defines the n-ary 1-cell assignment, $f \longmapsto Ff$, of F.

F on n-ary 2-cells Given an n-ary 2-cell $\theta : f \longrightarrow f'$ in $\mathsf{M}(\langle x \rangle\,; y)$, consider the n-ary 2-cell

$$\gamma\Big(1_{\eta_y}\,;\gamma\Big(\theta\,;1_{\eta_{\langle x \rangle}^{-1}}\Big)\Big) : GFf \longrightarrow GFf' \quad \text{in} \quad \mathsf{M}\big(GF\langle x \rangle\,; GFy\big),$$

which we visualize as the long composite below.

$$\begin{array}{ccc} \langle x \rangle & \xleftarrow[\cong]{\eta_{\langle x \rangle}^{-1}} & GF\langle x \rangle \\ f \;\overset{\theta}{\Rightarrow}\; f' & & GFf \;\overset{G(F\theta)}{\Rightarrow}\; GFf' \\ y & \xrightarrow[\cong]{\eta_y} & GFy \end{array}$$

By the isomorphism G in (3.51), there exists a unique n-ary 2-cell

$$Ff \xrightarrow{F\theta} Ff' \quad \text{in} \quad \mathsf{N}\big(F\langle x \rangle\,; Fy\big)$$

such that

$$G(F\theta) = \gamma\Big(1_{\eta_y}\,;\gamma\Big(\theta\,;1_{\eta_{\langle x \rangle}^{-1}}\Big)\Big). \tag{3.53}$$

This defines the n-ary 2-cell assignment, $\theta \longmapsto F\theta$, of F.

This finishes the definitions of F and η. Moreover, we make the same definitions in the non-symmetric case with G, F, and η non-symmetric. ◇

Explanation 3.54. Consider Definition 3.49.

1. This definition still makes sense even if $G : \mathsf{N} \longrightarrow \mathsf{M}$ is a non-symmetric Cat-multifunctor between non-symmetric Cat-multicategories because the symmetric group action on M and N and the symmetric group action axiom for G are not used.

2. The definitions (3.52) and (3.53) are modeled after, respectively, the object and morphism Cat-naturality conditions, (3.36) and (3.37), for $\eta : 1_{\mathsf{M}} \longrightarrow GF$. Once we prove that F is a (non-symmetric) Cat-multifunctor, these Cat-naturality conditions for η will then follow. ◇

To show that $F : \mathsf{M} \longrightarrow \mathsf{N}$ is a (non-symmetric) Cat-multifunctor, first we check its functoriality between multimorphism categories. The following lemma holds in both the symmetric and the non-symmetric cases.

Lemma 3.55. *In the context of Definition 3.49, for objects $\langle x\rangle = \langle x_j\rangle_{j=1}^n$ and y in M, the assignments*

$$\mathsf{M}(\langle x\rangle ; y) \xrightarrow{F} \mathsf{N}(F\langle x\rangle ; Fy)$$

form a functor.

Proof. Suppose $f : \langle x\rangle \longrightarrow y$ is an n-ary 1-cell in M. To show that F preserves the identity n-ary 2-cell $1_f : f \longrightarrow f$, we compute as follows.

$$\begin{aligned} GF1_f &= \gamma\big(1_{\eta_y} ; \gamma(1_f ; 1_{\eta^{-1}_{\langle x\rangle}})\big) && \text{by (3.53)}\\ &= 1_{\gamma(\eta_y ; \gamma(f ; \eta^{-1}_{\langle x\rangle}))} && \text{by functoriality of } \gamma \text{ in } \mathsf{M}\\ &= 1_{GFf} && \text{by (3.52)}\\ &= G1_{Ff} && \text{by functoriality of } G \text{ in (3.51)} \end{aligned}$$

By the injectivity of G in (3.51) on n-ary 2-cells, it follows that

$$F1_f = 1_{Ff} \quad \text{in} \quad \mathsf{N}\big(F\langle x\rangle ; Fy\big).$$

This shows that F preserves identity n-ary 2-cells.

To show that F preserves composition of n-ary 2-cells

$$f \xrightarrow{\theta} f' \xrightarrow{\theta'} f'' \quad \text{in} \quad \mathsf{M}\big(\langle x\rangle ; y\big),$$

we compute as follows.

$$\begin{aligned} &GF(\theta' \circ \theta)\\ &= \gamma\big(1_{\eta_y} ; \gamma(\theta' \circ \theta ; 1_{\eta^{-1}_{\langle x\rangle}})\big) && \text{by (3.53)}\\ &= \Big[\gamma\big(1_{\eta_y} ; \gamma(\theta' ; 1_{\eta^{-1}_{\langle x\rangle}})\big)\Big] \circ \Big[\gamma\big(1_{\eta_y} ; \gamma(\theta ; 1_{\eta^{-1}_{\langle x\rangle}})\big)\Big] && \text{by functoriality of } \gamma\\ &= (GF\theta') \circ (GF\theta) && \text{by (3.53)}\\ &= G\big(F\theta' \circ F\theta\big) && \text{by functoriality of } G \end{aligned}$$

By the injectivity of G in (3.51) on n-ary 2-cells, it follows that

$$F(\theta' \circ \theta) = F\theta' \circ F\theta \quad \text{in} \quad \mathsf{N}\big(F\langle x \rangle\,;\, Fy\big).$$

This proves that F is a functor on each multimorphism category. □

Lemma 3.56. *In the context of Definition 3.49,*

$$\mathsf{M} \xrightarrow{\;F\;} \mathsf{N}$$

is a Cat*-multifunctor. If G is a non-symmetric* Cat*-multifunctor, then F is a non-symmetric* Cat*-multifunctor.*

Proof. We must show that F preserves the colored units, composition, and, in the symmetric case only, symmetric group action as in Definition 3.19.

Step 1: Colored Units

For each object $x \in \mathsf{M}$, its colored unit $1_x : x \longrightarrow x$ is a 1-ary 1-cell in M. We compute as follows.

$$\begin{aligned} GF1_x &= \gamma\big(\eta_x\,;\, \gamma(1_x\,;\, \eta_x^{-1})\big) && \text{by (3.52)} \\ &= \gamma\big(\eta_x\,;\, \eta_x^{-1}\big) && \text{by left unity (3.9)} \\ &= 1_{GFx} && \text{by (3.50)} \\ &= G1_{Fx} && \text{by (3.21) for } G \end{aligned}$$

The injectivity of G in (3.51) on 1-ary 1-cells implies that

$$F1_x = 1_{Fx} \quad \text{in} \quad \mathsf{N}\big(Fx\,;\, Fx\big).$$

This shows that F preserves colored units in the sense of (3.21).

Step 2: Composition

To show that F preserves composition as in (3.22), consider objects

$$y, \quad \langle x \rangle = \langle x_j \rangle_{j=1}^{n}, \quad \text{and} \quad \langle w_j \rangle = \langle w_{ji} \rangle_{i=1}^{\ell_j}$$

in M for $j \in \{1, \ldots, n\}$ with $\langle w \rangle = \langle \langle w_j \rangle \rangle_{j=1}^{n}$. We must show that the following diagram is commutative.

$$\begin{array}{ccc} \mathsf{M}\big(\langle x \rangle\,;\, y\big) \times \prod_{j=1}^{n} \mathsf{M}\big(\langle w_j \rangle\,;\, x_j\big) & \xrightarrow{\;F \times \prod_j F\;} & \mathsf{N}\big(F\langle x \rangle\,;\, Fy\big) \times \prod_{j=1}^{n} \mathsf{N}\big(F\langle w_j \rangle\,;\, Fx_j\big) \\ \Big\downarrow{\scriptstyle \gamma} & & \Big\downarrow{\scriptstyle \gamma} \\ \mathsf{M}\big(\langle w \rangle\,;\, y\big) & \xrightarrow{\;F\;} & \mathsf{N}\big(F\langle w \rangle\,;\, Fy\big) \end{array} \tag{3.57}$$

Given

- an n-ary 1-cell $f : \langle x \rangle \longrightarrow y$ and
- an ℓ_j-ary 1-cell $f_j : \langle w_j \rangle \longrightarrow x_j$ for each $j \in \{1, \ldots, n\}$

in M, we compute as follows.

$$\begin{cases} \quad GF\gamma\big(f\,;\langle f_j\rangle\big) \\ = \gamma\Big(\eta_y\,;\gamma\big(\gamma(f\,;\langle f_j\rangle)\,;\eta^{-1}_{\langle w\rangle}\big)\Big) & \text{by (3.52)} \\ = \gamma\Big(\gamma\big(\eta_y\,;\gamma(f\,;\eta^{-1}_{\langle x\rangle})\big)\,;\big\langle\gamma\big(\eta_{x_j}\,;\gamma(f_j\,;\eta^{-1}_{\langle w_j\rangle})\big)\big\rangle\Big) & \text{by (3.7) and (3.8)} \\ = \gamma\big(GFf\,;GF\langle f_j\rangle\big) & \text{by (3.52)} \\ = G\gamma\big(Ff\,;F\langle f_j\rangle\big) & \text{by (3.22) for } G \end{cases} \tag{3.58}$$

The computation (3.58) and the injectivity of G on $(\ell_1 + \cdots + \ell_n)$-ary 1-cells imply

$$F\gamma\big(f\,;\langle f_j\rangle\big) = \gamma\big(Ff\,;F\langle f_j\rangle\big) \quad \text{in} \quad \mathsf{N}\big(F\langle w\rangle\,;Fy\big).$$

This proves the commutativity of (3.57) on objects.

Suppose given

- an n-ary 2-cell $\theta : f \longrightarrow f'$ in $\mathsf{M}(\langle x\rangle\,;y)$ and
- an ℓ_j-ary 2-cell $\theta_j : f_j \longrightarrow f'_j$ in $\mathsf{M}(\langle w_j\rangle\,;x_j)$ for each $j \in \{1,\ldots,n\}$.

Then

- the computation (3.58) with (f, f_j, η) and (3.52) replaced by, respectively, $(\theta, \theta_j, 1_\eta)$ and (3.53), and
- the injectivity of G on $(\ell_1 + \cdots + \ell_n)$-ary 2-cells

imply the commutativity of (3.57) on morphisms. This proves that F preserves composition.

The proof so far is still valid if G is a non-symmetric Cat-multifunctor. In this case, we have shown that F is a non-symmetric Cat-multifunctor.

Step 3: Symmetric Group Action

This case only applies if G is a Cat-multifunctor. With the notation above, F preserves the symmetric group action in the sense of (3.20) if, for each permutation $\sigma \in \Sigma_n$, the following diagram commutes.

$$\begin{array}{ccc} \mathsf{M}(\langle x\rangle\,;y) & \xrightarrow{\;F\;} & \mathsf{N}\big(F\langle x\rangle\,;Fy\big) \\ {\scriptstyle\sigma}\downarrow & & \downarrow{\scriptstyle\sigma} \\ \mathsf{M}(\langle x\rangle\sigma\,;y) & \xrightarrow{\;F\;} & \mathsf{N}\big(F\langle x\rangle\sigma\,;Fy\big) \end{array} \tag{3.59}$$

For an n-ary 1-cell $f : \langle x \rangle \longrightarrow y$ in M, we compute as follows.

$$\begin{cases} G\big((Ff)\sigma\big) & \\ = (GFf)\sigma & \text{by (3.20) for } G \\ = \gamma\big(\eta_y\,;\gamma(f\,;\eta^{-1}_{\langle x \rangle})\big)\sigma & \text{by (3.52)} \\ = \gamma\big(\eta_y\,;\gamma(f\,;\eta^{-1}_{\langle x \rangle})\sigma\big) & \text{by bottom eq. (3.12)} \\ = \gamma\big(\eta_y\,;\gamma(f\sigma\,;\eta^{-1}_{\langle x \rangle\sigma})\big) & \text{by top eq. (3.10)} \\ = GF(f\sigma) & \text{by (3.52)} \end{cases} \tag{3.60}$$

The injectivity of G on n-ary 1-cells implies

$$(Ff)\sigma = F(f\sigma).$$

This proves the commutativity of (3.59) on objects.

The commutativity of (3.59) on morphisms is proved using

- the computation (3.60) with f, η, and (3.52) replaced by, respectively, an n-ary 2-cell in $\mathsf{M}(\langle x \rangle\,; y)$, 1_η, and (3.53), and
- the injectivity of G on n-ary 2-cells.

This finishes the proof that F preserves the symmetric group action if G is a Cat-multifunctor. □

Recall from Explanation 3.33 the description of a (non-symmetric) Cat-multinatural transformation or isomorphism.

Lemma 3.61. *In the context of Definition 3.49,*

$$1_{\mathsf{M}} \xrightarrow{\ \eta\ } GF$$

is a Cat-*multinatural isomorphism. If G is a non-symmetric* Cat-*multifunctor, then η is a non-symmetric* Cat-*multinatural isomorphism.*

Proof. In Lemma 3.56 we showed that $F : \mathsf{M} \longrightarrow \mathsf{N}$ is a (non-symmetric) Cat-multifunctor. By definition (3.50) each component 1-ary 1-cell $\eta_x : x \longrightarrow GFx$ is an isomorphism in M. It remains to check the two Cat-naturality conditions, (3.36) and (3.37), for η.

For an n-ary 1-cell $f : \langle x \rangle \longrightarrow y$ in M with $\langle x \rangle = \langle x_j \rangle_{j=1}^n$, the following equalities prove the object Cat-naturality condition (3.36) for $\eta : 1_{\mathsf{M}} \longrightarrow GF$.

$$\begin{cases} \gamma\big(GFf\,;\eta_{\langle x \rangle}\big) & \\ = \gamma\Big(\gamma\big(\eta_y\,;\gamma(f\,;\eta^{-1}_{\langle x \rangle})\big)\,;\eta_{\langle x \rangle}\Big) & \text{by (3.52)} \\ = \gamma\Big(\eta_y\,;\gamma\big(f\,;\big\langle\gamma(\eta^{-1}_{x_j}\,;\eta_{x_j})\big\rangle\big)\Big) & \text{by (3.7)} \\ = \gamma\big(\eta_y\,;f\big) & \text{by (3.8)} \end{cases} \tag{3.62}$$

The morphism Cat-naturality condition (3.37) for η is proved using the computation (3.62) by replacing f, η, and (3.52) with, respectively, an n-ary 2-cell in $\mathsf{M}(\langle x \rangle\,; y)$, 1_η, and (3.53). □

Counit

Next we define the counit for (F, G). The formula (3.65) below comes from the right triangle identity (1.47) of an adjunction.

Definition 3.63. In the context of Definition 3.49, define the data of a Cat-multinatural isomorphism

$$FG \xrightarrow[\cong]{\varepsilon} 1_{\mathsf{N}}$$

as follows. For each object $z \in \mathsf{N}$, the (Gz)-component 1-ary 1-cell of η in M

$$Gz \xrightarrow[\cong]{\eta_{Gz}} GFGz$$

is an isomorphism (3.50), with inverse denoted by

$$GFGz \xrightarrow[\cong]{\eta_{Gz}^{-1}} Gz.$$

Since the multimorphism functor

$$\mathsf{N}\big(FGz\,;z\big) \xrightarrow[\cong]{G} \mathsf{M}\big(GFGz\,;Gz\big)$$

is an isomorphism, there exists a unique invertible 1-ary 1-cell in N

$$FGz \xrightarrow[\cong]{\varepsilon_z} z \tag{3.64}$$

such that

$$G\varepsilon_z = \eta_{Gz}^{-1}. \tag{3.65}$$

This defines the z-component of ε. ◇

Lemma 3.66. *In the context of Definitions 3.49 and 3.63,*

$$FG \xrightarrow{\varepsilon} 1_{\mathsf{N}}$$

is a Cat*-multinatural isomorphism. If G is a non-symmetric* Cat*-multifunctor, then ε is a non-symmetric* Cat*-multinatural isomorphism.*

Proof. Since each component of ε is an isomorphism (3.64), it suffices to check the two Cat-naturality conditions, (3.36) and (3.37), for ε. Suppose $g : \langle w \rangle \longrightarrow z$ is an n-ary 1-cell in N with $\langle w \rangle = \langle w_j \rangle_{j=1}^n$. The injectivity of G on n-ary 1-cells and the following equalities prove the object Cat-naturality condition (3.36) for $\varepsilon : FG \longrightarrow 1_{\mathsf{N}}$.

$$\left\{\begin{aligned}
&G\gamma\big(g\,;\varepsilon_{\langle w \rangle}\big) && \\
&= \gamma\big(Gg\,;G\varepsilon_{\langle w \rangle}\big) && \text{by (3.22) for } G\\
&= \gamma\big(Gg\,;\eta_{G\langle w \rangle}^{-1}\big) && \text{by (3.65)}\\
&= \gamma\big(\eta_{Gz}^{-1}\,;GFGg\big) && \text{by (3.7) through (3.9) and (3.62)}\\
&= \gamma\big(G\varepsilon_z\,;GFGg\big) && \text{by (3.65)}\\
&= G\gamma\big(\varepsilon_z\,;FGg\big) && \text{by (3.22) for } G
\end{aligned}\right. \tag{3.67}$$

The morphism Cat-naturality condition (3.37) for ε is proved using

- the injectivity of G on n-ary 2-cells and
- the computation (3.67) with g, η, ε, and (3.62) replaced by, respectively, an n-ary 2-cell in N, 1_η, 1_ε, and the morphism Cat-naturality condition (3.37) for η (Lemma 3.61).

This finishes the proof. □

Lemma 3.68. *In the context of Definitions 3.49 and 3.63, the triangle identities* (1.47) *hold for* $(F, G, \eta, \varepsilon)$.

Proof. The right triangle identity

$$\gamma\big(G\varepsilon_z\,;\,\eta_{Gz}\big) = 1_{Gz}$$

follows from the definition (3.65) of ε_z by applying $\gamma(-\,;\,\eta_{Gz})$ to both sides.

The left triangle identity follows from the following computation for each object $x \in \mathsf{M}$ and the injectivity of G on 1-ary 1-cells.

$$\begin{aligned}
&G\gamma\big(\varepsilon_{Fx}\,;\,F\eta_x\big) \\
&= \gamma\big(G\varepsilon_{Fx}\,;\,GF\eta_x\big) && \text{by (3.22) for } G \\
&= \gamma\Big(\eta^{-1}_{GFx}\,;\,\gamma\big(\eta_{GFx}\,;\,\gamma(\eta_x\,;\,\eta_x^{-1})\big)\Big) && \text{by (3.52) and (3.65)} \\
&= 1_{GFX} && \text{by (3.8)} \\
&= G1_{Fx} && \text{by (3.21) for } G
\end{aligned}$$

This finishes the proof. □

Cat-Multiequivalences

We are now ready for the main result of this section, which is a practical characterization of a Cat-multiequivalence (Definition 3.42).

Theorem 3.69 (Multiequivalence). *Suppose*

$$\mathsf{N} \xrightarrow{\;G\;} \mathsf{M}$$

is a Cat-*multifunctor between* Cat-*multicategories. Then the following three statements are equivalent.*

1. *G is the right adjoint of an adjoint* Cat-*multiequivalence* $(F, G, \eta, \varepsilon)$.
2. *G is a* Cat-*multiequivalence.*
3. *G is*
   - *essentially surjective on objects and*
   - *an isomorphism on each multimorphism category.*

Moreover, the non-symmetric variant is also true, where each of M, N, G, *and* $(F, G, \eta, \varepsilon)$ *is non-symmetric.*

Proof. We consider each implication as follows.

- (1) $\Rightarrow$ (2) holds by Definitions 1.46 and 3.42.
- (2) $\Rightarrow$ (3) holds by Lemma 3.45.
- (3) $\Rightarrow$ (1) holds by Lemmas 3.56, 3.61, 3.66, and 3.68.

This completes the proof. □

Recall from Definition 1.49 the notions of (adjoint) 2-equivalences. Restricting the proofs of the lemmas referenced in the proof of Theorem 3.69 to the 1-ary case recovers the following characterization of a 2-equivalence [JY21, 7.5.8].

Theorem 3.70 (2-equivalence)**.** *Suppose*

$$\mathsf{C} \xrightarrow{G} \mathsf{B}$$

is a 2-functor between 2-categories. Then the following three statements are equivalent.

1. *G is the right adjoint of an adjoint 2-equivalence $(F, G, \eta, \varepsilon)$.*
2. *G is a 2-equivalence.*
3. *G is*
   - *essentially surjective on objects and*
   - *an isomorphism on each hom category.*

In [JY21, 7.5.8] this theorem is obtained by restricting the proof of an analogous characterization of a biequivalence between bicategories [JY21, 7.4.1]. In contrast, here we obtain Theorem 3.70 by restricting the proof of Theorem 3.69, which is a characterization of a Cat-multiequivalence. See Question A.1 for a possible generalization involving a lax variant of a Cat-multicategory.

4

Pseudo Symmetry

In this chapter we define *pseudo symmetric* Cat-*multifunctors* (Definition 4.1) between Cat-multicategories. They are weaker variants of Cat-multifunctors that preserve the symmetric group action up to coherent isomorphisms. The corresponding notion of a *pseudo symmetric* Cat-*multinatural transformation* is in Definition 4.19. These notions are the 1-cells and 2-cells of a 2-category $\mathsf{Cat}\text{-}\mathsf{Multicat}^{\mathsf{ps}}$ with small Cat-multicategories as objects (Theorem 4.34). Theorem 4.64 characterizes pseudo symmetric Cat-multiequivalences in terms of essential surjectivity on objects and invertibility on multimorphism categories. This is the pseudo symmetric analogue of Theorem 3.69. Below is a summary of some of the definitions and results in this chapter.

- Pseudo symmetric Cat-multifunctors. Definition 4.1.
- Pseudo symmetric Cat-multinatural transformations. Definition 4.19.
- 2-category $\mathsf{Cat}\text{-}\mathsf{Multicat}^{\mathsf{ps}}$. Theorem 4.34.
- Pseudo symmetric Cat-multiequivalences. Definition 4.37.
- Characterization of pseudo symmetric Cat-multiequivalences. Theorem 4.64.

Pseudo symmetric Cat-multifunctors are utilized in subsequent chapters as follows.

The Grothendieck Construction and Inverse K-Theory

The prime example of a pseudo symmetric Cat-multifunctor is the Grothendieck construction

$$\mathcal{D}\mathsf{Cat} \xrightarrow{\int_{\mathcal{D}}} \mathsf{PermCat}^{\mathsf{sg}}$$

on any small tight bipermutative category $(\mathcal{D}, \oplus, \otimes)$ (Theorem 6.110).

- The domain Cat-multicategory $\mathcal{D}\mathsf{Cat}$ (Theorem 5.71) has additive symmetric monoidal functors $\mathcal{D}^{\oplus} \longrightarrow \mathsf{Cat}$ as objects. Its multimorphism categories (Definition 5.58) are defined using the full extent of the bipermutative structure of $\mathcal{D}$ and Laplaza's Coherence Theorem 2.90.

DOI: 10.1201/9781003451235-4

- The codomain Cat-multicategory $\mathsf{PermCat}^{\mathsf{sg}}$ has small permutative categories as objects (Theorem 6.31).

An important consequence is that inverse K-theory $\int_{\mathcal{A}} \circ A$, which involves the Grothendieck construction $\int_{\mathcal{A}}$ on the small tight bipermutative category $\mathcal{A}$ in Example 2.63, is a pseudo symmetric Cat-multifunctor but *not* a Cat-multifunctor (Theorem 10.12).

Taking the projection down to $\mathcal{D}$ into account, the Grothendieck construction above lifts to a non-symmetric Cat-multifunctor (7.65)

$$\mathcal{D}\mathsf{Cat} \xrightarrow{\int_{\mathcal{D}}} \mathsf{PFib}^{\mathcal{D}}.$$

The codomain $\mathsf{PFib}^{\mathcal{D}}$ is a non-symmetric Cat-multicategory with small permutative opfibrations over $\mathcal{D}$ as objects (Theorem 7.51). While these objects only use the additive structure $\mathcal{D}^{\oplus}$, the multimorphism categories of $\mathsf{PFib}^{\mathcal{D}}$ (Definition 7.34) make full use of the bipermutative structure of $\mathcal{D}$. The characterization of a Cat-multiequivalence in Theorem 3.69 is used to show that this lifted Grothendieck construction $\int_{\mathcal{D}}$ is a non-symmetric Cat-multiequivalence (Theorem 8.76).

Organization

In Section 4.1 we define *pseudo symmetric* Cat-*multifunctors*. These weaker variants of Cat-multifunctors are only non-strict with respect to the symmetric group action. They are required to *strictly* preserve the colored units and the Cat-multicategorical composition. We define them this way because the Grothendieck construction $\int_{\mathcal{D}}$ has precisely these properties (Theorem 6.110). See Explanation 4.35 and Question A.1 regarding further lax variants of Cat-multicategories and Cat-multifunctors.

In Section 4.2 we define *pseudo symmetric* Cat-*multinatural transformations* between pseudo symmetric Cat-multifunctors. They have the same data as a Cat-multinatural transformation, namely, component 1-ary 1-cells, and satisfy the Cat-naturality conditions (3.36) and (3.37). Moreover, each pseudo symmetric Cat-multinatural transformation is compatible with the pseudo symmetry isomorphisms of its domain and codomain. There is a 2-category $\mathsf{Cat}\text{-}\mathsf{Multicat}^{\mathsf{ps}}$ with pseudo symmetric Cat-multifunctors as 1-cells and pseudo symmetric Cat-multinatural transformations as 2-cells (Theorem 4.34).

In Section 4.3 we extend Theorem 3.69 to the pseudo symmetric context. Theorem 4.64 proves that a pseudo symmetric Cat-multifunctor is a pseudo symmetric Cat-*multiequivalence* if and only if it is essentially surjective on objects and an isomorphism on each multimorphism category.

We remind the reader of our left normalized bracketing Convention 1.24.

4.1 Pseudo Symmetric Cat-Multifunctors

In this section we define a variant of a Cat-multifunctor $F : \mathsf{M} \longrightarrow \mathsf{N}$ between Cat-multicategories, called a *pseudo symmetric* Cat-multifunctor, that preserves the symmetric group action (3.20) up to coherent natural isomorphisms. Pseudo symmetric Cat-multifunctors are closed under composition (Lemma 4.16), which is strictly associative and unital. Transformations between them are discussed in Section 4.2.

Our main use of pseudo symmetric Cat-multifunctors is Theorem 6.110. This result shows that, for each small tight bipermutative category $\mathcal{D}$, the Grothendieck construction $\int_{\mathcal{D}}$ is a pseudo symmetric Cat-multifunctor that is generally not a Cat-multifunctor. As a practical consequence, inverse K-theory is a pseudo symmetric Cat-multifunctor but not a Cat-multifunctor (Theorem 10.12).

Definition 4.1. Suppose M and N are Cat-multicategories. A *pseudo symmetric* Cat-*multifunctor*

$$\mathsf{M} \xrightarrow{(F, \{F_{\sigma,\langle c\rangle,c'}\})} \mathsf{N}$$

consists of

- an object assignment $F : \operatorname{Ob}\mathsf{M} \longrightarrow \operatorname{Ob}\mathsf{N}$;
- for each $(\langle c\rangle ; c') \in \mathsf{Prof}(\mathsf{M}) \times \operatorname{Ob}\mathsf{M}$ with $\langle c\rangle = \langle c_j\rangle_{j=1}^n$, a component functor

  $$\mathsf{M}(\langle c\rangle ; c') \xrightarrow{F} \mathsf{N}(F\langle c\rangle ; Fc')$$

  with $F\langle c\rangle = \langle Fc_j\rangle_{j=1}^n$; and
- for each permutation $\sigma \in \Sigma_n$ and object $(\langle c\rangle ; c')$ as above, a natural isomorphism

  $$\begin{array}{ccc} \mathsf{M}(\langle c\rangle ; c') & \xrightarrow{F} & \mathsf{N}(F\langle c\rangle ; Fc') \\ \sigma\big\downarrow & \overset{F_{\sigma,\langle c\rangle,c'}}{\underset{\cong}{\Longrightarrow}} & \big\downarrow\sigma \\ \mathsf{M}(\langle c\rangle\sigma ; c') & \xrightarrow[F]{} & \mathsf{N}(F\langle c\rangle\sigma ; Fc') \end{array} \tag{4.2}$$

  called the *pseudo symmetry isomorphism.*

We abbreviate $F_{\sigma,\langle c\rangle,c'}$ to F_σ when the domain object $(\langle c\rangle ; c')$ can be inferred from the context. The above data are subject to

- the unit axiom (3.21),
- the composition axiom (3.22), and
- the four axioms (4.3) through (4.6) below with $(\langle c\rangle ; c')$ as above.

Unit Permutation For the identity permutation $\mathrm{id}_n \in \Sigma_n$, the following equality holds.

$$F_{\mathrm{id}_n,\langle c\rangle,c'} = 1_F \tag{4.3}$$

Product Permutation For $\sigma, \tau \in \Sigma_n$ the following pasting diagram equality holds.

$$\begin{array}{ccc} \mathsf{M}(\langle c\rangle; c') & \xrightarrow{F} & \mathsf{N}(F\langle c\rangle; Fc') \\ \sigma\big\downarrow & \overset{F_\sigma}{\Longrightarrow} & \big\downarrow\sigma \\ \mathsf{M}(\langle c\rangle\sigma; c') & \xrightarrow{F} & \mathsf{N}(F\langle c\rangle\sigma; Fc') \\ \tau\big\downarrow & \overset{F_\tau}{\Longrightarrow} & \big\downarrow\tau \\ \mathsf{M}(\langle c\rangle\sigma\tau; c') & \xrightarrow{F} & \mathsf{N}(F\langle c\rangle\sigma\tau; Fc') \end{array} \quad = \quad \begin{array}{ccc} \mathsf{M}(\langle c\rangle; c') & \xrightarrow{F} & \mathsf{N}(F\langle c\rangle; Fc') \\ \sigma\tau\Bigg\downarrow & \overset{F_{\sigma\tau}}{\Longrightarrow} & \Bigg\downarrow\sigma\tau \\ \mathsf{M}(\langle c\rangle\sigma\tau; c') & \xrightarrow{F} & \mathsf{N}(F\langle c\rangle\sigma\tau; Fc') \end{array} \tag{4.4}$$

Top Equivariance In the context of (3.10) we use the following shorter notation.

$$\begin{cases} \langle\langle c_j\rangle\rangle_{j=1}^n = (\langle c_1\rangle, \ldots, \langle c_n\rangle) & \langle\langle c_{\sigma(j)}\rangle\rangle_{j=1}^n = (\langle c_{\sigma(1)}\rangle, \ldots, \langle c_{\sigma(n)}\rangle) \\ \langle F\langle c_j\rangle\rangle_{j=1}^n = (F\langle c_1\rangle, \ldots, F\langle c_n\rangle) & \langle F\langle c_{\sigma(j)}\rangle\rangle_{j=1}^n = (F\langle c_{\sigma(1)}\rangle, \ldots, F\langle c_{\sigma(n)}\rangle) \\ \overline{\sigma} = \sigma\langle k_{\sigma(1)}, \ldots, k_{\sigma(n)}\rangle & \end{cases}$$

Then the following pasting diagram equality holds.

$$\begin{array}{ccc} \mathsf{M}(\langle c'\rangle; c'') \times \prod_{j=1}^n \mathsf{M}(\langle c_j\rangle; c_j') & \xrightarrow[\prod_j F]{F\times} & \mathsf{N}(F\langle c'\rangle; Fc'') \times \prod_{j=1}^n \mathsf{N}(F\langle c_j\rangle; Fc_j') \\ \gamma\big\downarrow & = & \big\downarrow\gamma \\ \mathsf{M}(\langle\langle c_j\rangle\rangle_{j=1}^n; c'') & \xrightarrow{F} & \mathsf{N}(\langle F\langle c_j\rangle\rangle_{j=1}^n; Fc'') \\ \overline{\sigma}\big\downarrow & \overset{F_{\overline{\sigma}}}{\Longrightarrow} & \big\downarrow\overline{\sigma} \\ \mathsf{M}(\langle\langle c_{\sigma(j)}\rangle\rangle_{j=1}^n; c'') & \xrightarrow{F} & \mathsf{N}(\langle F\langle c_{\sigma(j)}\rangle\rangle_{j=1}^n; Fc'') \end{array} = \begin{array}{ccc} \mathsf{M}(\langle c'\rangle; c'') \times \prod_{j=1}^n \mathsf{M}(\langle c_j\rangle; c_j') & \xrightarrow[\prod_j F]{F\times} & \mathsf{N}(F\langle c'\rangle; Fc'') \times \prod_{j=1}^n \mathsf{N}(F\langle c_j\rangle; Fc_j') \\ (\sigma,\sigma^{-1})\big\downarrow & \overset{F_\sigma \times 1}{\Longrightarrow} & \big\downarrow(\sigma,\sigma^{-1}) \\ \mathsf{M}(\langle c'\rangle\sigma; c'') \times \prod_{j=1}^n \mathsf{M}(\langle c_{\sigma(j)}\rangle; c_{\sigma(j)}') & \xrightarrow[\prod_j F]{F\times} & \mathsf{N}(F\langle c'\rangle\sigma; Fc'') \times \prod_{j=1}^n \mathsf{N}(F\langle c_{\sigma(j)}\rangle; Fc_{\sigma(j)}') \\ \gamma\big\downarrow & = & \big\downarrow\gamma \\ \mathsf{M}(\langle\langle c_{\sigma(j)}\rangle\rangle_{j=1}^n; c'') & \xrightarrow{F} & \mathsf{N}(\langle F\langle c_{\sigma(j)}\rangle\rangle_{j=1}^n; Fc'') \end{array} \tag{4.5}$$

Bottom Equivariance In the context of (3.12), we use the following shorter notation.

$$\begin{cases} \langle\langle c_j\rangle\tau_j\rangle_{j=1}^n = (\langle c_1\rangle\tau_1, \ldots, \langle c_n\rangle\tau_n) & \tau^\times = \tau_1 \times \cdots \times \tau_n \\ \langle F\langle c_j\rangle\tau_j\rangle_{j=1}^n = (F\langle c_1\rangle\tau_1, \ldots, F\langle c_n\rangle\tau_n) & \end{cases}$$

Then the following pasting diagram equality holds.

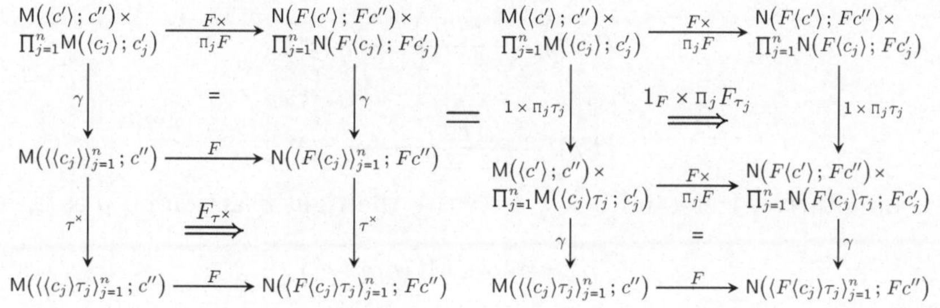

(4.6)

This finishes the definition of a pseudo symmetric Cat-multifunctor.

Moreover, the *identity pseudo symmetric* Cat-*multifunctor* of M, denoted 1_{M}, consists of

- the identity object assignment on $\mathsf{Ob}\,\mathsf{M}$,
- identity component functors, and
- identity natural transformations as the pseudo symmetry isomorphisms.

This finishes the definition of 1_{M}. ◇

Example 4.7 (Cat-Multifunctors)**.** The domain and codomain of the pseudo symmetry isomorphism in (4.2) form the boundary of the symmetric group action axiom (3.20). Therefore, there is a canonical bijection between

- Cat-multifunctors $F : \mathsf{M} \longrightarrow \mathsf{N}$ and
- pseudo symmetric Cat-multifunctors $(F, \{F_\sigma\}) : \mathsf{M} \longrightarrow \mathsf{N}$ with each pseudo symmetry isomorphism F_σ given by the identity natural transformation.

In this monograph, when a Cat-multifunctor F is regarded as a pseudo symmetric Cat-multifunctor, we always mean taking each F_σ as the identity natural transformation. See Question A.2 for a possible variation. ◇

Example 4.8 (Non-Symmetric Cat-Multifunctors)**.** Each pseudo symmetric Cat-multifunctor $(F, \{F_\sigma\}) : \mathsf{M} \longrightarrow \mathsf{N}$ yields a non-symmetric Cat-multifunctor $F : \mathsf{M} \longrightarrow \mathsf{N}$ by forgetting the pseudo symmetry isomorphisms $\{F_\sigma\}$. ◇

Explanation 4.9 (Pseudo Symmetry Data)**.** The pseudo symmetry isomorphism F_σ in (4.2) consists of, for each n-ary 1-cell $p \in \mathsf{M}(\langle c\rangle\,;c')$, an invertible n-ary 2-cell

$$F(p^\sigma) \xrightarrow[\cong]{F_{\sigma;p}} (Fp)^\sigma \quad \text{in} \quad \mathsf{N}\big(F\langle c\rangle\sigma\,;Fc'\big)$$

such that, for each n-ary 2-cell $f : p \longrightarrow q$ in $\mathsf{M}(\langle c\rangle\,;\,c')$, the naturality diagram

$$\begin{array}{ccc} F(p^\sigma) & \xrightarrow{F(f^\sigma)} & F(q^\sigma) \\ {\scriptstyle F_{\sigma;p}}\downarrow & & \downarrow{\scriptstyle F_{\sigma;q}} \\ (Fp)^\sigma & \xrightarrow{(Ff)^\sigma} & (Fq)^\sigma \end{array}$$

is commutative. Here and below we write the right σ-action on p as

$$p^\sigma = p\cdot\sigma \in \mathsf{M}\big(\langle c\rangle\sigma\,;\,c'\big)$$

and the p-component of F_σ as $F_{\sigma;p}$ or $F_{\sigma,p}$. ◇

Explanation 4.10 (Pseudo Symmetry Axioms)**.** Consider Definition 4.1 of a pseudo symmetric Cat-multifunctor $(F,\{F_\sigma\})$.

1. The unit axiom (3.21) and the composition axiom (3.22) are assumed to hold strictly. These two axioms do not involve the symmetric group action and the pseudo symmetry isomorphisms.

2. The unit permutation axiom (4.3) and the product permutation axiom (4.4) are modeled after the symmetric group action axiom (3.6) of a Cat-multicategory. The product permutation axiom (4.4) means that, for each n-ary 1-cell $p \in \mathsf{M}(\langle c\rangle\,;\,c')$, the diagram of n-ary 2-cells

$$\begin{array}{ccc} F(p^{\sigma\tau}) & \xrightarrow{F_{\sigma\tau;p}} & (Fp)^{\sigma\tau} \\ & {\scriptstyle F_{\tau;p^\sigma}}\searrow \quad \nearrow{\scriptstyle (F_{\sigma;p})^\tau} & \\ & (F(p^\sigma))^\tau & \end{array}$$

in $\mathsf{N}\big(F\langle c\rangle\sigma\tau\,;\,Fc'\big)$ is commutative.

3. By the unit permutation axiom (4.3), the pseudo symmetry isomorphism $F_{\sigma,\langle c\rangle,c'}$ is the identity natural transformation if $\langle c\rangle$ is either the empty sequence or has length 1. It follows that the restriction of a pseudo symmetric Cat-multifunctor to the underlying 1-ary 2-categories is a 2-functor. See Proposition 1.35, Definition 1.36, and Example 3.18.

4. In the top and bottom equivariance axioms, (4.5) and (4.6), the four regions labeled = are commutative by the composition axiom (3.22), which is assumed to hold for a pseudo symmetric Cat-multifunctor. The top and bottom equivariance axioms of a pseudo symmetric Cat-multifunctor are modeled after, respectively, the top and bottom equivariance axioms of a Cat-multicategory, (3.10) and (3.12).

The top equivariance axiom (4.5) means that, for n-ary 1-cells

$$p \in \mathsf{M}\big(\langle c'\rangle\,;\,c''\big) \quad\text{and}\quad p_j \in \mathsf{M}\big(\langle c_j\rangle\,;\,c_j'\big),$$

the following equality of n-ary 2-cells holds in $\mathsf{N}\big(\langle F\langle c_{\sigma(j)}\rangle\rangle_{j=1}^{n}\,;\,Fc''\big)$.

$$F_{\overline{\sigma};\gamma\left(p\,;\,\langle p_j\rangle_{j=1}^{n}\right)} = \gamma\left(F_{\sigma;p}\,;\,\langle 1_{Fp_{\sigma(j)}}\rangle_{j=1}^{n}\right) \tag{4.11}$$

The bottom equivariance axiom (4.6) means the equality of n-ary 2-cells

$$F_{\tau^{\times};\gamma\left(p\,;\,\langle p_j\rangle_{j=1}^{n}\right)} = \gamma\left(1_{Fp}\,;\,\langle F_{\tau_j;p_j}\rangle_{j=1}^{n}\right) \tag{4.12}$$

in $\mathsf{N}\big(\langle F\langle c_j\rangle\tau_j\rangle_{j=1}^{n}\,;\,Fc''\big)$. ◇

Composition of Pseudo Symmetric Cat-Multifunctors

Definition 4.13. Suppose given pseudo symmetric Cat-multifunctors

$$\mathsf{M} \xrightarrow{(F\,,\{F_{\sigma,\langle c\rangle,c'}\})} \mathsf{N} \xrightarrow{(G\,,\{G_{\sigma,\langle d\rangle,d'}\})} \mathsf{P}.$$

Their *composite*

$$\mathsf{M} \xrightarrow{(GF\,,\{GF_{\sigma,\langle c\rangle,c'}\})} \mathsf{P}$$

consists of

- the composite function $GF : \operatorname{Ob}\mathsf{M} \longrightarrow \operatorname{Ob}\mathsf{P}$;
- for each $(\langle c\rangle\,;\,c') \in \mathsf{Prof}(\mathsf{M}) \times \operatorname{Ob}\mathsf{M}$ the composite component functor

  $$\mathsf{M}\big(\langle c\rangle\,;\,c'\big) \xrightarrow{F} \mathsf{N}\big(F\langle c\rangle\,;\,Fc'\big) \xrightarrow{G} \mathsf{P}\big(GF\langle c\rangle\,;\,GFc'\big);$$

  and
- for each permutation $\sigma \in \Sigma_n$ and object $(\langle c\rangle\,;\,c')$ as above, the natural isomorphism $GF_{\sigma,\langle c\rangle,c'}$ given by the following pasting of pseudo symmetry isomorphisms.

$$\begin{array}{ccccc}
\mathsf{M}\big(\langle c\rangle\,;\,c'\big) & \xrightarrow{F} & \mathsf{N}\big(F\langle c\rangle\,;\,Fc'\big) & \xrightarrow{G} & \mathsf{P}\big(GF\langle c\rangle\,;\,GFc'\big)\\
\sigma\big\downarrow & \underset{\cong}{\overset{F_{\sigma,\langle c\rangle,c'}}{\Longrightarrow}} & \big\downarrow\sigma & \underset{\cong}{\overset{G_{\sigma,F\langle c\rangle,Fc'}}{\Longrightarrow}} & \big\downarrow\sigma\\
\mathsf{M}\big(\langle c\rangle\sigma\,;\,c'\big) & \xrightarrow[F]{} & \mathsf{N}\big(F\langle c\rangle\sigma\,;\,Fc'\big) & \xrightarrow[G]{} & \mathsf{P}\big(GF\langle c\rangle\sigma\,;\,GFc'\big)
\end{array} \tag{4.14}$$

This finishes the definition of the composite $\big(GF, \{GF_{\sigma,\langle c\rangle,c'}\}\big)$. ◇

Explanation 4.15. For each n-ary 1-cell $p \in \mathsf{M}\big(\langle c\rangle\,;\,c'\big)$, the p-component of GF_{σ} in (4.14) is given by the categorical composite

$$\begin{array}{ccc}
GF(p^{\sigma}) & \xrightarrow{(GF)_{\sigma;p}} & (GFp)^{\sigma}\\
{\scriptstyle G(F_{\sigma;p})}\searrow & & \nearrow{\scriptstyle G_{\sigma;\,Fp}}\\
& G\left((Fp)^{\sigma}\right) &
\end{array}$$

in $\mathsf{P}\big(GF\langle c\rangle\sigma\,;\,GFc'\big)$. ◇

Lemma 4.16. *In the context of Definition 4.13, the following statements hold.*

1. *The composite* $(GF, \{GF_{\sigma,\langle c\rangle,c'}\})$ *is a pseudo symmetric* Cat*-multifunctor.*
2. *Composition of pseudo symmetric* Cat*-multifunctors is associative and unital with respect to identity pseudo symmetric* Cat*-multifunctors.*

Proof. For assertion 1, first note that the composite GF satisfies the unit axiom (3.21) and the composition axiom (3.22) because each of F and G satisfies these axioms, which do not involve the symmetric group action. The unit permutation axiom (4.3) is true for GF because the pasting (4.14) of two identity natural transformations is an identity natural transformation.

Each of the other three axioms, (4.4) through (4.6), for GF is obtained by pasting together the respective pasting diagrams for F and G. For example, for the product permutation axiom (4.4) for GF, the left-hand side is the following pasting diagram.

$$\begin{array}{ccccc}
\mathsf{M}(\langle c\rangle; c') & \xrightarrow{F} & \mathsf{N}(F\langle c\rangle; Fc') & \xrightarrow{G} & \mathsf{P}(GF\langle c\rangle; GFc') \\
\sigma\downarrow & \overset{F_\sigma}{\Longrightarrow} & \downarrow\sigma & \overset{G_\sigma}{\Longrightarrow} & \downarrow\sigma \\
\mathsf{M}(\langle c\rangle\sigma; c') & \xrightarrow{F} & \mathsf{N}(F\langle c\rangle\sigma; Fc') & \xrightarrow{G} & \mathsf{P}(GF\langle c\rangle\sigma; GFc') \\
\tau\downarrow & \overset{F_\tau}{\Longrightarrow} & \downarrow\tau & \overset{G_\tau}{\Longrightarrow} & \downarrow\tau \\
\mathsf{M}(\langle c\rangle\sigma\tau; c') & \xrightarrow{F} & \mathsf{N}(F\langle c\rangle\sigma\tau; Fc') & \xrightarrow{G} & \mathsf{P}(GF\langle c\rangle\sigma\tau; GFc')
\end{array} \tag{4.17}$$

By the product permutation axiom (4.4) for each of F and G, the previous pasting diagram is equal to the following pasting, which is $GF_{\sigma\tau}$.

$$\begin{array}{ccccc}
\mathsf{M}(\langle c\rangle; c') & \xrightarrow{F} & \mathsf{N}(F\langle c\rangle; Fc') & \xrightarrow{G} & \mathsf{P}(GF\langle c\rangle; GFc') \\
\sigma\tau\downarrow & \overset{F_{\sigma\tau}}{\Longrightarrow} & \downarrow\sigma\tau & \overset{G_{\sigma\tau}}{\Longrightarrow} & \downarrow\sigma\tau \\
\mathsf{M}(\langle c\rangle\sigma\tau; c') & \xrightarrow{F} & \mathsf{N}(F\langle c\rangle\sigma\tau; Fc') & \xrightarrow{G} & \mathsf{P}(GF\langle c\rangle\sigma\tau; GFc')
\end{array} \tag{4.18}$$

This proves the axiom (4.4) for GF. The top and bottom equivariance axioms, (4.5) and (4.6), for GF are verified in the same way. This proves assertion 1.

Assertion 2 is true because

- composition of functions on objects,
- composition of component functors, and
- pasting of natural isomorphisms (4.14)

are all associative and unital. □

4.2 Pseudo Symmetric Cat-Multinatural Transformations

In this section we discuss transformations between pseudo symmetric Cat-multifunctors.

- These transformations are defined in Definition 4.19.
- Lemmas 4.28 and 4.29 show that identities, vertical composition, and horizontal composition of these transformations are well defined.
- Theorem 4.34 shows that there is a 2-category $\mathsf{Cat}\text{-}\mathsf{Multicat}^{\mathsf{ps}}$ with small Cat-multicategories as objects, pseudo symmetric Cat-multifunctors as 1-cells, and pseudo symmetric Cat-multinatural transformations as 2-cells.

A pseudo symmetric Cat-multifunctor (Definition 4.1) has not only an object assignment and multimorphism functors but also pseudo symmetry isomorphisms. Thus a transformation between pseudo symmetric Cat-multifunctors should also preserve the pseudo symmetry isomorphisms as in the next definition.

Definition 4.19. Suppose given pseudo symmetric Cat-multifunctors as follows.

$$\mathsf{M} \underset{(G\,,\{G_{\sigma,\langle c\rangle,c'}\})}{\overset{(F\,,\{F_{\sigma,\langle c\rangle,c'}\})}{\rightrightarrows}} \mathsf{N}$$

A *pseudo symmetric* Cat-*multinatural transformation* $\theta : F \longrightarrow G$ consists of a component 1-ary 1-cell

$$Fc \xrightarrow{\theta_c} Gc \qquad \text{in } \mathsf{N} \text{ for each } c \in \operatorname{Ob}\mathsf{M}. \tag{4.20}$$

These data are required to satisfy

- the object Cat-naturality condition (3.36),
- the morphism Cat-naturality condition (3.37), and
- the *pseudo symmetry preservation axiom* (4.21) below.

For each n-ary 1-cell $p \in \mathsf{M}(\langle c\rangle\,; c')$ and permutation $\sigma \in \Sigma_n$, the following equality of n-ary 2-cells holds in $\mathsf{N}\big(F\langle c\rangle\sigma\,; Gc'\big)$.

$$\gamma\big(1_{\theta_{c'}}\,; F_{\sigma;p}\big) = \gamma\big(G_{\sigma;p}\,; 1_{\theta_{\langle c\rangle\sigma}}\big) \tag{4.21}$$

Here $F_{\sigma;p}$ denotes the p-component of F_σ and likewise for $G_{\sigma;p}$.

A *pseudo symmetric* Cat-*multinatural isomorphism* is a pseudo symmetric Cat-multinatural transformation with each component 1-ary 1-cell invertible with respect to γ in N.

Moreover, for pseudo symmetric Cat-multinatural transformations between pseudo symmetric Cat-multifunctors, we define

- *vertical composition* as in (3.26),
- *horizontal composition* as in (3.27), and
- the *identity pseudo symmetric* Cat-*multinatural transformation* 1_F as consisting of the component colored units (3.34)

$$(1_F)_c = 1_{Fc} : Fc \longrightarrow Fc.$$

Vertical composition and identities are well defined by Lemma 4.28 below. Horizontal composition is well defined by Lemma 4.29 below. This finishes the definition. ◇

Explanation 4.22. Similar to (3.40), the pseudo symmetry preservation axiom (4.21) says that the following two multicategorical composites in N are equal, with $(-)^\sigma$ denoting the right σ-action.

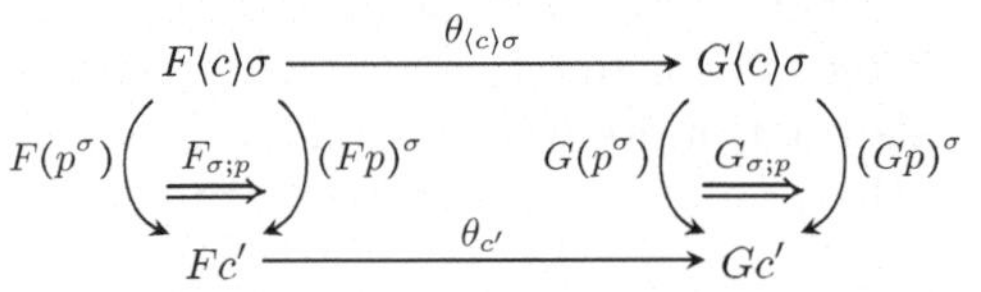

(4.23)

The domain n-ary 1-cells of the two n-ary 2-cell composites in (4.23),

$$\gamma\big(G(p^\sigma)\,;\,\theta_{\langle c\rangle\sigma}\big) = \gamma\big(\theta_{c'}\,;\,F(p^\sigma)\big), \tag{4.24}$$

are equal by the object Cat-naturality condition (3.36) applied to p^σ. The codomain n-ary 1-cells of the two n-ary 2-cell composites in (4.23) are equal by the following equalities.

$$\begin{cases} \gamma\big((Gp)^\sigma\,;\,\theta_{\langle c\rangle\sigma}\big) = \gamma\big(Gp\,;\,\theta_{\langle c\rangle}\big)\cdot\sigma & \text{by (3.10)} \\ \qquad = \gamma\big(\theta_{c'}\,;\,Fp\big)\cdot\sigma & \text{by (3.36)} \\ \qquad = \gamma\big(\theta_{c'}\,;\,(Fp)^\sigma\big) & \text{by (3.12)} \end{cases} \tag{4.25}$$

A pseudo symmetric Cat-multinatural transformation has the same definition as a Cat-multinatural transformation (Explanation 3.33) with one additional axiom (4.21) but *no* extra data. ◇

Example 4.26 (Cat-Multinatural Transformations)**.** Suppose $\theta : F \longrightarrow G$ is a Cat-multinatural transformation (Explanation 3.33) between Cat-multifunctors $F, G : \mathsf{M} \longrightarrow \mathsf{N}$. We regard each of F and G as a pseudo symmetric Cat-multifunctor with identity pseudo symmetry isomorphisms (Example 4.7). Then θ is a pseudo symmetric Cat-multinatural transformation. Its pseudo symmetry preservation axiom (4.21) follows from the object Cat-naturality condition (3.36) as in (4.24) and the functoriality of γ. ◇

Example 4.27 (Non-Symmetric Cat-Multinatural Transformations)**.** Suppose $\theta : F \longrightarrow G$ is a pseudo symmetric Cat-multinatural transformation between pseudo symmetric Cat-multifunctors. Then θ yields a non-symmetric

Cat-multinatural transformation between the underlying non-symmetric Cat-multifunctors of F and G (Example 4.8), since θ satisfies the Cat-naturality conditions (3.36) and (3.37). $\diamond$

Lemma 4.28. *In Definition 4.19,*

- *identity pseudo symmetric* Cat-*multinatural transformations and*
- *vertical composites of pseudo symmetric* Cat-*multinatural transformations*

are well defined.

Proof. The Cat-naturality conditions (3.36) and (3.37) do not involve the symmetric group action. They are satisfied by the identities and vertical composites as part of the 2-category structure on $\mathsf{Cat}\text{-}\mathsf{Multicat}^{\mathsf{ns}}$ (Theorem 3.28).

It remains to verify the pseudo symmetry preservation axiom (4.21). For the identity 1_F, the axiom (4.21) holds by the right and left unity axioms, (3.8) and (3.9), since each component of 1_F is a colored unit.

Consider pseudo symmetric Cat-multinatural transformations

$$F \xrightarrow{\ \theta\ } G \xrightarrow{\ \psi\ } H.$$

The axiom (4.21) for the vertical composite $\psi\theta$ (3.26) is obtained by pasting (4.23) and the one for ψ together as follows.

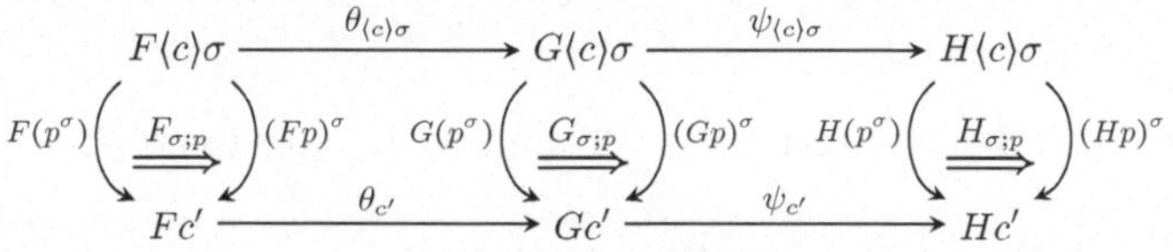

This is valid by the associativity axiom (3.7) in N and the pseudo symmetry preservation axiom (4.21) for θ and ψ. $\square$

Lemma 4.29. *In Definition 4.19, horizontal composites of pseudo symmetric* Cat-*multinatural transformations are well defined.*

Proof. Consider the horizontal composite $\theta' * \theta$ (3.27) of the following pseudo symmetric Cat-multinatural transformations.

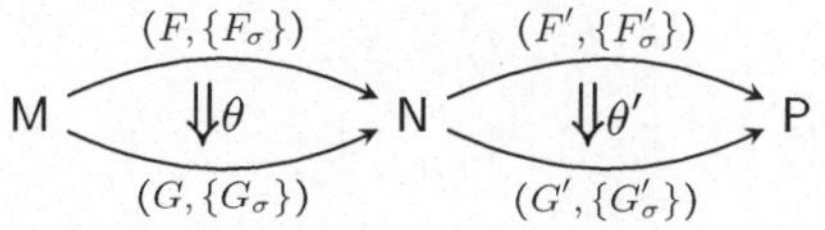

The Cat-naturality conditions (3.36) and (3.37) do not involve the symmetric group action and are satisfied by the horizontal composite as part of the 2-category structure on $\mathsf{Cat}\text{-}\mathsf{Multicat}^{\mathsf{ns}}$ (Theorem 3.28). It remains to verify the pseudo symmetry preservation axiom (4.21), which is the following equality in $\mathsf{P}\big(F'F\langle c\rangle\sigma\,;\,G'Gc'\big)$ for each n-ary 1-cell $p \in \mathsf{M}(\langle c\rangle\,;\,c')$ and permutation $\sigma \in \Sigma_n$.

$$\gamma\big(1_{(\theta'*\theta)_{c'}}\,;\,(F'F)_{\sigma;p}\big) = \gamma\big((G'G)_{\sigma;p}\,;\,1_{(\theta'*\theta)_{\langle c\rangle\sigma}}\big) \tag{4.30}$$

To prove (4.30), we use

- the second expression in (3.41) for the horizontal composite

$$(\theta' * \theta)_{c'} = \gamma\big(G'\theta_{c'}\,;\,\theta'_{Fc'}\big)$$

 and likewise for $(\theta' * \theta)_{c_{\sigma(j)}}$ for $j \in \{1,\ldots,n\}$;

- the definition (4.14) of the pseudo symmetry isomorphism

$$(F'F)_{\sigma;p} = F'_{\sigma;Fp} \circ F'(F_{\sigma;p})$$

 and likewise for $(G'G)_{\sigma;p}$; and

- the functoriality and associativity (3.7) of γ in P.

Then the desired equality (4.30) becomes the following equality.

$$\begin{aligned}&\gamma\Big(1_{G'\theta_{c'}}\,;\,\gamma\Big(1_{\theta'_{Fc'}}\,;\,F'_{\sigma;Fp}\Big)\Big)\circ\gamma\Big(1_{G'\theta_{c'}}\,;\,\gamma\Big(1_{\theta'_{Fc'}}\,;\,F'(F_{\sigma;p})\Big)\Big)\\ &= \gamma\Big(\gamma\big(G'_{\sigma;Gp}\,;\,1_{G'\theta_{\langle c\rangle\sigma}}\big)\,;\,1_{\theta'_{F\langle c\rangle\sigma}}\Big)\circ\gamma\Big(\gamma\big(G'(G_{\sigma;p})\,;\,1_{G'\theta_{\langle c\rangle\sigma}}\big)\,;\,1_{\theta'_{F\langle c\rangle\sigma}}\Big)\end{aligned} \tag{4.31}$$

We prove (4.31) by showing that the left, respectively, right, factors on the two sides are equal using

- the Cat-naturality conditions (3.36) and (3.37) for θ and θ',
- the associativity (3.7) of γ, and
- the fact that G', as a pseudo symmetric Cat-multifunctor, strictly preserves the multicategorical composition (3.22).

The equality of the right factors in (4.31) are proved as follows.

$$\begin{aligned}&\gamma\Big(1_{G'\theta_{c'}}\,;\,\gamma\Big(1_{\theta'_{Fc'}}\,;\,F'(F_{\sigma;p})\Big)\Big)\\ &= \gamma\Big(1_{G'\theta_{c'}}\,;\,\gamma\Big(G'(F_{\sigma;p})\,;\,1_{\theta'_{F\langle c\rangle\sigma}}\Big)\Big) && \text{by (3.37) for } \theta'\\ &= \gamma\Big(\gamma\big(1_{G'\theta_{c'}}\,;\,G'(F_{\sigma;p})\big)\,;\,1_{\theta'_{F\langle c\rangle\sigma}}\Big) && \text{by (3.7)}\\ &= \gamma\Big(G'\gamma\big(1_{\theta_{c'}}\,;\,F_{\sigma;p}\big)\,;\,1_{\theta'_{F\langle c\rangle\sigma}}\Big) && \text{by (3.22)}\\ &= \gamma\Big(G'\gamma\big(G_{\sigma;p}\,;\,1_{\theta_{\langle c\rangle\sigma}}\big)\,;\,1_{\theta'_{F\langle c\rangle\sigma}}\Big) && \text{by (4.21)}\\ &= \gamma\Big(\gamma\big(G'(G_{\sigma;p})\,;\,1_{G'\theta_{\langle c\rangle\sigma}}\big)\,;\,1_{\theta'_{F\langle c\rangle\sigma}}\Big) && \text{by (3.22)}\end{aligned}$$

The equality of the left factors in (4.31) are proved as follows.

$$\begin{aligned}
&\gamma\Big(1_{G'\theta_{c'}}\,;\gamma\Big(1_{\theta'_{Fc'}}\,;F'_{\sigma;Fp}\Big)\Big)\\
&=\gamma\Big(1_{G'\theta_{c'}}\,;\gamma\Big(G'_{\sigma;Fp}\,;1_{\theta'_{F\langle c\rangle\sigma}}\Big)\Big) && \text{by (4.21) for } \theta'\\
&=\gamma\Big(\gamma\big(1_{G'\theta_{c'}}\,;G'_{\sigma;Fp}\big)\,;1_{\theta'_{F\langle c\rangle\sigma}}\Big) && \text{by (3.7)}\\
&=\gamma\Big(\gamma\big(G'_{\sigma;Gp}\,;1_{G'\theta_{\langle c\rangle\sigma}}\big)\,;1_{\theta'_{F\langle c\rangle\sigma}}\Big) && \text{by (4.32) below}
\end{aligned}$$

The last equality follows from the following computation, which uses the top and bottom equivariance axioms, (4.11) and (4.12), for G'.

$$\begin{cases}
\gamma\big(1_{G'\theta_{c'}}\,;G'_{\sigma;\,Fp}\big)\\
=G'_{\sigma;\,\gamma(\theta_{c'}\,;\,Fp)} & \text{by (4.12) for } (\theta_{c'},Fp)\\
=G'_{\sigma;\,\gamma(Gp\,;\,\theta_{\langle c\rangle})} & \text{by (3.36)}\\
=\gamma\big(G'_{\sigma;\,Gp}\,;1_{G'\theta_{\langle c\rangle\sigma}}\big) & \text{by (4.11) for } (Gp,\theta_{\langle c\rangle})
\end{cases} \tag{4.32}$$

This finishes the proof of the desired equality (4.31). □

Explanation 4.33. The computation (4.32) asserts that the following two composite n-ary 2-cells in P are equal.

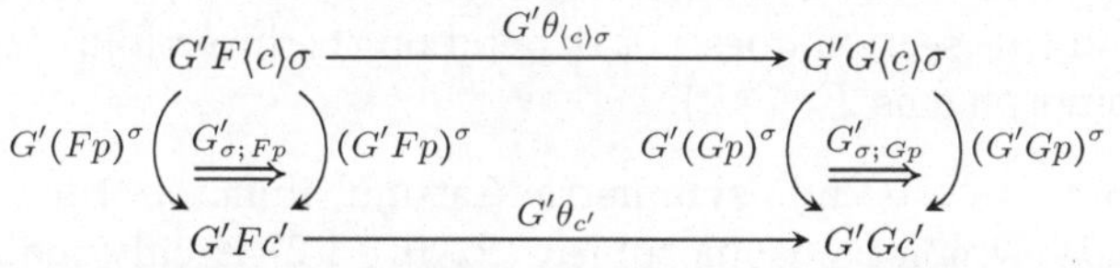

The top and bottom equivariance axioms, (4.11) and (4.12), of a pseudo symmetric Cat-multifunctor are necessary to prove this equality. ◇

Theorem 4.34. *There is a 2-category* Cat-Multicat^{ps} *consisting of the following data.*

- *Its objects are small* Cat-*multicategories.*
- *For small* Cat-*multicategories* M *and* N*, the hom category*

  Cat-Multicat^{ps}(M, N)

  has

  – *pseudo symmetric* Cat-*multifunctors* M ⟶ N *(Definition 4.1) as 1-cells and*

  – *pseudo symmetric* Cat-*multinatural transformations (Definition 4.19) as 2-cells.*

- *For pseudo symmetric* Cat*-multinatural transformations, vertical composition, horizontal composition, and identity 2-cells are given as in Definition 4.19.*
- *The identity 1-cell* 1_{M} *is the identity pseudo symmetric* Cat*-multifunctor* 1_{M} *(Definition 4.1).*
- *Horizontal composition of 1-cells is the composition of pseudo symmetric* Cat*-multifunctors (Definition 4.13).*

Proof. Consider axioms 1 through 4 in Definition 1.32 of a 2-category for $\mathsf{Cat}\text{-}\mathsf{Multicat}^{\mathsf{ps}}$.

- By Lemma 4.16 the horizontal composition of 1-cells is well defined, and the 1-cell axiom 3 holds.
- Vertical and horizontal compositions of 2-cells and identity 2-cells are well defined Lemmas 4.28 and 4.29.
- The 2-cell axioms 1, 2, and 4 hold as they do in the 2-category $\mathsf{Cat}\text{-}\mathsf{Multicat}^{\mathsf{ns}}$ (Theorem 3.28) because the symmetric group action is not involved.

This finishes the proof. □

Explanation 4.35 (Lax Symmetric Cat-Multifunctors)**.** The discussion in Section 4.1 and this section does *not* depend on the invertibility of the pseudo symmetry isomorphisms F_σ (4.2).

- Definition 4.1 of a pseudo symmetric Cat-multifunctor still makes sense if each pseudo symmetry isomorphism F_σ in (4.2) is only required to be a natural *transformation*, instead of a natural isomorphism. We call this lax variant a *lax symmetric* Cat*-multifunctor* and F_σ a *lax symmetry.*
- Composites of lax symmetric Cat-multifunctors are well defined, with the composite lax symmetry given by the pasting (4.14).
- The lax variants of Lemmas 4.16, 4.28, and 4.29, Definition 4.19, and Theorem 4.34 also hold.

Therefore, there is a 2-category

$$\mathsf{Cat}\text{-}\mathsf{Multicat}^{\mathsf{lax}}$$

with

- small Cat-multicategories as objects,
- lax symmetric Cat-multifunctors as 1-cells,
- lax symmetric Cat-multinatural transformations as 2-cells (as in Definition 4.19), and

- identity 1- and 2-cells, vertical composition, and horizontal composition given as in Lemmas 4.16, 4.28, and 4.29.

It is related to the 2-categories in Theorems 3.28 and 4.34 via the following 2-functors.

$$\begin{array}{ccc} \mathsf{Cat\text{-}Multicat} & & \mathsf{Cat\text{-}Multicat}^{\mathsf{ns}} \\ \big\downarrow{\scriptstyle i_1} & & \big\uparrow{\scriptstyle j} \\ \mathsf{Cat\text{-}Multicat}^{\mathsf{ps}} & \xrightarrow{\;i_2\;} & \mathsf{Cat\text{-}Multicat}^{\mathsf{lax}} \end{array} \tag{4.36}$$

Each of these 2-functors is the identity on

- objects, which are small Cat-multicategories, and
- 2-cells (Examples 4.26 and 4.27).

On 1-cells they are defined as follows.

- The assignment i_1 sends a Cat-multifunctor F to the pseudo symmetric Cat-multifunctor $(F, \{\mathrm{Id}\})$ (Example 4.7).
- The assignment i_2 is the inclusion of pseudo symmetric Cat-multifunctors into lax symmetric Cat-multifunctors.
- The assignment j sends a lax symmetric Cat-multifunctor $(F, \{F_\sigma\})$ to its underlying non-symmetric Cat-multifunctor F (Example 4.8).

See Questions A.1 and A.2 for related discussion. ◇

4.3 Pseudo Symmetric Cat-Multiequivalences

In this section we show that the characterization of Cat-multiequivalences in Theorem 3.69 also holds for the 2-category $\mathsf{Cat\text{-}Multicat}^{\mathsf{ps}}$ in Theorem 4.34. See Theorem 4.64. In other words, a pseudo symmetric Cat-multifunctor is a pseudo symmetric Cat-multiequivalence if and only if it is essentially surjective on objects and an isomorphism on each multimorphism category. For the *if* direction, most of the work involves defining the pseudo symmetry isomorphisms for the adjoint and showing that the (co)unit preserves them.

Properties of Pseudo Symmetric Cat-Multiequivalences

The following notions are restrictions of those in Definition 1.46 to the 2-category $\mathsf{Cat\text{-}Multicat}^{\mathsf{ps}}$ and analogues of those in Definition 3.42.

Definition 4.37. We define the following.

- A *pseudo symmetric* Cat-*multiequivalence* is an equivalence in the 2-category $\mathsf{Cat\text{-}Multicat}^{\mathsf{ps}}$.

- An *adjoint pseudo symmetric* Cat-*multiequivalence* is an adjoint equivalence in the 2-category $\mathsf{Cat}\text{-}\mathsf{Multicat}^{\mathsf{ps}}$. ◇

Explanation 4.38. In Definition 4.37 the use of the 2-category $\mathsf{Cat}\text{-}\mathsf{Multicat}^{\mathsf{ps}}$ is simply a matter of conceptual convenience. Indeed, a pseudo symmetric Cat-multifunctor (Definition 4.1)

$$\mathsf{M} \xrightarrow{(F,\{F_\sigma\})} \mathsf{N}$$

is a *pseudo symmetric* Cat-*multiequivalence* if there exist

- a pseudo symmetric Cat-multifunctor $(G, \{G_\sigma\}) : \mathsf{N} \longrightarrow \mathsf{M}$ and
- pseudo symmetric Cat-multinatural isomorphisms (Definition 4.19)

$$1_{\mathsf{M}} \xrightarrow[\cong]{\eta} GF \quad \text{and} \quad FG \xrightarrow[\cong]{\varepsilon} 1_{\mathsf{N}}. \tag{4.39}$$

The quadruple $(F, G, \eta, \varepsilon)$ is an *adjoint pseudo symmetric* Cat-*multiequivalence* if, moreover, the two triangle identities (1.47) are satisfied. ◇

Example 4.40 (Cat-Multiequivalences)**.** By the 2-functor

$$\mathsf{Cat}\text{-}\mathsf{Multicat} \xhookrightarrow{i_1} \mathsf{Cat}\text{-}\mathsf{Multicat}^{\mathsf{ps}}$$

in (4.36), a Cat-multiequivalence $F : \mathsf{M} \longrightarrow \mathsf{N}$ yields a pseudo symmetric Cat-multiequivalence $(F, \{\mathrm{Id}\})$ when it is equipped with identity pseudo symmetry isomorphisms. The adjoint variant is also true by the 2-functoriality of i_1. ◇

Example 4.41 (Non-Symmetric Cat-Multiequivalences)**.** By the 2-functor

$$\mathsf{Cat}\text{-}\mathsf{Multicat}^{\mathsf{ps}} \xrightarrow{j \circ i_2} \mathsf{Cat}\text{-}\mathsf{Multicat}^{\mathsf{ns}}$$

in (4.36), a pseudo symmetric Cat-multiequivalence $(F, \{F_\sigma\})$ yields a non-symmetric Cat-multiequivalence F by forgetting the pseudo symmetry isomorphisms. The adjoint variant is also true by the 2-functoriality of $j \circ i_2$. ◇

Lemma 4.42. *Each pseudo symmetric* Cat-*multiequivalence is*

- *essentially surjective on objects and*
- *an isomorphism on each multimorphism category.*

Proof. By Example 4.41 each pseudo symmetric Cat-multiequivalence $(F, \{F_\sigma\})$ yields a non-symmetric Cat-multiequivalence F, which has the two stated properties by Lemma 3.45. □

Adjoint, Unit, and Counit

The main part of Theorem 4.64 is the converse of Lemma 4.42. We proceed as in Section 3.4. Given a pseudo symmetric Cat-multifunctor that satisfies the two properties in Lemma 4.42, we define its adjoint, unit, and counit as in the non-symmetric case as follows.

Definition 4.43. Suppose given a pseudo symmetric Cat-multifunctor (Definition 4.1)

$$\mathsf{N} \xrightarrow{(G,\{G_\sigma\})} \mathsf{M}$$

that is

- essentially surjective on objects and
- an isomorphism on each multimorphism category.

Define

- a non-symmetric Cat-multifunctor $F : \mathsf{M} \longrightarrow \mathsf{N}$ as in Definition 3.49;
- a non-symmetric Cat-multinatural isomorphism $\eta : 1_{\mathsf{M}} \xrightarrow{\cong} GF$ as in (3.50); and
- a non-symmetric Cat-multinatural isomorphism $\varepsilon : FG \xrightarrow{\cong} 1_{\mathsf{N}}$ as in Definition 3.63.

By Lemmas 3.56, 3.61, and 3.66, respectively, F, η, and ε are well defined. Moreover, by Lemma 3.68 the quadruple $(F, G, \eta, \varepsilon)$ satisfies the triangle identities (1.47). ◇

In the context of Definition 4.43, to finish the proof of Theorem 4.64, we proceed as follows.

1. Define the pseudo symmetry isomorphisms F_σ for F (Definition 4.48).
2. Show that $(F, \{F_\sigma\}) : \mathsf{M} \longrightarrow \mathsf{N}$ is a pseudo symmetric Cat-multifunctor (Lemma 4.55).
3. Show that each of η and ε in Definition 4.43 is a pseudo symmetric Cat-multinatural isomorphism (Lemmas 4.56 and 4.58).
4. Conclude that the quadruple $(F, G, \eta, \varepsilon)$ is an adjoint pseudo symmetric Cat-multiequivalence (Theorem 4.64).

The following preliminary observation is needed to define F_σ. Recall that $(-)^\sigma$ denotes the right σ-action.

Lemma 4.44. *In the context of Definition 4.43, suppose $\sigma \in \Sigma_n$ is a permutation and*

$$p \in \mathsf{M}(\langle c \rangle\,; c') \quad \textit{and} \quad q \in \mathsf{N}(\langle d \rangle\,; d')$$

are n-ary 1-cells. Then the following equalities hold.

$$\begin{cases} GF(p^\sigma) = (GFp)^\sigma \in \mathsf{M}\big(GF\langle c\rangle\sigma\,;\, GFc'\big) \\ FG(q^\sigma) = (FGq)^\sigma \in \mathsf{N}\big(FG\langle d\rangle\sigma\,;\, FGd'\big) \end{cases} \tag{4.45}$$

Proof. To prove the first equality in (4.45), we compute as follows using the object Cat-naturality of η (3.62) and the top and bottom equivariance axioms, (3.10) and (3.12), in M.

$$\begin{cases} \gamma\big(GF(p^\sigma)\,;\, \eta_{\langle c\rangle\sigma}\big) & \\ = \gamma\big(\eta_{c'}\,;\, p^\sigma\big) & \text{by (3.62)} \\ = \gamma\big(\eta_{c'}\,;\, p\big)\cdot\sigma & \text{by (3.12) in } \mathsf{M} \\ = \gamma\big(GFp\,;\, \eta_{\langle c\rangle}\big)\cdot\sigma & \text{by (3.62)} \\ = \gamma\big((GFp)^\sigma\,;\, \eta_{\langle c\rangle\sigma}\big) & \text{by (3.10) in } \mathsf{M} \end{cases} \tag{4.46}$$

Since each component 1-ary 1-cell η_{c_j} is invertible by (3.50), we obtain the first equality in (4.45) by

- applying $\gamma\big(-\,;\, \eta^{-1}_{\langle c\rangle\sigma}\big)$ to (4.46) and
- using associativity (3.7) and right unity (3.8) in M.

The second equality in (4.45) is proved similarly by first obtaining the following equalities using the object Cat-naturality of ε (3.67) and the top and bottom equivariance axioms in N.

$$\begin{cases} \gamma\big(\varepsilon_{d'}\,;\, FG(q^\sigma)\big) & \\ = \gamma\big(q^\sigma\,;\, \varepsilon_{\langle d\rangle\sigma}\big) & \text{by (3.67)} \\ = \gamma\big(q\,;\, \varepsilon_{\langle d\rangle}\big)\cdot\sigma & \text{by (3.10) in } \mathsf{N} \\ = \gamma\big(\varepsilon_{d'}\,;\, FGq\big)\cdot\sigma & \text{by (3.67)} \\ = \gamma\big(\varepsilon_{d'}\,;\, (FGq)^\sigma\big) & \text{by (3.12) in } \mathsf{N} \end{cases} \tag{4.47}$$

Since the component 1-ary 1-cell $\varepsilon_{d'}$ is invertible by (3.64), we obtain the second equality in (4.45) by

- applying $\gamma\big(\varepsilon^{-1}_{d'}\,;\, -\big)$ to (4.47) and
- using associativity (3.7) and left unity (3.9) in N.

This finishes the proof. □

Definition 4.48. In the context of Definition 4.43 and Lemma 4.44, define the data of a natural isomorphism

$$\begin{array}{ccc} \mathsf{M}\big(\langle c\rangle\,;\, c'\big) & \xrightarrow{\quad F \quad} & \mathsf{N}\big(F\langle c\rangle\,;\, Fc'\big) \\ {\scriptstyle\sigma}\big\downarrow & \overset{F_\sigma}{\underset{\cong}{\Longrightarrow}} & \big\downarrow{\scriptstyle\sigma} \\ \mathsf{M}\big(\langle c\rangle\sigma\,;\, c'\big) & \xrightarrow[\quad F \quad]{} & \mathsf{N}\big(F\langle c\rangle\sigma\,;\, Fc'\big) \end{array}$$

as follows. For each n-ary 1-cell $p \in \mathsf{M}(\langle c\rangle\,;\,c')$, the (Fp)-component of the pseudo symmetry isomorphism G_σ has an inverse n-ary 2-cell

$$GF(p^\sigma) = (GFp)^\sigma \xrightarrow[\cong]{G^{-1}_{\sigma;Fp}} G(Fp)^\sigma \quad \text{in} \quad \mathsf{M}\big(GF\langle c\rangle\sigma\,;\,GFc'\big)$$

with the equality from (4.45). Since the multimorphism functor

$$\mathsf{N}\big(F\langle c\rangle\sigma\,;\,Fc'\big) \xrightarrow[\cong]{G} \mathsf{M}\big(GF\langle c\rangle\sigma\,;\,GFc'\big) \tag{4.49}$$

is an isomorphism, there exists a unique n-ary 2-cell

$$F(p^\sigma) \xrightarrow[\cong]{F_{\sigma;p}} (Fp)^\sigma \quad \text{in} \quad \mathsf{N}\big(F\langle c\rangle\sigma\,;\,Fc'\big) \tag{4.50}$$

such that

$$G(F_{\sigma;p}) = G^{-1}_{\sigma;Fp}. \tag{4.51}$$

We define $F_{\sigma;p}$ as the p-component of F_σ. Moreover, $F_{\sigma;p}$ is an invertible n-ary 2-cell by the fact that G in (4.49) is an isomorphism. The naturality of F_σ follows from the naturality of G_σ and the isomorphism G (4.49). ◇

Explanation 4.52. The definition (4.51) of F_σ means that the pasting

$$\begin{array}{ccccc}
\mathsf{M}(\langle c\rangle\,;\,c') & \xrightarrow{F} & \mathsf{N}\big(F\langle c\rangle\,;\,Fc'\big) & \xrightarrow{G} & \mathsf{M}\big(GF\langle c\rangle\,;\,GFc'\big) \\
\sigma\big\downarrow & \overset{F_\sigma}{\underset{\cong}{\Longrightarrow}} & \big\downarrow\sigma & \overset{G_\sigma}{\underset{\cong}{\Longrightarrow}} & \big\downarrow\sigma \\
\mathsf{M}(\langle c\rangle\sigma\,;\,c') & \xrightarrow[F]{} & \mathsf{N}\big(F\langle c\rangle\sigma\,;\,Fc'\big) & \xrightarrow[G]{} & \mathsf{M}\big(GF\langle c\rangle\sigma\,;\,GFc'\big)
\end{array} \tag{4.53}$$

is equal to the identity natural transformation

$$1_{GF(-)^\sigma} = 1_{(GF(-))^\sigma}$$

with the equality from (4.45). By the invertibility of the functor G (4.49) and the pseudo symmetry isomorphism G_σ, the equality

$$G_{\sigma;F(-)} \circ G(F_{\sigma;-}) = 1_{GF(-)^\sigma} \tag{4.54}$$

uniquely determines F_σ. Moreover, the pasting (4.53) is the pseudo symmetry isomorphism $(GF)_\sigma$ in (4.14) *if* we already know that $(F,\{F_\sigma\})$ is a pseudo symmetric Cat-multifunctor. ◇

Lemma 4.55. *In the context of Definitions 4.43 and 4.48,*

$$\mathsf{M} \xrightarrow{(F,\{F_\sigma\})} \mathsf{N}$$

is a pseudo symmetric Cat-*multifunctor.*

Proof. By Lemma 3.56 F preserves the colored units (3.21) and composition (3.22). It remains to check the axioms (4.3) through (4.6) for $(F, \{F_\sigma\})$.

Unit Permutation (4.3). This axiom says that

$$F_{\mathrm{id}_n} = 1_F.$$

To prove this equality, consider an n-ary 1-cell $p \in \mathsf{M}(\langle c \rangle ; c')$. The unit permutation axiom (4.3) for $(G, \{G_\sigma\})$ yields the equalities of n-ary 2-cells

$$G^{-1}_{\mathrm{id}_n;Fp} = 1^{-1}_{GFp} = 1_{GFp}.$$

With G the isomorphism in (4.49), we apply G^{-1} to the definition (4.51) of F_σ to obtain the following equalities.

$$\begin{aligned} F_{\mathrm{id}_n;p} &= G^{-1}\left(G^{-1}_{\mathrm{id}_n;Fp}\right) \\ &= G^{-1}\left(1_{GFp}\right) \\ &= 1_{Fp} \end{aligned}$$

This proves the unit permutation axiom for F.

Product Permutation (4.4). By

- the invertibility of the functor G (4.49),
- the invertibility of each pseudo symmetry isomorphism G_σ, and
- the product permutation axiom (4.4) for $(G, \{G_\sigma\})$,

it suffices to show that the two sides of (4.4) pasted with the corresponding sides for $(G, \{G_\sigma\})$ are equal. The two resulting pastings are (4.17) and (4.18). Each of these pastings is equal to the identity natural transformation by (4.54) for σ, τ, and $\sigma\tau$.

Top and Bottom Equivariance. The axioms (4.5) and (4.6) for $(F, \{F_\sigma\})$ are proved in the same way. In each case, we

- paste each side of the desired equality with the corresponding pasting for $(G, \{G_\sigma\})$ and
- observe that both sides are equal to the identity by (4.54).

This proves that $(F, \{F_\sigma\})$ is a pseudo symmetric Cat-multifunctor. □

Lemma 4.56. *In the context of Definitions 4.43 and 4.48,*

$$1_{\mathsf{M}} \xrightarrow{\eta} GF$$

is a pseudo symmetric Cat*-multinatural isomorphism.*

Proof. By Lemma 3.61 η is a non-symmetric Cat-multinatural isomorphism. It remains to check the pseudo symmetry preservation axiom (4.21). The axiom (4.21) for η says that, for each n-ary 1-cell $p \in \mathsf{M}(\langle c \rangle\,;c')$, the following two composite n-ary 2-cells in M are equal.

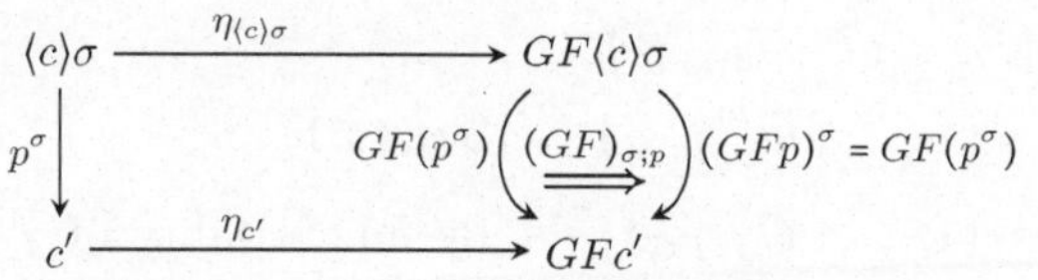

(4.57)

By Explanation 4.52 and Lemma 4.55, there are equalities

$$(GF)_\sigma = 1_{GF(-)^\sigma} = 1_{(GF(-))^\sigma}.$$

Thus, the two composites in (4.57) are equal by the object Cat-naturality condition for η, which is proved in (3.62). □

Lemma 4.58. *In the context of Definitions 4.43 and 4.48,*

$$FG \xrightarrow{\;\varepsilon\;} 1_{\mathsf{N}}$$

is a pseudo symmetric Cat-*multinatural isomorphism.*

Proof. By Lemma 3.66 ε is a non-symmetric Cat-multinatural isomorphism. It remains to check the pseudo symmetry preservation axiom (4.21). By the definition (4.14) of $(FG)_\sigma$, the axiom (4.21) for ε says that, for each n-ary 1-cell $q \in \mathsf{N}(\langle d \rangle\,; d')$, the following two composite n-ary 2-cells in N are equal, with the unlabeled vertical arrow given by $F(Gq)^\sigma$.

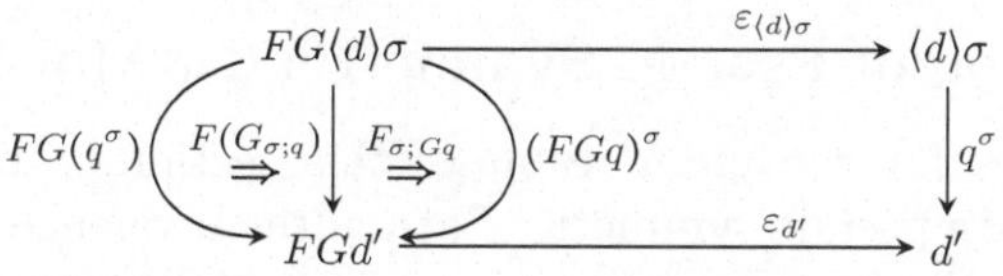

(4.59)

It suffices to show that the two composites in (4.59) have the same image under the isomorphism

$$\mathsf{N}\big(FG\langle d \rangle\sigma\,; d'\big) \xrightarrow[\cong]{\;G\;} \mathsf{M}\big(GFG\langle d \rangle\sigma\,; Gd'\big).$$

By the functoriality of γ, this means the following equality of n-ary 2-cells in M.

$$G\big[\gamma\big(1_{\varepsilon_{d'}}\,; F_{\sigma;\,Gq}\big) \circ \gamma\big(1_{\varepsilon_{d'}}\,; F(G_{\sigma;q})\big)\big] = G1_{\gamma(q^\sigma\,;\,\varepsilon_{\langle d \rangle\sigma})} \tag{4.60}$$

By the functoriality and the composition axiom (3.22) for G, the equality (4.60) becomes the following equality.

$$\gamma\big(1_{G\varepsilon_{d'}}\,; G(F_{\sigma;\,Gq})\big) \circ \gamma\big(1_{G\varepsilon_{d'}}\,; GF(G_{\sigma;q})\big) = 1_{\gamma(G(q^\sigma)\,;\,G\varepsilon_{\langle d \rangle\sigma})} \tag{4.61}$$

We prove (4.61) by computing the two terms on the left-hand side and observing that they are inverses of each other.

For the leftmost term in (4.61), we compute as follows using, in particular, the top and bottom equivariance axioms, (4.11) and (4.12), for G.

$$\begin{cases} \gamma\big(1_{G\varepsilon_{d'}}\,;\, G(F_{\sigma;\,Gq})\big) & \\ = \gamma\big(1_{G\varepsilon_{d'}}\,;\, G^{-1}_{\sigma;\,FGq}\big) & \text{by (4.51)} \\ = \gamma\big(1_{G\varepsilon_{d'}}\,;\, G_{\sigma;\,FGq}\big)^{-1} & \text{by functoriality of } \gamma \\ = \big(G_{\sigma;\,\gamma(\varepsilon_{d'}\,;\,FGq)}\big)^{-1} & \text{by (4.12) for } (\varepsilon_{d'}, FGq) \\ = \big(G_{\sigma;\,\gamma(q\,;\,\varepsilon_{\langle d\rangle})}\big)^{-1} & \text{by (3.67)} \\ = \gamma\big(G_{\sigma;q}\,;\, 1_{G\varepsilon_{\langle d\rangle\sigma}}\big)^{-1} & \text{by (4.11) for } (q, \varepsilon_{\langle d\rangle}) \end{cases} \tag{4.62}$$

For the second term in (4.61), we compute as follows.

$$\begin{cases} \gamma\big(1_{G\varepsilon_{d'}}\,;\, GF(G_{\sigma;q})\big) & \\ = \gamma\Big(1_{\eta^{-1}_{Gd'}}\,;\, \gamma\Big(1_{\eta_{Gd'}}\,;\, \gamma\Big(G_{\sigma;q}\,;\, 1_{\eta^{-1}_{G\langle d\rangle\sigma}}\Big)\Big)\Big) & \text{by (3.53) and (3.65)} \\ = \gamma\Big(1_{\gamma(\eta^{-1}_{Gd'}\,;\,\eta_{Gd'})}\,;\, \gamma\Big(G_{\sigma;q}\,;\, 1_{\eta^{-1}_{G\langle d\rangle\sigma}}\Big)\Big) & \text{by (3.7) and functoriality} \\ = \gamma\Big(G_{\sigma;q}\,;\, 1_{\eta^{-1}_{G\langle d\rangle\sigma}}\Big) & \text{by (3.9)} \\ = \gamma\big(G_{\sigma;q}\,;\, 1_{G\varepsilon_{\langle d\rangle\sigma}}\big) & \text{by (3.65)} \end{cases} \tag{4.63}$$

Since the last terms in (4.62) and (4.63) are inverses of each other, we have proved the desired equality (4.61). □

Characterization of Pseudo Symmetric Cat-Multiequivalences

We are now ready for the main result of this section, which is a practical characterization of a pseudo symmetric Cat-multiequivalence (Definition 4.37). The next result is the pseudo symmetric variant of Theorem 3.69.

Theorem 4.64. *Suppose*

$$\mathsf{N} \xrightarrow{(G,\{G_\sigma\})} \mathsf{M}$$

is a pseudo symmetric Cat*-multifunctor between* Cat*-multicategories. Then the following three statements are equivalent.*

1. *G is the right adjoint of an adjoint pseudo symmetric* Cat*-multiequivalence $(F, G, \eta, \varepsilon)$.*
2. *G is a pseudo symmetric* Cat*-multiequivalence.*
3. *G is*

- *essentially surjective on objects and*
- *an isomorphism on each multimorphism category.*

Proof. We consider each implication as follows.

- (1) $\Rightarrow$ (2) holds by Definitions 1.46 and 4.37.
- (2) $\Rightarrow$ (3) holds by Lemma 4.42.
- (3) $\Rightarrow$ (1) holds by Lemmas 3.68, 4.55, 4.56, and 4.58.

This completes the proof. □

Part II

Grothendieck Multiequivalence from Bipermutative-Indexed Categories to Permutative Opfibrations

5

Enriched Multicategories of Indexed Categories

Recall that $(\mathsf{Cat}, \times, \mathbf{1}, [\,,\,])$ is the complete and cocomplete symmetric monoidal closed category of small categories and functors with

- the Cartesian product as the monoidal product and
- diagram category as the closed structure (Example 1.16).

Suppose $\mathcal{D}$ is a small tight bipermutative category (Definition 2.41). The main purpose of this chapter is to construct a Cat-multicategory $\mathcal{D}\mathsf{Cat}$, with *additive* symmetric monoidal functors as objects, that fully incorporates the bipermutative structure of $\mathcal{D}$ (Theorem 5.71). We emphasize that the Cat-multicategory structure on $\mathcal{D}\mathsf{Cat}$ is generally *not* induced by a symmetric monoidal structure.

The Grothendieck Construction and Inverse K-Theory

In later chapters we utilize the Cat-multicategory $\mathcal{D}\mathsf{Cat}$ as follows.

1. In Chapter 6 $\mathcal{D}\mathsf{Cat}$ is the domain of the Grothendieck construction

$$\mathcal{D}\mathsf{Cat} \xrightarrow{\int_{\mathcal{D}}} \mathsf{PermCat}^{\mathsf{sg}}.$$

   The codomain $\mathsf{PermCat}^{\mathsf{sg}}$ is a Cat-multicategory with small permutative categories as objects. Using the Cat-multicategory structures on $\mathcal{D}\mathsf{Cat}$ and $\mathsf{PermCat}^{\mathsf{sg}}$, we show that $\int_{\mathcal{D}}$ is a *pseudo symmetric* Cat-*multifunctor* (Definition 4.1).

2. In Chapter 7 we lift $\int_{\mathcal{D}}$ to a non-symmetric Cat-multifunctor

$$\mathcal{D}\mathsf{Cat} \xrightarrow{\int_{\mathcal{D}}} \mathsf{PFib}^{\mathcal{D}}.$$

   The codomain $\mathsf{PFib}^{\mathcal{D}}$ is a non-symmetric Cat-multicategory with small permutative opfibrations over $\mathcal{D}$ as objects.

3. In Chapter 8 we prove that the lifted $\int_{\mathcal{D}}$ in Chapter 7 is a non-symmetric Cat-*multiequivalence*. This means that the lifted Grothendieck construction $\int_{\mathcal{D}}$ is

DOI: 10.1201/9781003451235-5

- essentially surjective on objects and
- an isomorphism on each multimorphism category.

4. In Chapters 9 and 10 we apply these results to the indexing category $\mathcal{A}$ (Example 2.63) for inverse K-theory $\int_{\mathcal{A}} A(-)$. The pseudo symmetric Cat-multifunctoriality of the Grothendieck construction $\int_{\mathcal{A}}$ implies that inverse K-theory is a pseudo symmetric Cat-multifunctor but *not* a Cat-multifunctor.

Organization

In Section 5.1 we first consider the diagram category $\mathcal{D}$-Cat when $(\mathcal{D}, \otimes, 1)$ is merely a permutative category (Definition 1.11). We recall the important fact that the symmetric monoidal closed structure on the diagram category $\mathcal{D}$-V induces a V-multicategory structure on $\mathcal{D}$-V. Specializing to the case V = Cat, we obtain a Cat-multicategory structure on $\mathcal{D}$-Cat. The rest of this section contains a detailed description of the Cat-multicategory $\mathcal{D}$-Cat. To emphasize the monoidal structure in use, we also denote $\mathcal{D}$-Cat by $\mathcal{D}^{\otimes}$-Cat in subsequent sections.

With $(\mathcal{D}, \oplus, \otimes)$ a small tight bipermutative category, in Section 5.2 we define some of the data of $\mathcal{D}$Cat.

- Its objects are *additive* symmetric monoidal functors, which are symmetric monoidal functors relative to the additive structure $\mathcal{D}^{\oplus} = (\mathcal{D}, \oplus, 0)$ of $\mathcal{D}$. This is in contrast to $\mathcal{D}^{\otimes}$-Cat, whose objects are functors $\mathcal{D} \longrightarrow$ Cat with no extra structure assumed.

- In each multimorphism category in $\mathcal{D}$Cat, the objects are *additive* natural transformations. These are natural transformations that satisfy a unity axiom and an additivity axiom that generalize those of monoidal natural transformations; see Explanation 5.48 (4).

- The morphisms between them are *additive* modifications. These are modifications that satisfy a unity axiom and an additivity axiom.

Additive natural transformations and additive modifications make use of both the additive structure $\mathcal{D}^{\oplus}$ and the multiplicative structure $\mathcal{D}^{\otimes}$ of $\mathcal{D}$. Moreover, their additivity axioms involve Laplaza coherence isomorphisms in $\mathcal{D}$; see (5.46) and (5.55). This section ends with the verification that each positive arity multimorphism category in $\mathcal{D}$Cat is a subcategory of the corresponding multimorphism category of $\mathcal{D}^{\otimes}$-Cat; see Lemma 5.60.

In Section 5.3 we define the Cat-multicategory structure on $\mathcal{D}$Cat as the one inherited from $\mathcal{D}^{\otimes}$-Cat in an appropriate sense. Most of the work involves checking that the Cat-multicategory structure on $\mathcal{D}^{\otimes}$-Cat restricts to additive natural transformations and additive modifications. In other words, we need to check that their unity and additivity axioms are preserved by the symmetric group action and composition in $\mathcal{D}^{\otimes}$-Cat.

Summary

There are two important differences between $\mathcal{D}$Cat and $\mathcal{D}^{\otimes}$-Cat.

1. The Cat-multicategory structure on $\mathcal{D}^{\otimes}$-Cat is induced by its symmetric monoidal closed structure given by the Day convolution with respect to the multiplicative structure $\otimes$ in $\mathcal{D}$. On the other hand, the Cat-multicategory structure on $\mathcal{D}$Cat does *not* arise from a symmetric monoidal structure. See Explanation 5.64.

2. The Cat-multicategory structure on $\mathcal{D}$Cat is defined using the one on $\mathcal{D}^{\otimes}$-Cat. However, $\mathcal{D}$Cat is *not* a sub-Cat-multicategory of $\mathcal{D}^{\otimes}$-Cat. See Explanation 5.73.

A summary of the Cat-multicategories $\mathcal{D}^{\otimes}$-Cat and $\mathcal{D}$Cat are given below.

- $\mathcal{D}^{\otimes}$-Cat (Explanation 5.2).
  - It is induced by a symmetric monoidal structure (Theorem 5.1).
  - Its n-ary 1-cells are natural transformations (5.5).
  - Its n-ary 2-cells are modifications (5.10).
- $\mathcal{D}$Cat (Theorem 5.71).
  - It is *not* induced by a symmetric monoidal structure (Explanation 5.64).
  - Its n-ary 1-cells are additive natural transformations (Definition 5.39).
  - Its n-ary 2-cells are additive modifications (Definition 5.52).

We remind the reader of our left normalized bracketing Convention 1.24 for iterated monoidal product.

5.1 Enriched Multicategories of Permutative-Indexed Categories

Recall the diagram category $\mathcal{D}$-C (Definition 1.15), which has $\mathcal{D}$-diagrams in C as objects and natural transformations as morphisms, and the notion of a V-multicategory (Definition 3.2). For a small permutative category $\mathcal{D}$, in Explanations 1.56 and 1.66 we described the symmetric monoidal structure on the diagram category $\mathcal{D}$-Cat, which has $\mathcal{D}$-indexed categories as objects, and its enriched, tensored, and cotensored structures over Cat. In this section we describe in detail the Cat-enriched multicategory structure on $\mathcal{D}$-Cat. The following theorem is an instance of [JY∞, 6.3.6] and guarantees the existence of the Cat-multicategory $\mathcal{D}$-Cat.

Theorem 5.1. *Suppose* V *is a symmetric monoidal closed category that is complete and cocomplete. Suppose* $\mathcal{D}$ *is a small symmetric monoidal category. Then the symmetric monoidal closed category* $\mathcal{D}\text{-}\mathsf{V}$ *in Theorem 1.63 has the structure of a* V*-multicategory with the same underlying* V*-category as that of* $\mathcal{D}\text{-}\mathsf{V}$.

Now we specialize to the case $\mathsf{V} = \mathsf{Cat}$, which is the main case of interest in this work.

Explanation 5.2 (Cat-Multicategory of Permutative-Indexed Categories)**.** In the context of Theorem 5.1, suppose

- $\mathsf{V} = (\mathsf{Cat}, \times, \mathbf{1}, [,])$ is the symmetric monoidal closed category of small categories and functors with the Cartesian product as the monoidal product and diagram category as the closed structure (Example 1.16) and
- $(\mathcal{D}, \otimes, 1, \beta)$ is a small permutative category (Definition 1.11).

By Theorem 5.1 $\mathcal{D}\text{-}\mathsf{Cat}$ has the structure of a Cat-multicategory, which we also denote by $\mathcal{D}\text{-}\mathsf{Cat}$, with the same underlying 2-category as that of $\mathcal{D}\text{-}\mathsf{Cat}$. Here we describe this Cat-multicategory structure in detail.

Objects The objects of the Cat-multicategory $\mathcal{D}\text{-}\mathsf{Cat}$ are $\mathcal{D}$-indexed categories (Definition 1.15), that is, functors $\mathcal{D} \longrightarrow \mathsf{Cat}$.

Multimorphism Categories in Arity 0 Suppose $Z : \mathcal{D} \longrightarrow \mathsf{Cat}$ is a $\mathcal{D}$-indexed category. The *0-ary multimorphism category* is given by

$$(\mathcal{D}\text{-}\mathsf{Cat})\big(\langle\rangle\,;\,Z\big) = (\mathcal{D}\text{-}\mathsf{Cat})(J\,,\,Z)$$

with $\langle\rangle$ the empty sequence and

$$J = \mathcal{D}(1,-) : \mathcal{D} \longrightarrow \mathsf{Cat}$$

the unit diagram in (1.69). As discussed in Explanation 1.66, the category $(\mathcal{D}\text{-}\mathsf{Cat})(J\,,\,Z)$ has

- natural transformations $J \longrightarrow Z$ as objects,
- modifications as morphisms,
- identity modifications as identities, and
- composition given by vertical composition of modifications.

By the Bicategorical Yoneda Lemma [JY21, 8.3.28] applied to the current 2-categorical context, there is a canonical isomorphism of categories

$$(\mathcal{D}\text{-}\mathsf{Cat})\big(\langle\rangle\,;\,Z\big) = (\mathcal{D}\text{-}\mathsf{Cat})(J\,,\,Z) \cong Z1 \tag{5.3}$$

given by the assignments

$$\begin{aligned} \phi &\longmapsto \phi_1(1_1) \\ \Phi &\longmapsto (\Phi_1)_{1_1} \end{aligned}$$

for each

- natural transformation $\phi : J \longrightarrow Z$ and
- modification $\Phi : \phi \longrightarrow \phi'$ with $\phi, \phi' : J \longrightarrow Z$ natural transformations.

Multimorphism Categories in Positive Arity Suppose given $\mathcal{D}$-indexed categories

$$Z, X_1, \ldots, X_n : \mathcal{D} \longrightarrow \mathsf{Cat}$$

with $n > 0$ and $\langle X \rangle = (X_1, \ldots, X_n)$. The *n-ary multimorphism category* is given by

$$(\mathcal{D}\text{-}\mathsf{Cat})\big(\langle X \rangle ; Z\big) = (\mathcal{D}\text{-}\mathsf{Cat})\big(\otimes_{j=1}^n X_j \,, Z\big). \tag{5.4}$$

For $n \geq 2$ the $\mathcal{D}$-indexed category $\otimes_{j=1}^n X_j$ is the iterated Day convolution in Explanation 1.59 with $\mathsf{V} = \mathsf{Cat}$. As discussed in Explanation 1.66, the category $(\mathcal{D}\text{-}\mathsf{Cat})\big(\otimes_{j=1}^n X_j \,, Z\big)$ has

- natural transformations $\otimes_{j=1}^n X_j \longrightarrow Z$ as objects,
- modifications as morphisms,
- identity modifications as identities, and
- composition given by vertical composition of modifications.

Unraveling the iterated Day convolution (1.60), we explicitly describe these natural transformations and modifications below.

A *natural transformation*

$$\otimes_{j=1}^n X_j \xrightarrow{\;\phi\;} Z \tag{5.5}$$

consists of, for each n-tuple of objects

$$\langle a \rangle = \langle a_j \rangle_{j=1}^n \in \mathcal{D}^n \quad \text{and} \quad a = \otimes_{j=1}^n a_j \in \mathcal{D}, \tag{5.6}$$

a component functor

$$\textstyle\prod_{j=1}^n X_j a_j \xrightarrow{\;\phi_{\langle a \rangle}\;} Za. \tag{5.7}$$

For each n-tuple of morphisms $\langle f_j : a_j \longrightarrow a'_j \rangle_{j=1}^n$ in $\mathcal{D}$, denote

$$\langle a' \rangle = \langle a'_j \rangle_{j=1}^n \,, \quad a' = \otimes_{j=1}^n a'_j \,, \quad \text{and} \quad f = \otimes_{j=1}^n f_j : a \longrightarrow a'. \tag{5.8}$$

The following *naturality diagram* of functors is required to commute.

$$\begin{array}{ccc} \prod_{j=1}^n X_j a_j & \xrightarrow{\;\phi_{\langle a \rangle}\;} & Za \\ {\scriptstyle \prod_{j=1}^n X_j f_j} \big\downarrow & & \big\downarrow {\scriptstyle Zf} \\ \prod_{j=1}^n X_j a'_j & \xrightarrow{\;\phi_{\langle a' \rangle}\;} & Za' \end{array} \tag{5.9}$$

A *modification* $\Phi : \phi \longrightarrow \varphi$ in $(\mathcal{D}\text{-}\mathsf{Cat})\big(\bigotimes_{j=1}^{n} X_j\,,\, Z\big)$ as in

$$\bigotimes_{j=1}^{n} X_j \;\overset{\phi}{\underset{\varphi}{\Downarrow\Phi}}\; Z \tag{5.10}$$

consists of a component natural transformation

$$\prod_{j=1}^{n} X_j a_j \;\overset{\phi_{\langle a\rangle}}{\underset{\varphi_{\langle a\rangle}}{\Downarrow\Phi_{\langle a\rangle}}}\; Za \tag{5.11}$$

for each n-tuple of objects $\langle a\rangle = \{a_j\}_{j=1}^{n}$ in $\mathcal{D}$ as in (5.6). The *modification axiom* in this case is equivalent to, in the context of (5.6) and (5.8), the equality of the following two whiskered natural transformations.

$$\begin{array}{ccc} \prod_{j=1}^{n} X_j a_j & \overset{\phi_{\langle a\rangle}}{\underset{\varphi_{\langle a\rangle}}{\Downarrow\Phi_{\langle a\rangle}}} & Za \\ {\scriptstyle \prod_{j=1}^{n} X_j f_j}\downarrow & & \downarrow{\scriptstyle Zf} \\ \prod_{j=1}^{n} X_j a'_j & \overset{\phi_{\langle a'\rangle}}{\underset{\varphi_{\langle a'\rangle}}{\Downarrow\Phi_{\langle a'\rangle}}} & Za' \end{array} \tag{5.12}$$

This finishes the description of the category $(\mathcal{D}\text{-}\mathsf{Cat})(\langle X\rangle\,;\, Z)$ in (5.4).

Colored Units For a $\mathcal{D}$-indexed category $Z : \mathcal{D} \longrightarrow \mathsf{Cat}$, the Z*-colored unit*

$$Z \xrightarrow{\;1_Z\;} Z \tag{5.13}$$

is the identity natural transformation.

Symmetric Group Action For a permutation $\sigma \in \Sigma_n$, the *symmetric group action*

$$(\mathcal{D}\text{-}\mathsf{Cat})\big(\langle X\rangle\,;\, Z\big) \xrightarrow[\cong]{\;\sigma\;} (\mathcal{D}\text{-}\mathsf{Cat})\big(\langle X\rangle\sigma\,;\, Z\big) \tag{5.14}$$

is the functor that sends a natural transformation $\phi : \bigotimes_{j=1}^{n} X_j \longrightarrow Z$ as in (5.5) to the natural transformation

$$\bigotimes_{j=1}^{n} X_{\sigma(j)} \xrightarrow{\;\phi^{\sigma}\;} Z.$$

For an n-tuple of objects $\langle a\rangle = \langle a_j\rangle_{j=1}^{n}$ in $\mathcal{D}$ as in (5.6), ϕ^{σ} has the following component composite functor.

$$\begin{array}{ccc} \prod_{j=1}^{n} X_{\sigma(j)} a_j & \xrightarrow{\;\phi^{\sigma}_{\langle a\rangle}\;} & Za \\ {\scriptstyle \sigma}\downarrow{\scriptstyle \cong} & & {\scriptstyle \cong}\uparrow{\scriptstyle Z(\sigma^{-1})} \\ \prod_{j=1}^{n} X_j a_{\sigma^{-1}(j)} & \xrightarrow{\;\phi_{\sigma\langle a\rangle}\;} & Za_{\sigma^{-1}} \end{array} \tag{5.15}$$

The arrows in (5.15) are given as follows.

- The left vertical arrow permutes the n factors according to $\sigma \in \Sigma_n$.
- The bottom horizontal arrow $\phi_{\sigma\langle a\rangle}$ is the component functor of ϕ as in (5.7) for the n-tuple of objects

$$\sigma\langle a\rangle = \langle a_{\sigma^{-1}(j)}\rangle_{j=1}^n \in \mathcal{D}^n \quad \text{with} \quad a_{\sigma^{-1}} = \bigotimes_{j=1}^n a_{\sigma^{-1}(j)} \in \mathcal{D}. \tag{5.16}$$

- The right vertical arrow $Z(\sigma^{-1})$ is the image under Z of the unique coherence isomorphism [ML98, XI.1]

$$a_{\sigma^{-1}} = \bigotimes_{j=1}^n a_{\sigma^{-1}(j)} \xrightarrow[\cong]{\sigma^{-1}} \bigotimes_{j=1}^n a_j = a \quad \text{in} \quad (\mathcal{D}, \otimes, \beta) \tag{5.17}$$

  that permutes the n factors according to $\sigma^{-1} \in \Sigma_n$.

This finishes the description of the symmetric group action (5.14) on natural transformations.

For a modification $\Phi : \phi \longrightarrow \varphi$ as in (5.10), its image under the symmetric group action σ in (5.14) is the following modification.

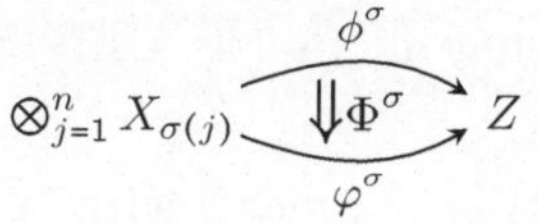

For an n-tuple of objects $\langle a\rangle = \langle a_j\rangle_{j=1}^n$ in $\mathcal{D}$ as in (5.6), Φ^σ has the component natural transformation $\Phi^\sigma_{\langle a\rangle}$ given by the following whiskering.

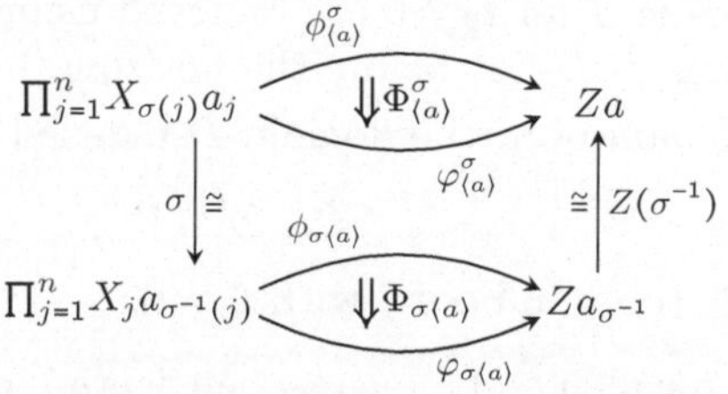

(5.18)

In (5.18) $\Phi_{\sigma\langle a\rangle}$ is the component natural transformation of Φ as in (5.11) for the n-tuple of objects $\sigma\langle a\rangle$ in (5.16). This finishes the description of the symmetric group action (5.14) on the Cat-multicategory $\mathcal{D}$-Cat.

Composition Suppose given $\mathcal{D}$-indexed categories

$$Z\,,\, \langle X_j\rangle_{j=1}^n\,,\, \langle\langle W_{ji}\rangle_{i=1}^{\ell_j}\rangle_{j=1}^n : \mathcal{D} \longrightarrow \mathsf{Cat}$$

with $n > 0$, $\langle X\rangle = (X_1, \ldots, X_n)$,

$$\langle W_j\rangle = \big(W_{j1}, \ldots, W_{j\ell_j}\big), \quad \text{and} \quad \langle W\rangle = \big(\langle W_1\rangle, \ldots, \langle W_n\rangle\big).$$

The *multicategorical composition*

$$(\mathcal{D}\text{-}\mathsf{Cat})\big(\langle X\rangle\,;\,Z\big) \times \textstyle\prod_{j=1}^{n}(\mathcal{D}\text{-}\mathsf{Cat})\big(\langle W_j\rangle\,;\,X_j\big) \xrightarrow{\ \gamma\ } (\mathcal{D}\text{-}\mathsf{Cat})\big(\langle W\rangle\,;\,Z\big) \tag{5.19}$$

is the following composite functor.

$$\begin{array}{ccc} \mathcal{D}\text{-}\mathsf{Cat}\big(\bigotimes_{j=1}^{n} X_j\,,\,Z\big) \times \prod_{j=1}^{n} \mathcal{D}\text{-}\mathsf{Cat}\big(\bigotimes_{i=1}^{\ell_j} W_{ji}\,,\,X_j\big) & \xrightarrow{\ \gamma\ } & \mathcal{D}\text{-}\mathsf{Cat}\big(\bigotimes_{j=1}^{n}\bigotimes_{i=1}^{\ell_j} W_{ji}\,,\,Z\big) \\ {\scriptstyle 1\times\otimes}\Big\downarrow & \nearrow{\scriptstyle m} & \\ \mathcal{D}\text{-}\mathsf{Cat}\big(\bigotimes_{j=1}^{n} X_j\,,\,Z\big) \times \mathcal{D}\text{-}\mathsf{Cat}\big(\bigotimes_{j=1}^{n}\bigotimes_{i=1}^{\ell_j} W_{ji}\,,\,\bigotimes_{j=1}^{n} X_j\big) & & \end{array} \tag{5.20}$$

- In (5.20) the arrow labeled m is the horizontal composition in the 2-category $\mathcal{D}$-Cat (Explanation 1.66).
- In the left vertical arrow in (5.20), $\otimes$ is the Day convolution in
  - (1.51) for $\mathcal{D}$-indexed categories,
  - (1.52) for natural transformations, and
  - (1.70) for modifications.

The diagram (5.20) assumes that each $\ell_j > 0$.

If $\ell_j = 0$ for some but not all $j \in \{1, \ldots, n\}$, then one makes the following adjustments in (5.20).

- Since $\langle W_j\rangle = \langle\rangle$ for each index j with $\ell_j = 0$, along the left vertical arrow in (5.20) the $\mathcal{D}$-indexed category $\bigotimes_{i=1}^{\ell_j} W_{ji}$ should be interpreted as the unit diagram $J = \mathcal{D}(1, -)$ in (1.69), which is the monoidal unit in $\mathcal{D}$-Cat with respect to the Day convolution.
- These copies of J for $\ell_j = 0$ are removed using the left and right unit isomorphisms in $\mathcal{D}$-Cat before the horizontal composition m.
- In the codomain of γ, the domain $\mathcal{D}$-indexed category

  $$\bigotimes_{j=1}^{n}\bigotimes_{i=1}^{\ell_j} W_{ji}$$

  should omit those indices j with $\ell_j = 0$.

Finally, if $\ell_j = 0$ for all $j \in \{1, \ldots, n\}$, then there is one further adjustment. Since $\langle W\rangle$ is also the empty sequence $\langle\rangle$, in the codomain of γ, the domain $\mathcal{D}$-indexed category should be J.

This completes the description of the Cat-multicategory $\mathcal{D}$-Cat when $(\mathcal{D}, \otimes, 1, \beta)$ is a small permutative category. ◇

Explanation 5.21 (Composition in $\mathcal{D}$-Cat)**.** The composition γ in (5.20) can be described more explicitly as follows. Suppose given natural transformations

$$\bigotimes_{j=1}^{n} X_j \xrightarrow{\ \phi\ } Z \quad \text{and} \quad \bigotimes_{i=1}^{\ell_j} W_{ji} \xrightarrow{\ \phi_j\ } X_j \tag{5.22}$$

for $1 \leq j \leq n$ with each $\ell_j > 0$. Their composite is the natural transformation

$$\bigotimes_{j=1}^{n} \bigotimes_{i=1}^{\ell_j} W_{ji} \xrightarrow{\gamma(\phi;\langle \phi_j \rangle)} Z \tag{5.23}$$

defined as follows. Suppose given objects $a_{ji} \in \mathcal{D}$ for $1 \leq j \leq n$ and $1 \leq i \leq \ell_j$, along with the following notation.

$$\begin{cases} \langle a_j \rangle = (a_{j1}, \ldots, a_{j\ell_j}) \in \mathcal{D}^{\ell_j} & a_j = \bigotimes_{i=1}^{\ell_j} a_{ji} \in \mathcal{D} \\ \langle a \rangle = (a_1, \ldots, a_n) \in \mathcal{D}^n & a = \bigotimes_{j=1}^{n} a_j \in \mathcal{D} \end{cases} \tag{5.24}$$

The corresponding component functor of $\gamma(\phi; \langle \phi_j \rangle)$ in (5.23) is the following composite.

$$\begin{array}{ccc} \prod_{j=1}^{n} \prod_{i=1}^{\ell_j} W_{ji} a_{ji} & \xrightarrow{\gamma(\phi;\langle \phi_j \rangle)_{\langle \langle a_{ji} \rangle_i \rangle_j}} & Za \\ & {\scriptstyle \prod_j (\phi_j)_{\langle a_j \rangle}} \searrow \quad \nearrow {\scriptstyle \phi_{\langle a \rangle}} & \\ & \prod_{j=1}^{n} X_j a_j & \end{array} \tag{5.25}$$

Here $\phi_{\langle a \rangle}$ and $(\phi_j)_{\langle a_j \rangle}$ are the indicated component functors of, respective, ϕ and ϕ_j as in (5.7).

Next, suppose given modifications Φ and Φ_j for $1 \leq j \leq n$ as follows.

$$\bigotimes_{j=1}^{n} X_j \underset{\varphi}{\overset{\phi}{\rightrightarrows}} Z \;\; (\Downarrow \Phi) \qquad \bigotimes_{i=1}^{\ell_j} W_{ji} \underset{\varphi_j}{\overset{\phi_j}{\rightrightarrows}} X_j \;\; (\Downarrow \Phi_j) \tag{5.26}$$

Their composite modification

$$\bigotimes_{j=1}^{n} \bigotimes_{i=1}^{\ell_j} W_{ji} \underset{\gamma(\varphi;\langle \varphi_j \rangle)}{\overset{\gamma(\phi;\langle \phi_j \rangle)}{\rightrightarrows}} Z \;\; (\Downarrow \gamma(\Phi; \langle \Phi_j \rangle))$$

has the following component horizontal composite natural transformation.

$$\prod_{j=1}^{n} \prod_{i=1}^{\ell_j} W_{ji} a_{ji} \underset{\prod_j (\varphi_j)_{\langle a_j \rangle}}{\overset{\prod_j (\phi_j)_{\langle a_j \rangle}}{\rightrightarrows}} \prod_{j=1}^{n} X_j a_j \underset{\varphi_{\langle a \rangle}}{\overset{\phi_{\langle a \rangle}}{\rightrightarrows}} Za \;\; (\Downarrow \textstyle\prod_j (\Phi_j)_{\langle a_j \rangle},\ \Downarrow \Phi_{\langle a \rangle}) \tag{5.27}$$

Here $\Phi_{\langle a \rangle}$ and $(\Phi_j)_{\langle a_j \rangle}$ are the indicated component natural transformations of, respectively, Φ and Φ_j as in (5.11).

The discussion above assumes that each $\ell_j > 0$. Suppose there is at least one index $j \in \{1, \ldots, n\}$ with $\ell_j = 0$. Then one makes the appropriate adjustments as discussed after (5.20), using the canonical isomorphism of categories in (5.3)

$$(\mathcal{D}\text{-}\mathsf{Cat})\big(\langle\rangle; X_j\big) = (\mathcal{D}\text{-}\mathsf{Cat})(J, X_j) \cong X_j 1.$$

For example, suppose $\ell_j = 0$ for all $j \in \{1, \ldots, n\}$. Then the composition in (5.20)

$$\mathcal{D}\text{-}\mathsf{Cat}\big(\otimes_{j=1}^n X_j \, , \, Z \big) \times \textstyle\prod_{j=1}^n X_j 1 \xrightarrow{\ \gamma\ } Z1$$

is given on objects by

$$\gamma\big(\phi; \langle x \rangle\big) = \phi_{\langle 1 \rangle} \langle x \rangle \in Z1$$

for

- each natural transformation $\phi : \otimes_{j=1}^n X_j \longrightarrow Z$ and
- objects $x_j \in X_j 1$ with $\langle x \rangle = (x_1, \ldots, x_n)$.

Here $\langle 1 \rangle$ is the n-tuple with copies of $1 \in \mathcal{D}$, and

$$\textstyle\prod_{j=1}^n X_j 1 \xrightarrow{\ \phi_{\langle 1 \rangle}\ } Z1$$

is the corresponding component functor of ϕ. This is well defined, since 1 is the strict monoidal unit in $\mathcal{D}$ and $\otimes_{j=1}^n 1 = 1$. ◇

5.2 Additive Natural Transformations and Modifications

Throughout this section we assume that

$$\big(\mathcal{D}, (\oplus, 0, \beta^{\oplus}), (\otimes, 1, \beta^{\otimes}), (\partial^l, \partial^r)\big) \tag{5.28}$$

is a small *tight* bipermutative category (Definition 2.41). The tightness condition means that the factorization morphisms ∂^l and ∂^r are natural isomorphisms. For example, $\mathsf{Fin}^{\mathsf{sk}}$, Σ, $\mathcal{F}$, and $\mathcal{A}$ in Section 2.3 are small tight bipermutative categories. As in Section 2.2, denote the additive structure of $\mathcal{D}$ by

$$\mathcal{D}^{\oplus} = (\mathcal{D}, \oplus, 0, \beta^{\oplus}) \tag{5.29}$$

and the multiplicative structure of $\mathcal{D}$ by

$$\mathcal{D}^{\otimes} = (\mathcal{D}, \otimes, 1, \beta^{\otimes}). \tag{5.30}$$

Each of $\mathcal{D}^{\oplus}$ and $\mathcal{D}^{\otimes}$ is a permutative category (Definition 1.11).

Our next objective is to construct a variant of the Cat-enriched multicategory $\mathcal{D}^{\otimes}$-Cat in Section 5.1, denoted $\mathcal{D}\mathsf{Cat}$, that incorporates the bipermutative structure of $\mathcal{D}$. In this section, we define its

- objects, which are called additive symmetric monoidal functors (Definition 5.31), and
- multimorphism categories in positive arity, with

– additive natural transformations (Definition 5.39) as objects and
– additive modifications (Definition 5.52) as morphisms.

In Section 5.3 we prove that $\mathcal{D}\mathsf{Cat}$ is a Cat-multicategory.

Objects: Additive Symmetric Monoidal Functors

We first define the objects in $\mathcal{D}\mathsf{Cat}$, which are symmetric monoidal functors with respect to the additive structure of $\mathcal{D}$.

Definition 5.31. An *additive symmetric monoidal functor* is a symmetric monoidal functor (Definition 1.17)

$$(X, X^2, X^0) : \mathcal{D}^{\oplus} \longrightarrow \mathsf{Cat}.$$

We call (X^2, X^0) the *additive structure* of X and suppress it from the notation if it is clear from the context. ◇

Explanation 5.32. An additive symmetric monoidal functor (X, X^2, X^0) is defined using the additive structure $\mathcal{D}^{\oplus} = (\mathcal{D}, \oplus, 0, \beta^{\oplus})$ of $\mathcal{D}$ in (5.29). Note that we do *not* require X^0 or X^2 to be invertible in general.

- The functor $X : \mathcal{D} \longrightarrow \mathsf{Cat}$ is a $\mathcal{D}$-indexed category.
- The *unit constraint* X^0 is a functor

  $$\mathbf{1} \xrightarrow{X^0} X0.$$

  So X^0 is determined by the *unit object*

  $$X^0 * \in X0 \tag{5.33}$$

  with $* \in \mathbf{1}$ the unique object and $0 \in \mathcal{D}$ the additive zero.
- The *monoidal constraint* X^2 is a natural transformation with a component functor

  $$Xa \times Xb \xrightarrow{X^2_{a,b}} X(a \oplus b) \tag{5.34}$$

  for each pair of objects $a, b \in \mathcal{D}$.

The data (X, X^2, X^0) satisfy the unity axiom (1.18), the associativity axiom (1.19), and the braiding axiom (1.20) for a symmetric monoidal functor. For future reference we state these axioms explicitly below.

Naturality The naturality of X^2 states that, for morphisms

$$f : a \longrightarrow b \quad \text{and} \quad f' : a' \longrightarrow b'$$

in $\mathcal{D}$, the following diagram commutes, with $(-)_* = X(-)$.

$$\begin{array}{ccc} Xa \times Xa' & \xrightarrow{X^2_{a,a'}} & X(a \oplus a') \\ {\scriptstyle f_* \times f'_*}\downarrow & & \downarrow{\scriptstyle (f \oplus f')_*} \\ Xb \times Xb' & \xrightarrow{X^2_{b,b'}} & X(b \oplus b') \end{array} \tag{5.35}$$

Unity The unity axiom (1.18) for (X, X^2, X^0) states that, for objects $a \in \mathcal{D}$ and $w \in Xa$ and morphisms $f \in Xa$, the following equalities hold in Xa.

$$\begin{cases} X^2_{0,a}\left(X^0 *, w\right) = w = X^2_{a,0}\left(w, X^0 *\right) \\ X^2_{0,a}\left(1_{X^0 *}, f\right) = f = X^2_{a,0}\left(f, 1_{X^0 *}\right) \end{cases} \tag{5.36}$$

Associativity The associativity axiom (1.19) for (X, X^2, X^0) states that, for objects $a, b, c \in \mathcal{D}$ and

$$(x, y, z) \in Xa \times Xb \times Xc$$

either all objects or all morphisms, the equality

$$X^2_{a\oplus b,c}\left(X^2_{a,b}(x,y), z\right) = X^2_{a,b\oplus c}\left(x, X^2_{b,c}(y,z)\right) \tag{5.37}$$

holds in $X(a \oplus b \oplus c)$.

Braiding The braiding axiom (1.20) for (X, X^2, X^0) states the equality

$$\beta^{\oplus}_* X^2_{a,b}(x,y) = X^2_{b,a}(y,x) \tag{5.38}$$

in $X(b \oplus a)$ with

- $\beta^{\oplus} : a \oplus b \longrightarrow b \oplus a$ the braiding in $\mathcal{D}$ and
- $\beta^{\oplus}_* = X\beta^{\oplus} : X(a \oplus b) \longrightarrow X(b \oplus a)$ its image under X. ◇

Multimorphism Categories: Additive Natural Transformations

Next we define the objects in the multimorphism categories in $\mathcal{D}\mathsf{Cat}$. These are subcategories of the corresponding multimorphism categories in $\mathcal{D}^{\otimes}\text{-}\mathsf{Cat}$ (Explanation 5.2), with $\mathcal{D}^{\otimes} = (\mathcal{D}, \otimes, 1, \beta^{\otimes})$ the *multiplicative* structure of $\mathcal{D}$ in (5.30). Recall from (5.4) that $(\mathcal{D}^{\otimes}\text{-}\mathsf{Cat})(\langle X \rangle; Z)$ has

- natural transformations $\bigotimes_{j=1}^n X_j \longrightarrow Z$ as objects (5.5),
- modifications as morphisms (5.10), and
- identities and composition given by those of modifications under vertical composition.

In each multimorphism category in $\mathcal{D}\mathsf{Cat}$, the objects are natural transformations in $\mathcal{D}^{\otimes}\text{-}\mathsf{Cat}$ satisfying two compatibility conditions, which are (5.43) and (5.45) below, with respect to the *additive* structures of the additive symmetric monoidal functors in the input and the output.

Definition 5.39. Suppose given additive symmetric monoidal functors

$$(Z, Z^2, Z^0) \quad \text{and} \quad \left\langle (X_j, X^2_j, X^0_j) \right\rangle_{j=1}^n : \mathcal{D}^{\oplus} \longrightarrow \mathsf{Cat}$$

with $n > 0$ as in Definition 5.31. An *additive natural transformation*

$$\big\langle (X_j, X_j^2, X_j^0) \big\rangle_{j=1}^n \xrightarrow{\ \phi\ } (Z, Z^2, Z^0) \tag{5.40}$$

is a natural transformation in $\mathcal{D}^{\otimes}$-Cat as in (5.5)

$$\textstyle\bigotimes_{j=1}^n X_j \xrightarrow{\ \phi\ } Z \tag{5.41}$$

that satisfies the following unity axiom (5.43) and additivity axiom (5.45).

Unity For each $i \in \{1, \ldots, n\}$ and objects $a_j \in \mathcal{D}$ for $i \neq j \in \{1, \ldots, n\}$, consider the component functor of ϕ as in (5.7)

$$X_1 a_1 \times \cdots \times X_i \mathsf{0} \times \cdots \times X_n a_n \xrightarrow{\phi_{(a_1, \ldots, \mathsf{0}, \ldots, a_n)}} Z\mathsf{0} \tag{5.42}$$

with

- i-th domain factor $X_i \mathsf{0}$ and
- j-th domain factor $X_j a_j$ for $j \neq i$.

For each $j \neq i$, suppose $x_j \in X_j a_j$ is an object, and $f_j \in X_j a_j$ is a morphism. Then the following object, respectively, morphism, equalities are required to hold in $Z\mathsf{0}$, where $\phi = \phi_{(a_1, \ldots, \mathsf{0}, \ldots, a_n)}$.

$$\begin{cases} \phi\big(x_1, \ldots, X_i^0 *, \ldots, x_n\big) = Z^0 * \\ \phi\big(f_1, \ldots, 1_{X_i^0 *}, \ldots, f_n\big) = 1_{Z^0 *} \end{cases} \tag{5.43}$$

Here $X_i^0 * \in X_i \mathsf{0}$ and $Z^0 * \in Z\mathsf{0}$ are the unit objects (5.33) of, respectively, X_i and Z.

Additivity Suppose given objects $\langle a \rangle = \langle a_j \rangle_{j=1}^n \in \mathcal{D}^n$ and $a_i' \in \mathcal{D}$ for some $i \in \{1, \ldots, n\}$, along with the following notation.

$$\begin{cases} a = a_1 \otimes \cdots \otimes a_i \otimes \cdots \otimes a_n & a_i'' = a_i \oplus a_i' \\ a' = a_1 \otimes \cdots \otimes a_i' \otimes \cdots \otimes a_n & \langle a' \rangle = (a_1, \ldots, a_i', \ldots, a_n) \\ a'' = a_1 \otimes \cdots \otimes a_i'' \otimes \cdots \otimes a_n & \langle a'' \rangle = (a_1, \ldots, a_i'', \ldots, a_n) \\ X\langle a \rangle = X_1 a_1 \times \cdots \times X_i a_i \times \cdots \times X_n a_n & \\ X\langle a' \rangle = X_1 a_1 \times \cdots \times X_i a_i' \times \cdots \times X_n a_n & \\ X\langle a'' \rangle = X_1 a_1 \times \cdots \times X_i a_i'' \times \cdots \times X_n a_n & \end{cases} \tag{5.44}$$

- In each of a, a', and a'', the j-th factor is a_j for $j \neq i$, and the i-th factor is, respectively, a_i, a_i', and a_i''.
- In each of $\langle a' \rangle$ and $\langle a'' \rangle$, the j-th entry is a_j for $j \neq i$, and the i-th entry is, respectively, a_i' and a_i''.

- In each of $X\langle a\rangle$, $X\langle a'\rangle$, and $X\langle a''\rangle$, the j-th factor is $X_j a_j$ for $j \neq i$, and the i-th factor is, respectively, $X_i a_i$, $X_i a_i'$, and $X_i a_i''$.

Then the following diagram of functors is required to commute.

$$\begin{array}{ccccc} & & X_1a_1 \times \cdots \times (X_ia_i \times X_ia_i') \times \cdots \times X_na_n & & \\ & \swarrow^{1 \times \cdots \times (X_i^2)_{a_i,a_i'} \times \cdots \times 1} & & \searrow^{\varrho} & \\ X\langle a''\rangle & & & & X\langle a\rangle \times X\langle a'\rangle \\ \downarrow{\scriptstyle \phi_{\langle a''\rangle}} & & & & \downarrow{\scriptstyle \phi_{\langle a\rangle} \times \phi_{\langle a'\rangle}} \\ Za'' & & & & Za \times Za' \\ & \searrow_{Z(\mathsf{L})}^{\cong} & & \swarrow_{Z^2_{a,a'}} & \\ & & Z(a \oplus a') & & \end{array} \tag{5.45}$$

The detail of (5.45) is as follows.

- At the top node of (5.45), the i-th factor is $X_ia_i \times X_ia_i'$, and the j-th factor is X_ja_j for $j \neq i$.
- $Z^2_{a,a'}$ and

$$X_ia_i \times X_ia_i' \xrightarrow{(X_i^2)_{a_i,a_i'}} X_i(a_i \oplus a_i') = X_ia_i''$$

  are the indicated component functors of the monoidal constraints (5.34) of, respectively, Z and X_i.
- Each of $\phi_{\langle a\rangle}$, $\phi_{\langle a'\rangle}$, and $\phi_{\langle a''\rangle}$ is the indicated component functor (5.7) of ϕ.
- Omitting the $\otimes$ symbol to save space, the morphism

$$\begin{array}{ccc} a'' & \xrightarrow[\cong]{\mathsf{L}} & a \oplus a' \\ \| & & \| \\ a_1 \cdots (a_i \oplus a_i') \cdots a_n & & (a_1 \cdots a_i \cdots a_n) \oplus (a_1 \cdots a_i' \cdots a_n) \end{array} \tag{5.46}$$

  in (5.45) is the unique Laplaza coherence isomorphism in the tight bipermutative category $\mathcal{D}$ (Example 2.91) that distributes over the sum.
- ϱ in (5.45) is the following composite functor.

$$\begin{array}{ccc} X_1a_1 \times \cdots \times (X_ia_i \times X_ia_i') \times \cdots \times X_na_n & & \\ \downarrow{\scriptstyle \varrho_1} & \searrow^{\varrho} & \\ & & X\langle a\rangle \times X\langle a'\rangle \\ & \nearrow_{\cong}^{\varrho_2} & \\ (X_1a_1)^2 \times \cdots \times (X_ia_i \times X_ia_i') \times \cdots \times (X_na_n)^2 & & \end{array} \tag{5.47}$$

- ϱ_1 is the product of
  (i) the diagonal functors $X_j a_j \longrightarrow (X_j a_j)^2$ for $j \neq i$ and
  (ii) the identity functor of $X_i a_i \times X_i a'_i$.
- ϱ_2 permutes the $2n$ factors according to the transpose permutation $\xi^{\otimes}_{2,n} \in \Sigma_{2n}$ in (2.52).

This finishes the definition of an additive natural transformation. An explicit form of the additivity axiom (5.45) is given in (5.51). $\diamond$

Explanation 5.48 (Additive Natural Transformations)**.** Consider Definition 5.39.

1. An additive natural transformation is a natural transformation in $\mathcal{D}^{\otimes}$-Cat with two additional properties—the unity axiom (5.43) and the additivity axiom (5.45)—but *no* extra structure.

2. The domain of the natural transformation $\phi : \bigotimes_{j=1}^n X_j \longrightarrow Z$ in (5.41) is the iterated Day convolution $\bigotimes_{j=1}^n X_j$ in (1.60). It is constructed using the *multiplicative* structure $\mathcal{D}^{\otimes} = (\mathcal{D}, \otimes, 1, \beta^{\otimes})$ and the underlying functors $X_j : \mathcal{D} \longrightarrow \mathsf{Cat}$.

3. The component functor ϕ in (5.42) is well defined because the iterated product in $\mathcal{D}$

$$\overbrace{a_1 \otimes \cdots \otimes a_{i-1}}^{\text{empty if } i=1} \otimes 0 \otimes \overbrace{a_{i+1} \otimes \cdots \otimes a_n}^{\text{empty if } i=n} = 0$$

   by the multiplicative zero axiom (2.31) of a ring, hence also bipermutative, category.

4. In the case $n = 1$, an additive natural transformation

$$X_1 \xrightarrow{\phi} Z$$

   is precisely a monoidal natural transformation (Definition 1.21). The unity axiom (5.43) and the additivity axiom (5.45) reduce to the diagrams in (1.22) when $n = 1$.

5. In (5.47) ϱ_2 is the unique coherence isomorphism in the symmetric monoidal category Cat that permutes the factors to $X\langle a\rangle \times X\langle a'\rangle$ in such a way that the relative order of the two factors in

   - each $(X_j a_j)^2$ for $j \neq i$ and
   - $X_i a_i \times X_i a'_i$

   remain unchanged.

6. The additivity axiom (5.45) is given explicitly as follows. Suppose given

$$\langle x\rangle = \langle x_j\rangle_{j=1}^n \in \textstyle\prod_{j=1}^n X_j a_j \quad \text{and} \quad x'_i \in X_i a'_i \tag{5.49}$$

that are either all objects or all morphisms, along with the following notation.

$$\begin{cases} x_i'' = (X_i^2)_{a_i,a_i'}(x_i, x_i') \in X_i a_i'' = X_i(a_i \oplus a_i') \\ \langle x' \rangle = (x_1, \ldots, x_i', \ldots, x_n) \in X\langle a' \rangle \\ \langle x'' \rangle = (x_1, \ldots, x_i'', \ldots, x_n) \in X\langle a'' \rangle \end{cases} \tag{5.50}$$

Then the additivity axiom (5.45) is the equality

$$\mathsf{L}_* \phi_{\langle a'' \rangle} \langle x'' \rangle = Z^2_{a,a'}\big(\phi_{\langle a \rangle}\langle x \rangle \,,\, \phi_{\langle a' \rangle}\langle x' \rangle\big) \tag{5.51}$$

in $Z(a \oplus a')$, with $\mathsf{L}_* = Z(\mathsf{L})$. ◇

Multimorphism Categories: Additive Modifications

Next we define the morphisms in the multimorphism categories in $\mathcal{D}\mathsf{Cat}$.

Definition 5.52. In the context of Definition 5.39, suppose ϕ and ψ as in

$$\big\langle (X_j, X_j^2, X_j^0) \big\rangle_{j=1}^n \overset{\phi}{\underset{\psi}{\rightrightarrows}} (Z, Z^2, Z^0)$$

are additive natural transformations. An *additive modification* $\Phi : \phi \longrightarrow \psi$, which is also denoted

$$\big\langle (X_j, X_j^2, X_j^0) \big\rangle_{j=1}^n \underset{\psi}{\overset{\phi}{\Downarrow \Phi}} (Z, Z^2, Z^0), \tag{5.53}$$

is a modification

$$\bigotimes_{j=1}^n X_j \underset{\psi}{\overset{\phi}{\Downarrow \Phi}} Z$$

as in (5.10) that satisfies the following unity and additivity axioms.

Unity In the context of (5.42) and (5.43), the component morphism of Φ in $Z0$

$$\begin{array}{c} \phi(x_1, \ldots, X_i^0 *, \ldots, x_n) = Z^0 * \\ \Big\downarrow \Phi_{(x_1, \ldots, X_i^0 *, \ldots, x_n)} \\ \psi(x_1, \ldots, X_i^0 *, \ldots, x_n) = Z^0 * \end{array} \tag{5.54}$$

is the identity morphism $1_{Z^0 *}$.

Additivity In the context of (5.44) and (5.45), the following two whiskered natural transformations are equal, where each $\Phi_?$ is the indicated component natural transformation of Φ as in (5.11).

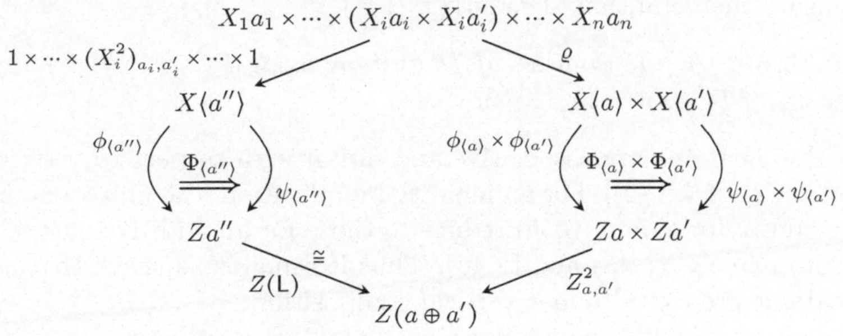

(5.55)

This finishes the definition of an additive modification. ◇

Explanation 5.56 (Additive Modifications)**.** Consider Definition 5.52.

1. The morphism in the unity axiom (5.54) for Φ is the component of the component natural transformation of Φ

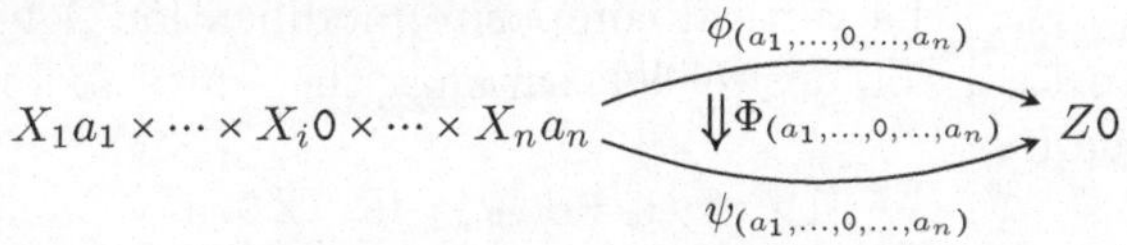

at the object $(x_1, \ldots, X_i^0 *, \ldots, x_n)$.

2. In the context of (5.50) and (5.51) with objects $x_j \in X_ja_j$ and $x'_i \in X_ia'_i$, the additivity axiom (5.55) for Φ is the morphism equality

$$\mathsf{L}_*\big(\Phi_{\langle a'' \rangle}\big)_{\langle x'' \rangle} = Z^2_{a,a'}\big((\Phi_{\langle a \rangle})_{\langle x \rangle}, (\Phi_{\langle a' \rangle})_{\langle x' \rangle}\big) \tag{5.57}$$

in $Z(a \oplus a')$. ◇

Multimorphism Categories

Next we define the multimorphism categories in $\mathcal{D}\mathsf{Cat}$ in positive arity and check that they are well defined. Recall the multimorphism category $\mathcal{D}^{\otimes}\text{-}\mathsf{Cat}(\langle X \rangle; Z)$ in (5.4).

Definition 5.58. In the context of Definition 5.39 with $\langle X \rangle = \langle X_j \rangle_{j=1}^n$, define the data of a subcategory

$$\mathcal{D}\mathsf{Cat}\big(\langle X \rangle; Z\big) \tag{5.59}$$

of $\mathcal{D}^{\otimes}\text{-}\mathsf{Cat}(\langle X \rangle; Z)$ as follows.

Objects An object is an *additive* natural transformation as in (5.40).

Morphisms A morphism is an *additive* modification as in (5.53).

Identities Identity morphisms are identity modifications.

Composition It is the vertical composition of modifications (1.44).

This finishes the definition of $\mathcal{D}\mathsf{Cat}(\langle X \rangle\,;\, Z)$. ◇

Lemma 5.60. *In the context of Definition 5.58, $\mathcal{D}\mathsf{Cat}(\langle X \rangle\,;\, Z)$ defines a subcategory of $\mathcal{D}^{\otimes}\text{-}\mathsf{Cat}(\langle X \rangle\,;\, Z)$ in* (5.4).

Proof. Modifications are associative and unital with respect to vertical composition [JY21, 4.4.8 (2)]. For an identity modification, the unity axiom (5.54) and the additivity axiom (5.55) reduce to those for an additive natural transformation, namely, (5.43) and (5.45). Thus it remains to check that additive modifications are closed under vertical composition.

Suppose given additive modifications Φ and Ψ as follows.

$$\big\langle (X_j, X_j^2, X_j^0)\big\rangle_{j=1}^{n} \xrightarrow[\omega]{\;\phi\; \Downarrow\Phi\;\; \psi\;\; \Downarrow\Psi\;} (Z, Z^2, Z^0)$$

We must show that the vertical composite modification $\Psi\Phi : \phi \longrightarrow \omega$ is additive. The vertical composite $\Psi\Phi$ satisfies the unity axiom (5.54) by the morphism equality

$$1_{Z^0 *} 1_{Z^0 *} = 1_{Z^0 *} \quad \text{in} \quad Z\mathsf{0}.$$

The following computation shows that the vertical composite $\Psi\Phi$ satisfies the additivity axiom (5.57).

$$\begin{aligned}
&\mathsf{L}_*\big((\Psi\Phi)_{\langle a''\rangle}\big)_{\langle x''\rangle}\\
&= \mathsf{L}_*\Big(\big(\Psi_{\langle a''\rangle}\big)_{\langle x''\rangle}\big(\Phi_{\langle a''\rangle}\big)_{\langle x''\rangle}\Big)\\
&= \Big(\mathsf{L}_*\big(\Psi_{\langle a''\rangle}\big)_{\langle x''\rangle}\Big)\Big(\mathsf{L}_*\big(\Phi_{\langle a''\rangle}\big)_{\langle x''\rangle}\Big)\\
&= \Big(Z^2_{a,a'}\big((\Psi_{\langle a\rangle})_{\langle x\rangle}, (\Psi_{\langle a'\rangle})_{\langle x'\rangle}\big)\Big)\Big(Z^2_{a,a'}\big((\Phi_{\langle a\rangle})_{\langle x\rangle}, (\Phi_{\langle a'\rangle})_{\langle x'\rangle}\big)\Big)\\
&= Z^2_{a,a'}\big((\Psi_{\langle a\rangle})_{\langle x\rangle}(\Phi_{\langle a\rangle})_{\langle x\rangle}, (\Psi_{\langle a'\rangle})_{\langle x'\rangle}(\Phi_{\langle a'\rangle})_{\langle x'\rangle}\big)\\
&= Z^2_{a,a'}\big(((\Psi\Phi)_{\langle a\rangle})_{\langle x\rangle}, ((\Psi\Phi)_{\langle a'\rangle})_{\langle x'\rangle}\big)
\end{aligned}$$

The above equalities hold for the following reasons:

- The first and the last equalities follow from the definitions of the vertical composite modification $\Psi\Phi$.
- The second equality follows from the functoriality of $\mathsf{L}_* = Z(\mathsf{L})$.
- The third equality follows from the additivity axiom (5.57) for Φ and Ψ.
- The fourth equality follows from the functoriality of $Z^2_{a,a'}$.

This proves that the vertical composite $\Psi\Phi$ is an additive modification. □

5.3 Enriched Multicategories of Bipermutative-Indexed Categories

We continue to assume that $\mathcal{D}$ is a small tight bipermutative category as in (5.28). In this section we construct the Cat-enriched multicategory $\mathcal{D}\mathsf{Cat}$ that incorporates the bipermutative structure of $\mathcal{D}$ (Theorem 5.71). While the Cat-multicategory structure on $\mathcal{D}\mathsf{Cat}$ is inherited from that on $\mathcal{D}^{\otimes}$-Cat in Explanation 5.2, there are several subtleties that one should be aware of:

1. The Cat-multicategory structure on $\mathcal{D}^{\otimes}$-Cat is induced by a symmetric monoidal closed structure (Theorem 5.1). This is *not* true for the Cat-multicategory $\mathcal{D}\mathsf{Cat}$. We discuss this point in Explanation 5.64.

2. Additive natural transformations (Definition 5.39) are natural transformations with extra properties. So we have to check that the Cat-multicategory structure on $\mathcal{D}^{\otimes}$-Cat preserves additive natural transformations. The same remark also applies to additive modifications (Definition 5.52), which are modifications with extra properties.

3. $\mathcal{D}\mathsf{Cat}$ is generally *not* a sub-Cat-multicategory of $\mathcal{D}^{\otimes}$-Cat because the Cat-multifunctor
$$\mathcal{D}\mathsf{Cat} \longrightarrow \mathcal{D}^{\otimes}\text{-}\mathsf{Cat}$$
is not injective on objects; see Explanation 5.73.

Definitions

Now we define the Cat-multicategory structure on $\mathcal{D}\mathsf{Cat}$.

Definition 5.61. For a small tight bipermutative category $\mathcal{D}$ as in (5.28), define the data of a Cat-multicategory $\mathcal{D}\mathsf{Cat}$ as follows.

Objects These are additive symmetric monoidal functors $\mathcal{D}^{\oplus} \longrightarrow \mathsf{Cat}$ (Definition 5.31).

Multimorphism Categories in Arity 0 For each additive symmetric monoidal functor
$$(Z, Z^2, Z^0) : \mathcal{D}^{\oplus} \longrightarrow \mathsf{Cat},$$
define the 0-ary multimorphism category
$$\mathcal{D}\mathsf{Cat}\big(\langle\rangle\,;\,Z\big) = Z1. \tag{5.62}$$

Multimorphism Categories in Positive Arity Suppose given additive symmetric monoidal functors
$$(Z, Z^2, Z^0) \quad\text{and}\quad \big\langle (X_j, X_j^2, X_j^0)\big\rangle_{j=1}^{n} : \mathcal{D}^{\oplus} \longrightarrow \mathsf{Cat} \tag{5.63}$$

with $n > 0$ and $\langle X\rangle = \langle X_j\rangle_{j=1}^n$. Define the n-ary multimorphism category

$$\mathcal{D}\mathsf{Cat}\big(\langle X\rangle\,;\,Z\big)$$

to be the category in (5.59).

Colored Units For an additive symmetric monoidal functor (Z, Z^2, Z^0), the Z-colored unit is the identity natural transformation $1_Z : Z \longrightarrow Z$.

Symmetric Group Action and Composition These are restrictions of those in the Cat-multicategory $\mathcal{D}^{\otimes}$-Cat—namely, (5.14) and (5.19)—using

- (5.3) and (5.62) to identify multimorphism categories with arity 0 input and
- Lemma 5.60 for multimorphism categories with positive arity input.

This finishes the definition of $\mathcal{D}\mathsf{Cat}$. ◇

Explanation 5.64 ($\mathcal{D}\mathsf{Cat}$ is Not Symmetric Monoidal)**.** Here we discuss a subtle difference between $\mathcal{D}^{\otimes}$-Cat and $\mathcal{D}\mathsf{Cat}$. On the one hand, the Cat-multicategory structure on $\mathcal{D}^{\otimes}$-Cat is induced by the symmetric monoidal closed structure on $\mathcal{D}^{\otimes}$-Cat (Explanation 5.2). Its monoidal unit is the unit diagram (1.69)

$$J = \mathcal{D}(\mathsf{1}, -) : \mathcal{D} \longrightarrow \mathsf{Cat}.$$

By the definition of the induced Cat-multicategory, for each $\mathcal{D}$-indexed category $Z : \mathcal{D} \longrightarrow \mathsf{Cat}$, the 0-ary multimorphism category in $\mathcal{D}^{\otimes}$-Cat is

$$(\mathcal{D}^{\otimes}\text{-}\mathsf{Cat})\big(\langle\rangle\,;\,Z\big) = (\mathcal{D}\text{-}\mathsf{Cat})(J\,,\,Z) \cong Z\mathsf{1},$$

with the canonical isomorphism given by the Bicategorical Yoneda Lemma.

On the other hand, in (5.62) we directly define the 0-ary multimorphism category $\mathcal{D}\mathsf{Cat}(\langle\rangle\,;\,Z)$ to be $Z\mathsf{1}$, without using J. The reason is that, in general, $J = \mathcal{D}(\mathsf{1}, -)$ does *not* admit the structure of a symmetric monoidal functor $\mathcal{D}^{\oplus} \longrightarrow \mathsf{Cat}$. Therefore, the Cat-multicategory $\mathcal{D}\mathsf{Cat}$, which we will verify shortly, is generally *not* induced by a restriction of the symmetric monoidal structure on $\mathcal{D}^{\otimes}$-Cat.

In more detail, by definition a monoidal functor structure on J has a unit constraint, which is a choice of a unit object (5.33)

$$J^0 * \in J\mathsf{0} = \mathcal{D}(\mathsf{1}, \mathsf{0}).$$

For a general bipermutative category $\mathcal{D}$, the morphism set $\mathcal{D}(\mathsf{1},\mathsf{0})$ may be empty, so such a unit object J^0* does not exist. Here are some concrete examples of small tight bipermutative categories where this morphism set is empty:

1. In $\mathsf{Fin}^{\mathsf{sk}}$ (Example 2.48) there is an empty morphism set

$$\mathsf{Fin}^{\mathsf{sk}}(\overline{1}, \overline{0}) = \mathsf{Set}\big(\{1\}\,,\,\varnothing\big) = \varnothing.$$

2. In Σ (Example 2.57) there is an empty morphism set
$$\Sigma(1,0) = \varnothing.$$

3. In $\mathcal{A}$ (Example 2.63) there is an empty morphism set
$$\mathcal{A}\big((1)\,,\,\langle\rangle\big) = \varnothing$$
because there are no functions $\overline{1} = \{1\} \longrightarrow \varnothing$.

In each of these examples, J cannot have a unit object and, therefore, does not admit the structure of a monoidal functor $\mathcal{D}^{\oplus} \longrightarrow \mathsf{Cat}$. $\diamond$

Symmetric Group Action

To show that $\mathcal{D}\mathsf{Cat}$ as in Definition 5.61 is a Cat-multicategory, we first check that the symmetric group action, which is a restriction of the symmetric group action in $\mathcal{D}^{\otimes}$-Cat, is well defined.

Lemma 5.65. *Additive natural transformations and additive modifications are closed under the symmetric group action* (5.14) *in* $\mathcal{D}^{\otimes}$-Cat.

Proof. Suppose given additive symmetric monoidal functors Z and X_j as in (5.63) with $n > 0$ and $\langle X\rangle = \langle X_j\rangle_{j=1}^n$. Suppose $\sigma \in \Sigma_n$ is a permutation. Given an additive natural transformation $\phi : \bigotimes_{j=1}^n X_j \longrightarrow Z$ as in (5.41), we must show that the natural transformation
$$\bigotimes_{j=1}^n X_{\sigma(j)} \xrightarrow{\phi^\sigma} Z$$
satisfies the unity axiom (5.43) and the additivity axiom (5.45) of an additive natural transformation. By definition (5.15), for objects $\langle a\rangle = \langle a_j\rangle_{j=1}^n$ in $\mathcal{D}$, the component functor $\phi^\sigma_{\langle a\rangle}$ is the following composite.
$$\begin{array}{ccc}
\prod_{j=1}^n X_{\sigma(j)} a_j & \xrightarrow{\phi^\sigma_{\langle a\rangle}} & Za \\
{\scriptstyle \sigma}\big\downarrow{\scriptstyle \cong} & & {\scriptstyle \cong}\big\uparrow{\scriptstyle Z(\sigma^{-1})} \\
\prod_{j=1}^n X_j a_{\sigma^{-1}(j)} & \xrightarrow{\phi_{\sigma\langle a\rangle}} & Za_{\sigma^{-1}}
\end{array}$$

For the unity axiom (5.43) for ϕ^σ, suppose $a_i = 0$ for some index $i \in \{1,\ldots,n\}$. The two equalities in the unity axiom hold for $\phi^\sigma_{\langle a\rangle}$ by

- the unity axiom (5.43) for $\phi_{\sigma\langle a\rangle}$;
- the uniqueness of Laplaza coherence isomorphisms (Theorem 2.90), which implies that the coherence isomorphism
$$a_{\sigma^{-1}} = \bigotimes_{j=1}^n a_{\sigma^{-1}(j)} = 0 \xrightarrow[\cong]{\sigma^{-1}} 0 = \bigotimes_{j=1}^n a_j = a$$
is the identity morphism 1_0 in $\mathcal{D}$; and
- the functoriality of Z, which implies that $Z(1_0) = 1_{Z0}$.

For the additivity axiom (5.45) for ϕ^σ, in addition to the notation in (5.16) and (5.44), we use the following notation.

$$\begin{cases} \sigma\langle a'\rangle = \big(a_{\sigma^{-1}(1)}, \ldots, a'_i, \ldots, a_{\sigma^{-1}(n)}\big) \in \mathcal{D}^n \\ a'_{\sigma^{-1}} = a_{\sigma^{-1}(1)} \otimes \cdots \otimes a'_i \otimes \cdots \otimes a_{\sigma^{-1}(n)} \in \mathcal{D} \\ \sigma\langle a''\rangle = \big(a_{\sigma^{-1}(1)}, \ldots, a''_i, \ldots, a_{\sigma^{-1}(n)}\big) \in \mathcal{D}^n \\ a''_{\sigma^{-1}} = a_{\sigma^{-1}(1)} \otimes \cdots \otimes a''_i \otimes \cdots \otimes a_{\sigma^{-1}(n)} \in \mathcal{D} \end{cases} \tag{5.66}$$

$$\begin{cases} X\sigma\langle a\rangle = X_1 a_{\sigma^{-1}(1)} \times \cdots \times X_{\sigma(i)} a_i \times \cdots \times X_n a_{\sigma^{-1}(n)} \\ X\sigma\langle a'\rangle = X_1 a_{\sigma^{-1}(1)} \times \cdots \times X_{\sigma(i)} a'_i \times \cdots \times X_n a_{\sigma^{-1}(n)} \\ X\sigma\langle a''\rangle = X_1 a_{\sigma^{-1}(1)} \times \cdots \times X_{\sigma(i)} a''_i \times \cdots \times X_n a_{\sigma^{-1}(n)} \end{cases}$$

$$\begin{cases} X^\sigma\langle a\rangle = X_{\sigma(1)} a_1 \times \cdots \times X_{\sigma(i)} a_i \times \cdots \times X_{\sigma(n)} a_n \\ X^\sigma\langle a'\rangle = X_{\sigma(1)} a_1 \times \cdots \times X_{\sigma(i)} a'_i \times \cdots \times X_{\sigma(n)} a_n \\ X^\sigma\langle a''\rangle = X_{\sigma(1)} a_1 \times \cdots \times X_{\sigma(i)} a''_i \times \cdots \times X_{\sigma(n)} a_n \end{cases}$$

$$\begin{cases} Y = X_{\sigma(1)} a_1 \times \cdots \times \big(X_{\sigma(i)} a_i \times X_{\sigma(i)} a'_i\big) \times \cdots \times X_{\sigma(n)} a_n \\ W = X_1 a_{\sigma^{-1}(1)} \times \cdots \times \big(X_{\sigma(i)} a_i \times X_{\sigma(i)} a'_i\big) \times \cdots \times X_n a_{\sigma^{-1}(n)} \end{cases}$$

With these notation, the additivity axiom (5.45) for ϕ^σ asserts the commutativity of the boundary of the following diagram of functors.

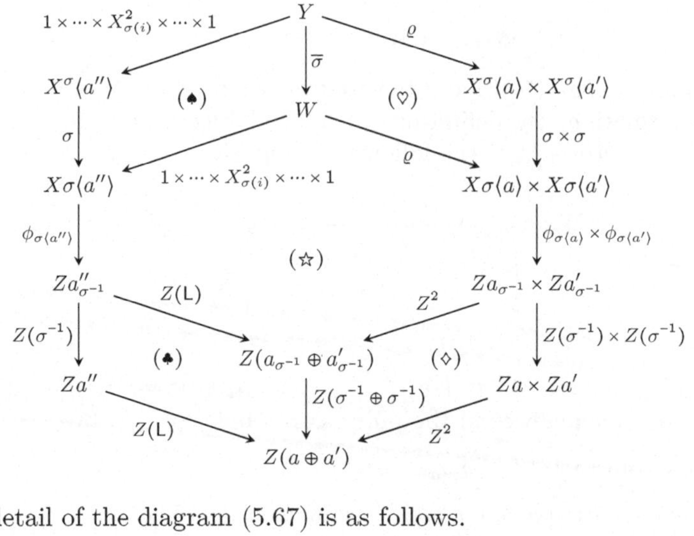

(5.67)

The detail of the diagram (5.67) is as follows.

- The isomorphism $\overline{\sigma} : Y \longrightarrow W$ is the block permutation induced by $\sigma \in \Sigma_n$ that regards the product $X_{\sigma(i)} a_i \times X_{\sigma(i)} a'_i$ in Y as a single block.
- The region $(\spadesuit)$ commutes by the naturality of the braiding in the symmetric monoidal category $(\mathsf{Cat}, \times)$.

- The region (♡) commutes because the two composites become equal after projecting onto any factor of the common codomain, $X\sigma\langle a\rangle \times X\sigma\langle a'\rangle$.
- The region (♣) commutes by the
  - functoriality of Z and
  - uniqueness of Laplaza coherence isomorphisms (Theorem 2.90).
- The region (◇) commutes by the naturality of the monoidal constraint Z^2 of the additive symmetric monoidal functor (Z, Z^2, Z^0).
- The hexagon (☆) commutes by the additivity axiom (5.45) for ϕ.

Thus ϕ^σ is an additivity natural transformation. This proves that additive natural transformations are closed under the symmetric group action (5.14) in $\mathcal{D}^\otimes$-Cat.

The proof that additive modifications are closed under the symmetric group action in $\mathcal{D}^\otimes$-Cat is obtained from the argument above by replacing ϕ with an additive modification Φ. This is valid for the following two reasons:

- Φ^σ (5.18) is defined in the same way as ϕ^σ, with $\Phi_{\sigma\langle a\rangle}$ replacing $\phi_{\sigma\langle a\rangle}$.
- The unity axiom (5.54) and the additivity axiom (5.55) for an additive modification are obtained from those for an additive natural transformation—namely, (5.43) and (5.45)—by replacing ϕ with Φ.

This finishes the proof. □

Composition

Next we check that the composition in $\mathcal{D}$Cat, which is a restriction of the composition in $\mathcal{D}^\otimes$-Cat, is well defined.

Lemma 5.68. *Additive natural transformations and additive modifications are closed under the composition* (5.19) *in* $\mathcal{D}^\otimes$-Cat.

Proof. Suppose given additive symmetric monoidal functors (Definition 5.31)

$$Z\,,\, \langle X_j\rangle_{j=1}^n\,,\, \langle\langle W_{ji}\rangle_{i=1}^{\ell_j}\rangle_{j=1}^n : \mathcal{D}^\oplus \longrightarrow \mathsf{Cat}$$

with $n > 0$, $\langle X\rangle = (X_1, \ldots, X_n)$,

$$\langle W_j\rangle = \big(W_{j1}, \ldots, W_{j\ell_j}\big), \quad \text{and} \quad \langle W\rangle = \big(\langle W_1\rangle, \ldots, \langle W_n\rangle\big).$$

Suppose given additive natural transformations (Definition 5.39)

$$\textstyle\bigotimes_{j=1}^n X_j \xrightarrow{\ \phi\ } Z \quad \text{and} \quad \bigotimes_{i=1}^{\ell_j} W_{ji} \xrightarrow{\ \phi_j\ } X_j$$

for $1 \leq j \leq n$. If some $\ell_j = 0$, then we make some appropriate adjustments in the argument below, as discussed in the last paragraph in Explanation 5.21. In the rest of this proof, we assume that each $\ell_j > 0$.

We must show that the composite natural transformation (5.23)

$$\bigotimes_{j=1}^n \bigotimes_{i=1}^{\ell_j} W_{ji} \xrightarrow{\gamma(\phi;\langle\phi_j\rangle)} Z$$

satisfies the unity axiom (5.43) and the additivity axiom (5.45). The unity axiom for $\gamma(\phi; \langle\phi_j\rangle)$ follows from

- the unity axiom for ϕ and ϕ_j and
- the definition (5.25) of each component functor of $\gamma(\phi; \langle\phi_j\rangle)$ as the composite

$$\prod_{j=1}^n \prod_{i=1}^{\ell_j} W_{ji} a_{ji} \xrightarrow{\prod_j (\phi_j)_{\langle a_j\rangle}} \prod_{j=1}^n X_j a_j \xrightarrow{\phi_{\langle a\rangle}} Za.$$

To check the additivity axiom (5.45) for $\gamma(\phi; \langle\phi_j\rangle)$, suppose given objects

- $a_{ji} \in \mathcal{D}$ for $j \in \{1, \ldots, n\}$ and $i \in \{1, \ldots, \ell_j\}$ and
- $a'_{hk} \in \mathcal{D}$ for one index $h \in \{1, \ldots, n\}$ and one index $k \in \{1, \ldots, \ell_h\}$,

along with the notation in (5.24) and below.

$$\begin{cases} a''_{hk} = a_{hk} \oplus a'_{hk} & \overline{a} = a_1 \otimes \cdots \otimes (a_h \oplus a'_h) \otimes \cdots \otimes a_n \\ a'_h = a_{h1} \otimes \cdots \otimes a'_{hk} \otimes \cdots \otimes a_{h\ell_h} & a' = a_1 \otimes \cdots \otimes a'_h \otimes \cdots \otimes a_n \\ a''_h = a_{h1} \otimes \cdots \otimes a''_{hk} \otimes \cdots \otimes a_{h\ell_h} & a'' = a_1 \otimes \cdots \otimes a''_h \otimes \cdots \otimes a_n \end{cases}$$

$$\begin{cases} A = \prod_{j=1}^{h-1} \prod_{i=1}^{\ell_j} W_{ji} a_{ji} & B_1 = \prod_{t=1}^{k-1} W_{ht} a_{ht} & \\ C = \prod_{j=h+1}^{n} \prod_{i=1}^{\ell_j} W_{ji} a_{ji} & B_2 = \prod_{t=k+1}^{\ell_h} W_{ht} a_{ht} & \\ B = W_{hk} a_{hk} & B' = W_{hk} a'_{hk} & B'' = W_{hk} a''_{hk} \end{cases}$$

$$\begin{cases} \phi^1 = \prod_{j=1}^{h-1} \phi_j & \phi^2 = \prod_{j=h+1}^{n} \phi_j & \widetilde{\phi} = \prod_{j=1}^{n} \phi_j = \phi^1 \times \phi_h \times \phi^2 \\ T = \prod_{j=1}^{h-1} X_j a_j & V = \prod_{j=h+1}^{n} X_j a_j & \overline{U} = X_h(a_h \oplus a'_h) \\ U = X_h a_h & U' = X_h a'_h & U'' = X_h a''_h \end{cases}$$

In ϕ^1, ϕ^2, and $\widetilde{\phi}$ above, the subscript in each component functor is omitted. To further simplify the diagram below, we omit the symbol $\times$ and use juxtaposition to denote products. With these notation and convention, the additivity axiom (5.45) for $\gamma(\phi; \langle\phi_j\rangle)$ asserts the commutativity of the boundary of the following diagram.

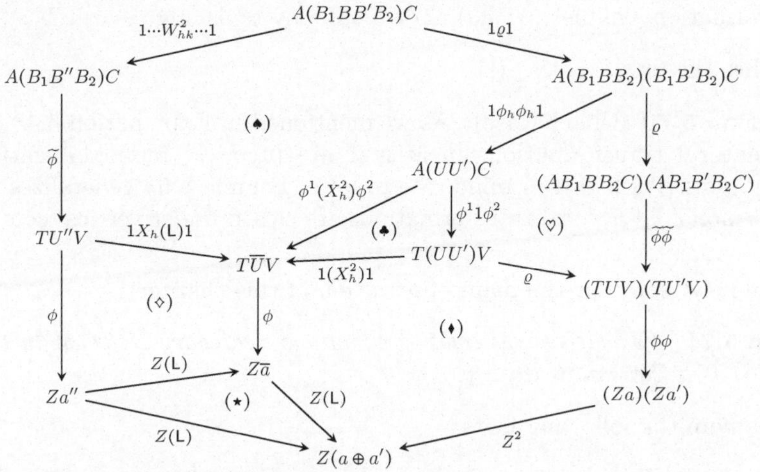

(5.69)

The detail of the diagram (5.69) is as follows.

- Each L is a Laplaza coherence isomorphism (Theorem 2.90).
- Each ϱ is a composite as in (5.47) that involves a product of diagonal functors on some of the factors and identity functors, followed by a permutation.
- The region (♠) commutes by the additivity axiom (5.45) for ϕ_h.
- The regions (♡) and (♣) commute by construction.
- The region (◇) commutes by the naturality (5.9) of ϕ.
- The region ($\star$) commutes by the
  - functoriality of Z and
  - uniqueness of Laplaza coherence isomorphisms (Theorem 2.90).
- The region (♦) commutes by the additivity axiom (5.45) for ϕ.

This proves that $\gamma(\phi; \langle\phi_j\rangle)$ satisfies the additivity axiom (5.45). We have shown that additive natural transformations are closed under the composition in $\mathcal{D}^{\otimes}$-Cat.

The proof that additive modifications are closed under the composition in $\mathcal{D}^{\otimes}$-Cat is obtained from the argument above by replacing ϕ and ϕ_j with, respectively, additive modifications $\Phi : \phi \longrightarrow \varphi$ and $\Phi_j : \phi_j \longrightarrow \varphi_j$. This is valid for the following two reasons:

- Each component natural transformation of $\gamma(\Phi; \langle\Phi_j\rangle)$ is the horizontal composite (5.27).

- The unity axiom (5.54) and the additivity axiom (5.55) for an additive modification are obtained from those for an additive natural transformation—namely, (5.43) and (5.45)—by replacing ϕ with Φ.

This finishes the proof. □

Explanation 5.70 (Unary Case). As we mentioned in Explanation 5.48 (4), additive natural transformations in arity 1 are precisely monoidal natural transformations. With this in mind, the proof of Lemma 5.68 generalizes the proof that monoidal natural transformations are closed under vertical composition. ◇

We are now ready for the main observation of this chapter.

Theorem 5.71. *For each small tight bipermutative category $\mathcal{D}$, $\mathcal{D}\mathsf{Cat}$ in Definition 5.61 is a* Cat*-multicategory.*

Proof. We note the following facts.

- By Lemma 5.60 each positive arity multimorphism category in $\mathcal{D}\mathsf{Cat}$ is a subcategory of the corresponding multimorphism category in $\mathcal{D}^{\otimes}\text{-}\mathsf{Cat}$.
- By (5.3) and (5.62) the arity 0 multimorphism categories in $\mathcal{D}\mathsf{Cat}$ and $\mathcal{D}^{\otimes}\text{-}\mathsf{Cat}$ are canonically isomorphic.
- The colored units in $\mathcal{D}^{\otimes}\text{-}\mathsf{Cat}$ and $\mathcal{D}\mathsf{Cat}$ are both given by identity natural transformations.
- By Lemmas 5.65 and 5.68, respectively, the symmetric group action and composition in $\mathcal{D}^{\otimes}\text{-}\mathsf{Cat}$ restrict to $\mathcal{D}\mathsf{Cat}$.

Thus the axioms of a Cat-multicategory (Definition 3.2) hold in $\mathcal{D}\mathsf{Cat}$ because they hold in $\mathcal{D}^{\otimes}\text{-}\mathsf{Cat}$ (Theorem 5.1). □

Since the Cat-multicategory structure on $\mathcal{D}\mathsf{Cat}$ (Theorem 5.71)—the colored units, the symmetric group action, and the composition—is inherited from that on $\mathcal{D}^{\otimes}\text{-}\mathsf{Cat}$ (Theorem 5.1), we obtain the following result. Recall from Definition 3.19 the notion of a V-multifunctor between V-multicategories.

Corollary 5.72. *For each small tight bipermutative category $\mathcal{D}$, there is a forgetful* Cat*-multifunctor*

$$\mathcal{D}\mathsf{Cat} \xrightarrow{\;U\;} \mathcal{D}^{\otimes}\text{-}\mathsf{Cat}$$

that is defined as follows.

- *U sends each additive symmetric monoidal functor (Z, Z^2, Z^0) to the underlying functor $Z : \mathcal{D} \longrightarrow \mathsf{Cat}$.*
- *U is the canonical isomorphism given by* (5.3) *and* (5.62) *for arity 0 multimorphism categories.*

- *U is the subcategory inclusion in Lemma 5.60 for positive arity multimorphism categories.*

Explanation 5.73 (Non-injectivity on Objects). For the Cat-multifunctor U in Corollary 5.72, each multimorphism functor is either an isomorphism or a subcategory inclusion. However, in general, U is *not* injective on objects because a $\mathcal{D}$-indexed category $Z : \mathcal{D} \longrightarrow \mathsf{Cat}$ may admit several different additive structures (Z^2, Z^0). Therefore, $\mathcal{D}\mathsf{Cat}$ is generally *not* a sub-Cat-multicategory of $\mathcal{D}^{\otimes}$-Cat. ◇

Example 5.74. Suppose $\mathcal{D}$ is any one of the small tight bipermutative categories

$$\mathsf{Fin}^{\mathsf{sk}}, \quad \Sigma, \quad \mathcal{F}, \quad \text{and} \quad \mathcal{A}$$

in, respectively, Examples 2.48, 2.57, 2.58, and 2.63. Then Theorem 5.71 and Corollary 5.72 apply to $\mathcal{D}\mathsf{Cat}$. In particular, there is a Cat-multicategory $\mathcal{A}\mathsf{Cat}$ whose objects are symmetric monoidal functors

$$\mathcal{A}^{\oplus} = \big(\mathcal{A}, \oplus, \langle\rangle, \beta^{\oplus}\big) \longrightarrow \big(\mathsf{Cat}, \times, \mathbf{1}, \xi\big).$$

As we will discuss in Chapters 9 and 10, $\mathcal{A}\mathsf{Cat}$ is an intermediate Cat-multicategory in inverse K-theory. ◇

6

The Grothendieck Construction is a Pseudo Symmetric Cat-*Multifunctor*

For a small tight bipermutative category $(\mathcal{D}, \oplus, \otimes)$, in Theorem 5.71 we observed that $\mathcal{D}\mathsf{Cat}$ is a Cat-multicategory, with additive symmetric monoidal functors $\mathcal{D}^{\oplus} \longrightarrow \mathsf{Cat}$ as objects (Definition 5.31). The main result of this chapter is Theorem 6.110. It shows that the Grothendieck construction of indexed categories extends to a *pseudo symmetric* Cat-multifunctor

$$\mathcal{D}\mathsf{Cat} \xrightarrow{\int_{\mathcal{D}}} \mathsf{PermCat}^{\mathsf{sg}}.$$

Furthermore, it is a Cat-multifunctor if the multiplicative braiding in $\mathcal{D}$ is the identity. Conversely, if $\int_{\mathcal{D}}$ is a Cat-multifunctor, then

$$x \otimes y = y \otimes x$$

for all objects $x, y \in \mathcal{D}$. The codomain $\mathsf{PermCat}^{\mathsf{sg}}$ (Section 6.1) is a Cat-multicategory with small permutative categories as objects.

Summary

Below is a summary of the Cat-multicategory $\mathsf{PermCat}^{\mathsf{sg}}$ and the pseudo symmetric Cat-multifunctor $\int_{\mathcal{D}}$.

- $\mathsf{PermCat}^{\mathsf{sg}}$ (Theorem 6.31).
  - Its objects are small permutative categories (Definition 1.11).
  - Its n-ary 1-cells are strong multilinear functors (Definition 6.4).
  - Its n-ary 2-cells are multilinear natural transformations (Definition 6.16).
  - Its colored units are identity 1-linear functors (Example 6.12).
  - Its composition is defined in (6.28).
  - Its symmetry is defined in (6.24).
  - It is *not* induced by a symmetric monoidal structure (Remark 6.32).

DOI: 10.1201/9781003451235-6

- $\int_{\mathcal{D}}$ (Theorem 6.110).
  - It has object assignment given in Lemma 6.42, n-ary 1-cell assignment given in Definitions 6.43 and 6.45, and n-ary 2-cell assignment given in Definition 6.61.
  - It preserves units (Lemma 6.73) and composition (Lemma 6.74).
  - Its pseudo symmetry isomorphisms are given in (6.86).

Pseudo Symmetry and Inverse K-Theory

Recall from Definition 4.1 that a pseudo symmetric Cat-multifunctor *strictly* preserves the colored units and multicategorical composition. It preserves the symmetric group action up to the pseudo symmetry isomorphisms, which satisfy the four coherence axioms (4.3) through (4.6). There is a canonical bijective correspondence between (i) Cat-multifunctors and (ii) pseudo symmetric Cat-multifunctors with identity pseudo symmetry isomorphisms (Example 4.7).

The components of the pseudo symmetry isomorphisms for the Grothendieck construction are defined in (6.89). They make full use of the multiplicative braiding $\beta^{\otimes}$ in $\mathcal{D}$, so they are almost never identities in practice. The Grothendieck construction $\int_{\mathcal{D}}$ is a Cat-multifunctor if and only if the multiplicative braiding $\beta^{\otimes}$ in $\mathcal{D}$ happens to be the identity natural transformation. For example, the pseudo symmetry isomorphisms for the Grothendieck construction are not identities for each of the small tight bipermutative categories $\mathsf{Fin}^{\mathsf{sk}}$, Σ, $\mathcal{F}$, and $\mathcal{A}$ in Section 2.3. In each of these cases, the Grothendieck construction $\int_{\mathcal{D}}$ is genuinely a pseudo symmetric Cat-multifunctor and *not* a Cat-multifunctor (Example 6.114).

An important consequence of Theorem 6.110 is that the inverse K-theory functor, from Γ-categories to small permutative categories, is a pseudo symmetric Cat-multifunctor but *not* a Cat-multifunctor. As we will discuss in Chapters 9 and 10, the inverse K-theory functor $\int_{\mathcal{A}} A(-)$ is a composite of two functors, the second of which is the Grothendieck construction $\int_{\mathcal{A}}$ with the indexing category $\mathcal{A}$ (Example 2.63). Since the multiplicative braiding in $\mathcal{A}$ is not the identity, the pseudo symmetry isomorphisms of $\int_{\mathcal{A}}$ are not identities.

Organization

In Section 6.1 we discuss the Cat-multicategory $\mathsf{PermCat}^{\mathsf{sg}}$. It is the codomain of the Grothendieck construction $\int_{\mathcal{D}}$.

- Its objects are small permutative categories.
- Its n-ary 1-cells—that is, objects in the multimorphism categories—are *strong* n-linear functors. The strong condition, which is what the

superscript sg refers to, means that each linearity constraint is a natural isomorphism.

- Its n-ary 2-cells are n-linear natural transformations.

In Section 6.2 we first review the Grothendieck construction of an indexed category. This construction associates to each functor $X : \mathcal{D} \longrightarrow \mathsf{Cat}$ a category $\int_{\mathcal{D}} X$, which is small if $\mathcal{D}$ is small. In the literature the Grothendieck construction is sometimes called the *wreath product* and denoted $\mathcal{D} \wr X$. Next we discuss the case with $\mathcal{D}$ a permutative category and X a symmetric monoidal functor. In this case the Grothendieck construction has the canonical structure of a permutative category. We emphasize that the symmetric monoidal functor X is *not* assumed to be strong. In other words, its unit and monoidal constraints are not required to be invertible. The construction $X \longmapsto \int_{\mathcal{D}} X$ is the object assignment of the Grothendieck construction.

In Section 6.3 we extend the Grothendieck construction $\int_{\mathcal{D}}$ to n-ary 1-cells in $\mathcal{D}\mathsf{Cat}$, which are additive natural transformations (Definition 5.39). For such an additive natural transformation ϕ, the Grothendieck construction $\int_{\mathcal{D}} \phi$ is a *strong* n-linear functor. The components of its linearity constraints $(\int_{\mathcal{D}} \phi)^2_i$ are defined in (6.56).

- As a morphism in a Grothendieck construction $\int_{\mathcal{D}} Z$, the first entry in each component of $(\int_{\mathcal{D}} \phi)^2_i$ is a morphism in the indexing small tight bipermutative category $\mathcal{D}$. In fact, it is defined as the inverse of a Laplaza coherence isomorphism (Example 2.91). Proofs regarding these linearity constraints often involve the uniqueness in Laplaza's Coherence Theorem 2.90. For example, in Lemma 6.57 we prove each of the multilinearity constraint axioms for $\int_{\mathcal{D}} \phi$ by applying Theorem 2.90.

- The second entry in each component of $(\int_{\mathcal{D}} \phi)^2_i$ is an identity morphism.

In Section 6.4 we extend the Grothendieck construction $\int_{\mathcal{D}}$ to n-ary 2-cells in $\mathcal{D}\mathsf{Cat}$, which are additive modifications (Definition 5.52). For such an additive modification Φ, the Grothendieck construction $\int_{\mathcal{D}} \Phi$ is an n-linear natural transformation. Each of its components is a morphism in a Grothendieck construction $\int_{\mathcal{D}} Z$; see (6.63).

- Its first entry is an identity morphism in $\mathcal{D}$.

- Its second entry is a component morphism of a component natural transformation of Φ.

Lemma 6.69 shows that the Grothendieck constructions of

- additive symmetric monoidal functors $\mathcal{D}^{\oplus} \longrightarrow \mathsf{Cat}$,

- additive natural transformations, and

- additive modifications

define functors from the multimorphism categories of $\mathcal{D}\mathsf{Cat}$ to those of $\mathsf{PermCat}^{\mathsf{sg}}$.

In Section 6.5 we show that the Grothendieck construction $\int_{\mathcal{D}}$ preserves the colored units and composition in the Cat-multicategories $\mathcal{D}\mathsf{Cat}$ and $\mathsf{PermCat}^{\mathsf{sg}}$. For additive natural transformations, the proof that $\int_{\mathcal{D}}$ preserves composition involves showing that the relevant linearity constraints are equal. This is once again an application of Laplaza's Coherence Theorem 2.90; see (6.83) and (6.84).

In Section 6.6 we first construct the pseudo symmetry isomorphisms $(\int_{\mathcal{D}})_\sigma$ for the Grothendieck construction. For an additive natural transformation ϕ, each component of $(\int_{\mathcal{D}})_{\sigma,\phi}$ is an isomorphism in a Grothendieck construction $\int_{\mathcal{D}} Z$; see (6.89).

- Its first entry is a coherence isomorphism in the multiplicative structure $\mathcal{D}^{\otimes}$ that permutes the tensor factors.
- Its second entry is an identity morphism.

In the rest of this section, we show that

- each component $(\int_{\mathcal{D}})_{\sigma,\phi}$ is a multilinear natural isomorphism (Lemmas 6.92 and 6.95) and
- it is a natural isomorphism in the ϕ variable (Lemma 6.100).

In Section 6.7 we finish the proof that the Grothendieck construction $\int_{\mathcal{D}}$ is a pseudo symmetric Cat-multifunctor by checking the four axioms in Definition 4.1. These proofs mostly involve unpacking the definitions of various constructions and do not involve Theorem 2.90. We provide all the explicit steps because we want to leave no doubt that our many constructions and definitions are correct. Moreover, $\int_{\mathcal{D}}$ is a Cat-multifunctor if the multiplicative braiding in $\mathcal{D}$ is the identity; a weaker converse is also true.

We remind the reader of our left normalized bracketing Convention 1.24 for iterated monoidal product.

6.1 Enriched Multicategory of Permutative Categories

In this section we review the Cat-enriched multicategory PermCat with

- small permutative categories (Definition 1.11) as objects,
- n-linear functors (Definition 6.4) as n-ary 1-cells, and
- n-linear natural transformations (Definition 6.16) as n-ary 2-cells.

The existence of the Cat-multicategory PermCat is stated in [EM06, page 174]. A detailed proof is given in [JY∞, Sections 6.5 and 6.6], from which the presentation below is adapted. Restricting to *strong* multilinear functors yields the Cat-enriched multicategory $\mathsf{PermCat}^{\mathsf{sg}}$ that is the codomain of the Grothendieck construction.

Multilinear Functors

The definitions of multilinear functors and multilinear natural transformations involve substitution of variables. We use the following notation to simplify relevant presentation.

Notation 6.1. Suppose

$$\langle x\rangle = (x_1, \ldots, x_n)$$

is an n-tuple of symbols, and x_i' is a symbol for some index $i \in \{1, \ldots, n\}$.

- We denote by

$$\langle x \circ_i x_i'\rangle = \langle x\rangle \circ_i x_i' = \big(\overbrace{x_1, \ldots, x_{i-1}}^{\text{empty if } i=1}, x_i', \overbrace{x_{i+1}, \ldots, x_n}^{\text{empty if } i=n}\big) \tag{6.2}$$

  the n-tuple obtained from $\langle x\rangle$ by replacing its i-th entry by x_i'.

- For $i \neq \ell \in \{1, \ldots, n\}$ and a symbol x_ℓ', we denote by

$$\langle x \circ_i x_i' \circ_\ell x_\ell'\rangle = \langle x\rangle \circ_i x_i' \circ_\ell x_\ell' \tag{6.3}$$

  the n-tuple obtained from $\langle x \circ_i x_i'\rangle$ by replacing its ℓ-th entry by x_ℓ'. ◇

The following definition of an n-linear functor is from [EM06, 3.2], as presented in [JY∞, 6.5.4]. For the rest of this section, suppose $\mathsf{C}_1, \ldots, \mathsf{C}_n$, and D are permutative categories (Definition 1.11). In each permutative category, the monoidal product, the monoidal unit, and the braiding are denoted, respectively, $\oplus$, e, and β.

Definition 6.4. An *n-linear functor*

$$\textstyle\prod_{j=1}^n \mathsf{C}_j \xrightarrow{(F, \langle F_j^2\rangle_{j=1}^n)} \mathsf{D} \tag{6.5}$$

consists of the following data.

- $F : \prod_{j=1}^n \mathsf{C}_j \longrightarrow \mathsf{D}$ is a functor.

- For each $j \in \{1, \ldots, n\}$, F_j^2 is a natural transformation, called the *j-th linearity constraint*, with component morphisms

$$F\langle X \circ_j X_j\rangle \oplus F\langle X \circ_j X_j'\rangle \xrightarrow{F_j^2} F\langle X \circ_j (X_j \oplus X_j')\rangle \in \mathsf{D} \tag{6.6}$$

  for objects $\langle X\rangle = \langle X_j\rangle \in \prod_{j=1}^n \mathsf{C}_j$ and $X_j' \in \mathsf{C}_j$.

The data above are subject to the following five axioms.

Unity F satisfies the following unity axioms for $j \in \{1, \ldots, n\}$, objects $\langle X \rangle$, and morphisms $\langle f \rangle$ in $\prod_{j=1}^{n} \mathsf{C}_j$.

$$\begin{cases} F\langle X \circ_j e \rangle = e \\ F\langle f \circ_j 1_e \rangle = 1_e \end{cases} \tag{6.7}$$

Constraint Unity

$$F_j^2 = 1 \qquad \text{if any } X_i = e \text{ or if } X_j' = e. \tag{6.8}$$

Constraint Associativity The following diagram commutes for each $i \in \{1, \ldots, n\}$ and objects $\langle X \rangle \in \prod_{j=1}^{n} \mathsf{C}_j$, with $X_i', X_i'' \in \mathsf{C}_i$.

$$\begin{array}{ccc} F\langle X \circ_i X_i \rangle \oplus F\langle X \circ_i X_i' \rangle \oplus F\langle X \circ_i X_i'' \rangle & \xrightarrow{1 \oplus F_i^2} & F\langle X \circ_i X_i \rangle \oplus F\langle X \circ_i (X_i' \oplus X_i'') \rangle \\ {\scriptstyle F_i^2 \oplus 1} \downarrow & & \downarrow {\scriptstyle F_i^2} \\ F\langle X \circ_i (X_i \oplus X_i') \rangle \oplus F\langle X \circ_i X_i'' \rangle & \xrightarrow{F_i^2} & F\langle X \circ_i (X_i \oplus X_i' \oplus X_i'') \rangle \end{array} \tag{6.9}$$

Constraint Symmetry The following diagram commutes for each $i \in \{1, \ldots, n\}$ and objects $\langle X \rangle \in \prod_{j=1}^{n} \mathsf{C}_j$, with $X_i' \in \mathsf{C}_i$.

$$\begin{array}{ccc} F\langle X \circ_i X_i \rangle \oplus F\langle X \circ_i X_i' \rangle & \xrightarrow{F_i^2} & F\langle X \circ_i (X_i \oplus X_i') \rangle \\ {\scriptstyle \beta} \downarrow & & \downarrow {\scriptstyle F\langle 1 \circ_i \beta \rangle} \\ F\langle X \circ_i X_i' \rangle \oplus F\langle X \circ_i X_i \rangle & \xrightarrow{F_i^2} & F\langle X \circ_i (X_i' \oplus X_i) \rangle \end{array} \tag{6.10}$$

Constraint 2-By-2 The following diagram commutes for each

$$i \neq k \in \{1, \ldots, n\}, \quad \langle X \rangle \in \textstyle\prod_{j=1}^{n} \mathsf{C}_j, \quad X_i' \in \mathsf{C}_i, \quad \text{and} \quad X_k' \in \mathsf{C}_k.$$

$$\begin{array}{ccc} & F\langle X \circ_i (X_i \oplus X_i') \circ_k X_k \rangle \oplus F\langle X \circ_i (X_i \oplus X_i') \circ_k X_k' \rangle & \\ {\scriptstyle F_i^2 \oplus F_i^2} \nearrow & & \searrow {\scriptstyle F_k^2} \\ \begin{array}{c} F\langle X \circ_i X_i \circ_k X_k \rangle \oplus F\langle X \circ_i X_i' \circ_k X_k \rangle \\ \oplus F\langle X \circ_i X_i \circ_k X_k' \rangle \oplus F\langle X \circ_i X_i' \circ_k X_k' \rangle \end{array} & & \\ {\scriptstyle 1 \oplus \beta \oplus 1} \downarrow & & F\langle X \circ_i (X_i \oplus X_i') \circ_k (X_k \oplus X_k') \rangle \\ \begin{array}{c} F\langle X \circ_i X_i \circ_k X_k \rangle \oplus F\langle X \circ_i X_i \circ_k X_k' \rangle \\ \oplus F\langle X \circ_i X_i' \circ_k X_k \rangle \oplus F\langle X \circ_i X_i' \circ_k X_k' \rangle \end{array} & & \\ {\scriptstyle F_k^2 \oplus F_k^2} \searrow & & \nearrow {\scriptstyle F_i^2} \\ & F\langle X \circ_i X_i \circ_k (X_k \oplus X_k') \rangle \oplus F\langle X \circ_i X_i' \circ_k (X_k \oplus X_k') \rangle & \end{array} \tag{6.11}$$

This finishes the definition of an n-linear functor.

Moreover, we define the following.

- If $n = 0$, then a *0-linear functor* is a choice of an object in D.
- An n-linear functor $(F, \langle F_j^2 \rangle)$ is *strong* if each linearity constraint F_j^2 is a natural isomorphism.
- A *multilinear functor* is an n-linear functor for some $n \geq 0$. ◇

Example 6.12 (Identities)**.** Each permutative category C has an *identity 1-linear functor*

$$\mathsf{C} \xrightarrow{1_{\mathsf{C}}} \mathsf{C}$$

consisting of

- the identity functor on C and
- the identity linearity constraint.

The latter is componentwise the identity morphism of $X \oplus X'$ for objects $X, X' \in \mathsf{C}$. Thus 1_{C} is a strong 1-linear functor. ◇

Example 6.13 (1-Linear Functors)**.** A 1-linear functor between permutative categories

$$\mathsf{C} \xrightarrow{(F,F^2)} \mathsf{D}$$

is precisely a strictly unital symmetric monoidal functor (Definition 1.17), since the constraint 2-by-2 axiom (6.11) is vacuous when $n = 1$. ◇

Example 6.14 (Bipermutative Categories)**.** Suppose given a *tight* bipermutative category (Definition 2.41)

$$\big(\mathsf{C}, (\oplus, 0, \beta^{\oplus}), (\otimes, 1, \beta^{\otimes}), \partial^l, \partial^r\big)$$

with additive structure $\mathsf{C}^{\oplus} = (\mathsf{C}, \oplus, 0, \beta^{\oplus})$. For each $n \geq 0$ there is a strong n-linear functor

$$\textstyle\prod_{j=1}^n \mathsf{C}^{\oplus} \xrightarrow{\big(\otimes\,,\,\langle \otimes_j^2 \rangle_{j=1}^n\big)} \mathsf{C}^{\oplus}$$

defined as follows.

- The functor

  $$(\mathsf{C}^{\oplus})^n \xrightarrow{\otimes} \mathsf{C}^{\oplus}$$

  is any one of the equal iterates of $\otimes : \mathsf{C}^2 \longrightarrow \mathsf{C}$. It is

  - the multiplicative unit $1 \in \mathsf{C}$ if $n = 0$ and
  - the identity functor 1_{C} if $n = 1$.

- For $n \geq 1$ and $j \in \{1, \ldots, n\}$, the j-th linearity constraint

$$\begin{array}{c}(X_1 \otimes \cdots \otimes X_j \otimes \cdots \otimes X_n) \oplus (X_1 \otimes \cdots \otimes X_j' \otimes \cdots \otimes X_n) \\ \Big\downarrow {\scriptstyle \otimes_j^2 = \mathsf{L}_j^{-1}} \\ X_1 \otimes \cdots \otimes (X_j \oplus X_j') \otimes \cdots \otimes X_n\end{array} \tag{6.15}$$

is the inverse of the unique Laplaza coherence isomorphism L_j (Example 2.91) that distributes over the sum $X_j \oplus X_j'$ in the j-th tensor factor. The naturality of $\otimes_j^2 = \mathsf{L}_j^{-1}$ follows from the naturality of Laplaza coherence isomorphisms.

The axioms of an n-linear functor hold as follows.

- The unity axiom (6.7) follows from the multiplicative zero axiom (2.31) in C.
- The four constraint axioms, (6.8) through (6.11), follow from the uniqueness in Laplaza's Coherence Theorem 2.90.

  For instance, if $n = 2$ then

- $\otimes_1^2 = \mathsf{L}_1^{-1}$ is the left factorization morphism ∂^l, and
- $\otimes_2^2 = \mathsf{L}_2^{-1}$ is the right factorization morphism ∂^r (2.30).

The four constraint axioms, (6.8) through (6.11), follow from, respectively, the ring category axioms (2.32), (2.35), (2.34), and (2.40) in C. ◇

Multilinear Natural Transformations

The following multilinear analogues of natural transformations are from [EM06, page 174], as presented in [JY∞, 6.5.11].

Definition 6.16. Suppose F and G are n-linear functors as displayed below.

$$\prod_{j=1}^n \mathsf{C}_j \overset{(F, \langle F_j^2 \rangle)}{\underset{(G, \langle G_j^2 \rangle)}{\Downarrow \theta}} \mathsf{D} \tag{6.17}$$

An *n-linear natural transformation* $\theta : F \longrightarrow G$ is a natural transformation of underlying functors that satisfies the following two axioms.

Unity

$$\theta_{\langle X \rangle} = 1_e \qquad \text{if any } X_i = e \in \mathsf{C}_i. \tag{6.18}$$

Constraint Compatibility The diagram

$$\begin{array}{ccc} F\langle X \circ_i X_i\rangle \oplus F\langle X \circ_i X_i'\rangle & \xrightarrow{F_i^2} & F\langle X \circ_i (X_i \oplus X_i')\rangle \\ {\scriptstyle \theta \oplus \theta}\downarrow & & \downarrow{\scriptstyle \theta} \\ G\langle X \circ_i X_i\rangle \oplus G\langle X \circ_i X_i'\rangle & \xrightarrow{G_i^2} & G\langle X \circ_i (X_i \oplus X_i')\rangle \end{array} \tag{6.19}$$

commutes for each $i \in \{1, \ldots, n\}$, objects $\langle X\rangle \in \prod_{j=1}^n \mathsf{C}_j$, and $X_i' \in \mathsf{C}_i$.

This finishes the definition of an n-linear natural transformation.

Moreover, we define the following.

- An *n-linear natural isomorphism* is an invertible n-linear natural transformation.
- A *0-linear natural transformation* is a morphism in D.
- A *multilinear natural transformation* is an n-linear natural transformation for some $n \geq 0$. ◇

Definition 6.20. For small permutative categories $\langle \mathsf{C}\rangle = \langle \mathsf{C}_j\rangle_{j=1}^n$ and D, define the category

$$\mathsf{PermCat}\big(\langle \mathsf{C}\rangle\, ; \mathsf{D}\big) = \mathsf{PermCat}\big(\mathsf{C}_1, \ldots, \mathsf{C}_n\, ; \mathsf{D}\big) \tag{6.21}$$

with

- n-linear functors $\prod_{j=1}^n \mathsf{C}_j \longrightarrow \mathsf{D}$ (Definition 6.4) as objects,
- n-linear natural transformations (Definition 6.16) as morphisms,
- identity natural transformations as identity morphisms, and
- vertical composition of natural transformations as composition.

This finishes the definition of the category $\mathsf{PermCat}(\langle \mathsf{C}\rangle\, ; \mathsf{D})$.

Moreover, we define the following.

- The *strong* variant

  $$\mathsf{PermCat}^{\mathsf{sg}}\big(\langle \mathsf{C}\rangle\, ; \mathsf{D}\big) = \mathsf{PermCat}^{\mathsf{sg}}\big(\mathsf{C}_1, \ldots, \mathsf{C}_n\, ; \mathsf{D}\big) \tag{6.22}$$

  is the full subcategory of $\mathsf{PermCat}(\langle \mathsf{C}\rangle\, ; \mathsf{D})$ with strong n-linear functors as objects.
- If $n = 0$ then we define

  $$\mathsf{PermCat}\big(\langle\rangle\, ; \mathsf{D}\big) = \mathsf{PermCat}^{\mathsf{sg}}\big(\langle\rangle\, ; \mathsf{D}\big) = \mathsf{D} \tag{6.23}$$

  as categories. ◇

In [JY∞, Sections 6.5 and 6.6] $\mathsf{PermCat}$ is denoted $\mathsf{PermCat}^{\mathsf{su}}$, with the superscript su referring to the strict unity conditions in Definitions 6.4 and 6.16. Next we describe the Cat-enriched multicategory structure on $\mathsf{PermCat}$, beginning with the symmetric group action.

Symmetric Group Action

Suppose given

- n-linear functors $F, G : \prod_{j=1}^{n} \mathsf{C}_j \longrightarrow \mathsf{D}$ and
- an n-linear natural transformation $\theta : F \longrightarrow G$ as in (6.17).

For a permutation $\sigma \in \Sigma_n$, the *symmetric group action functor*

$$\mathsf{PermCat}\big(\langle \mathsf{C} \rangle\, ;\, \mathsf{D}\big) \xrightarrow[\cong]{\sigma} \mathsf{PermCat}\big(\langle \mathsf{C} \rangle \sigma\, ;\, \mathsf{D}\big) \tag{6.24}$$

sends the n-linear functors and natural transformation in (6.17) to the following composite functors and whiskered natural transformation.

$$\prod_{j=1}^{n} \mathsf{C}_{\sigma(j)} \xrightarrow{\ \sigma\ } \prod_{j=1}^{n} \mathsf{C}_j \underset{(G, \langle G_j^2 \rangle)}{\overset{(F, \langle F_j^2 \rangle)}{\Downarrow \theta}} \mathsf{D} \tag{6.25}$$

For each $j \in \{1, \ldots, n\}$, the j-th linearity constraint of $F^{\sigma} = F \circ \sigma$ is the composite in D

$$\begin{array}{ccc} F^{\sigma}\langle A \rangle \oplus F^{\sigma}\langle A \circ_j A_j' \rangle & \xrightarrow{(F^{\sigma})_j^2} & F^{\sigma}\langle A \circ_j (A_j \oplus A_j') \rangle \\ \| & & \| \\ F\big(\sigma\langle A \rangle\big) \oplus F\big(\sigma\langle A \rangle \circ_{\sigma(j)} A_j'\big) & \xrightarrow{F^2_{\sigma(j)}} & F\big(\sigma\langle A \rangle \circ_{\sigma(j)} (A_j \oplus A_j')\big) \end{array} \tag{6.26}$$

for objects

$$\langle A \rangle = \langle A_j \rangle_{j=1}^{n} \in \textstyle\prod_{j=1}^{n} \mathsf{C}_{\sigma(j)} \quad \text{and} \quad A_j' \in \mathsf{C}_{\sigma(j)}.$$

If F is strong—that is, each F_j^2 is a natural isomorphism—then F^{σ} is also strong.

Composition

For the composition, suppose given, for each $j \in \{1, \ldots, n\}$,

- small permutative categories $\langle \mathsf{B}_j \rangle = \langle \mathsf{B}_{ji} \rangle_{i=1}^{k_j}$ with $k_j \geq 0$,
- k_j-linear functors $F_j', G_j' : \prod_{i=1}^{k_j} \mathsf{B}_{ji} \longrightarrow \mathsf{C}_j$, and
- a k_j-linear natural transformation $\theta_j : F_j' \longrightarrow G_j'$ as follows.

$$\textstyle\prod_{i=1}^{k_j} \mathsf{B}_{ji} \underset{G_j'}{\overset{F_j'}{\Downarrow \theta_j}} \mathsf{C}_j \tag{6.27}$$

With $\langle \mathsf{B} \rangle = \langle \langle \mathsf{B}_j \rangle \rangle_{j=1}^n$, the *multicategorical composition functor*

$$\mathsf{PermCat}\big(\langle \mathsf{C} \rangle ; \mathsf{D}\big) \times \textstyle\prod_{j=1}^n \mathsf{PermCat}\big(\langle \mathsf{B}_j \rangle ; \mathsf{C}_j\big) \xrightarrow{\ \gamma\ } \mathsf{PermCat}\big(\langle \mathsf{B} \rangle ; \mathsf{D}\big) \tag{6.28}$$

sends the data (6.17) and (6.27) to the following composite functors and horizontal composite natural transformations.

$$\prod_{j=1}^n \prod_{i=1}^{k_j} \mathsf{B}_{ji} \underset{\prod_j G'_j}{\overset{\prod_j F'_j}{\Downarrow \textstyle\prod_j \theta_j}} \prod_{j=1}^n \mathsf{C}_j \underset{G}{\overset{F}{\Downarrow \theta}} \mathsf{D} \tag{6.29}$$

If any $k_j = 0$, then the empty product $\prod_{i=1}^{k_j} \mathsf{B}_{ji}$ is, by definition, the terminal category $\mathbf{1}$. In this case,

- F'_j is determined by the object $F'_j * \in \mathsf{C}_j$, and
- θ_j is determined by the component morphism

$$(\theta_j)_* : F'_j * \longrightarrow G'_j * \quad \text{in} \quad \mathsf{C}_j.$$

In (6.29)

- the leftmost product should omit those indices j with $k_j = 0$, and
- $F'_j *$, $G'_j *$, and $(\theta_j)_*$ are used in C_j in the middle product.

If $k_j = 0$ for all $j \in \{1, \ldots, n\}$, then the leftmost product in (6.29) is $\mathbf{1}$.

Composite Linearity Constraints

To describe the linearity constraints of the composite functor $F \circ \prod_j F'_j$ in (6.29), suppose given objects

- $\langle W_j \rangle = \langle W_{ji} \rangle_{i=1}^{k_j} \in \prod_{i=1}^{k_j} \mathsf{B}_{ji}$ for each $j \in \{1, \ldots, n\}$,
- $\langle W \rangle = \langle \langle W_j \rangle \rangle_{j=1}^n \in \prod_{j=1}^n \prod_{i=1}^{k_j} \mathsf{B}_{ji}$,
- $W'_{ji} \in \mathsf{B}_{ji}$ for some choice of (j, i) with $\ell = k_1 + \cdots + k_{j-1} + i$, and
- $\langle F'W \rangle = \langle F'_j \langle W_j \rangle \rangle_{j=1}^n \in \prod_{j=1}^n \mathsf{C}_j$.

The following objects appear in (6.30) below.

$$\begin{aligned}
\langle W_j \circ_i W'_{ji} \rangle &= \big(\overbrace{W_{j1}, \ldots, W_{j,i-1}}^{\text{empty if } i=1}, W'_{ji}, \overbrace{W_{j,i+1}, \ldots, W_{jk_j}}^{\text{empty if } i=k_j} \big) \\
\big\langle W_j \circ_i (W_{ji} \oplus W'_{ji}) \big\rangle &= \big(\underbrace{W_{j1}, \ldots, W_{j,i-1}}_{\text{empty if } i=1}, W_{ji} \oplus W'_{ji}, \underbrace{W_{j,i+1}, \ldots, W_{jk_j}}_{\text{empty if } i=k_j} \big).
\end{aligned}$$

The *ℓ-th linearity constraint* $\big(F \circ \prod_j F'_j\big)^2_\ell$ is defined as the following composite morphism in D.

$$\begin{array}{ccc}
 & F\langle F'W\rangle \oplus F\big\langle F'W \circ_j F'_j\langle W_j \circ_i W'_{ji}\rangle\big\rangle & \\
\Vert & & \downarrow F^2_j \\
(F \circ \prod_j F'_j)\langle W\rangle \oplus (F \circ \prod_j F'_j)\langle W \circ_\ell W'_{ji}\rangle & & F\big\langle F'W \circ_j (F'_j\langle W_j\rangle \oplus F'_j\langle W_j \circ_i W'_{ji}\rangle)\big\rangle \\
\downarrow \big(F \circ \prod_j F'_j\big)^2_\ell & & \downarrow F\langle 1 \circ_j (F'_j)^2_i\rangle \\
(F \circ \prod_j F'_j)\big\langle W \circ_\ell (W_{ji} \oplus W'_{ji})\big\rangle & & \\
\Vert & & \\
 & F\big\langle F'W \circ_j F'_j\langle W_j \circ_i (W_{ji} \oplus W'_{ji})\rangle\big\rangle &
\end{array} \tag{6.30}$$

If F and each F'_j are strong—that is, each of their linearity constraints is a natural isomorphism—then $F \circ \prod_j F'_j$ is also strong.

A detailed proof of the following theorem for $\mathsf{PermCat}$ is given in [JY∞, Section 6.6]. That proof also applies to the strong variant $\mathsf{PermCat}^{\mathsf{sg}}$, with component categories defined in (6.22) and (6.23), by restricting to strong multilinear functors.

Theorem 6.31. *There is a* Cat*-multicategory* $\mathsf{PermCat}$ *with*

- *small permutative categories (Definition 1.11) as objects,*
- *multimorphism categories* $\mathsf{PermCat}(\langle \mathsf{C}\rangle\,;\mathsf{D})$ *as in* (6.21),
- *identity 1-linear functors (Example 6.12) as colored units,*
- *symmetric group action functor as in* (6.24), *and*
- *composition functor as in* (6.28).

Moreover, the strong variant $\mathsf{PermCat}^{\mathsf{sg}}$ *is also a* Cat*-multicategory.*

Remark 6.32. The Cat-multicategory structure on $\mathsf{PermCat}$ and $\mathsf{PermCat}^{\mathsf{sg}}$ are *not* induced by some symmetric monoidal structure on the Cat-multicategory of small pointed multicategories. This is explained in detail in [JY∞, 5.7.23 and 10.2.17]. The root cause for the failure of $\mathsf{PermCat}$ and $\mathsf{PermCat}^{\mathsf{sg}}$ to admit a monoidal structure is the non-existence of a suitable monoidal unit. This is similar to the reason for the non-existence of a monoidal structure on $\mathcal{D}\mathsf{Cat}$ (Explanation 5.64). $\diamond$

6.2 The Grothendieck Construction of Symmetric Monoidal Functors

Now we begin to prove that the Grothendieck construction is a pseudo symmetric Cat-multifunctor when the indexing category is a small tight bipermutative category. As the first step, in this section we review

- the Grothendieck construction $\int_{\mathcal{D}} X$ of an indexed category $X : \mathcal{D} \longrightarrow \mathsf{Cat}$ (Definition 1.15) and
- its permutative structure when $(\mathcal{D}, \oplus, 0, \beta^{\oplus})$ is a small permutative category with X a symmetric monoidal functor.

The assignment $X \longmapsto \int_{\mathcal{D}} X$ in the permutative case is the object assignment of the pseudo symmetric Cat-multifunctor $\int_{\mathcal{D}}$. The Grothendieck constructions of additive natural transformations and additive modifications, along with the pseudo symmetric Cat-multifunctoriality, are discussed in subsequent sections.

The Grothendieck Construction as a Category

A detailed reference for the Grothendieck construction is [JY21, Ch. 10], which discusses the more general situation when X is a lax functor, or a pseudofunctor for some results. That reference actually discusses the *contravariant* Grothendieck construction for $\mathcal{D}^{\mathsf{op}} \longrightarrow \mathsf{Cat}$. All the results there apply here as well by considering the opposite category of $\mathcal{D}^{\mathsf{op}}$, that is, $\mathcal{D}$. When we cite results in [JY21, Ch. 10], we are referring to the *covariant* version.

Definition 6.33. Suppose $\mathcal{D}$ is a small category, and $X : \mathcal{D} \longrightarrow \mathsf{Cat}$ is a $\mathcal{D}$-indexed category. The *Grothendieck construction* of X, denoted $\int_{\mathcal{D}} X$, is the category defined by the following data.

Objects An *object* in $\int_{\mathcal{D}} X$ is a pair of objects

$$(a, x) \in \mathcal{D} \times Xa.$$

Morphisms A *morphism*

$$(a, x) \xrightarrow{(f,p)} (b, y) \quad \text{in} \quad \int_{\mathcal{D}} X$$

consists of

- a morphism $f : a \longrightarrow b$ in $\mathcal{D}$ and
- a morphism $p : f_* x \longrightarrow y$ in Xb with the functor

$$f_* = Xf : Xa \longrightarrow Xb.$$

Identities For each object (a,x) in $\int_{\mathcal{D}} X$, the *identity morphism* consists of a pair of identity morphisms:

$$1_{(a,x)} = (1_a, 1_x). \tag{6.34}$$

Composition For another morphism

$$(b,y) \xrightarrow{(g,q)} (c,z) \quad \text{in} \quad \textstyle\int_{\mathcal{D}} X$$

the *composite* of

$$(a,x) \xrightarrow{(f,p)} (b,y) \xrightarrow{(g,q)} (c,z)$$

is given by

$$(g,q) \circ (f,p) = \big(gf, q \circ g_* p\big). \tag{6.35}$$

This finishes the definition of $\int_{\mathcal{D}} X$. ◇

The reader should not confuse the notation $\int_{\mathcal{D}} X$ for the Grothendieck construction with the coend in (1.51) or the end in (1.54). The following observation is a special case of [JY21, 10.1.8].

Lemma 6.36. *In the context of Definition 6.33, $\int_{\mathcal{D}} X$ is a small category.*

The Grothendieck Construction as a Permutative Category

Next we discuss the permutative variant of the Grothendieck construction. A reference is [GJO17, 4.29], which discusses the more general situation when the codomain is 2Cat instead of Cat. Recall from Definition 1.17 the notion of a symmetric monoidal functor between symmetric monoidal categories.

Definition 6.37. Suppose $(\mathcal{D}, \oplus, 0, \beta^{\oplus})$ is a small permutative category, and

$$(X, X^2, X^0) : \mathcal{D} \longrightarrow \mathsf{Cat}$$

is a symmetric monoidal functor. Define the data of a permutative category on the Grothendieck construction $\int_{\mathcal{D}} X$ as follows.

Monoidal Unit It is the object

$$(0, X^0 *) \in \textstyle\int_{\mathcal{D}} X \tag{6.38}$$

where $X^0 * \in X0$ is the unit object (5.33).

Monoidal Product on Objects For objects (a,x) and (a',x') in $\int_{\mathcal{D}} X$, their monoidal product is defined as the object

$$(a,x) \square (a',x') = \big(a \oplus a', X^2_{a,a'}(x,x')\big) \quad \text{in} \quad \textstyle\int_{\mathcal{D}} X \tag{6.39}$$

with

$$Xa \times Xa' \xrightarrow{X^2_{a,a'}} X(a \oplus a')$$

the (a,a')-component functor of the monoidal constraint X^2 (5.34).

Monoidal Product on Morphisms For morphisms

$$(a,x) \xrightarrow{(f,p)} (b,y) \quad \text{and} \quad (a',x') \xrightarrow{(f',p')} (b',y') \quad \text{in} \quad \textstyle\int_{\mathcal{D}} X$$

their monoidal product is defined as the following morphism.

$$\begin{array}{ccc} (a,x) \square (a',x') & \xrightarrow{(f,p)\square(f',p')} & (b,y) \square (b',y') \\ \| & & \| \\ \big(a \oplus a', X^2_{a,a'}(x,x')\big) & \xrightarrow{\big(f\oplus f',\, X^2_{b,b'}(p,p')\big)} & \big(b \oplus b', X^2_{b,b'}(y,y')\big) \end{array} \tag{6.40}$$

Braiding The component of the braiding $\beta^{\square}$ for objects (a,x) and (a',x') in $\int_{\mathcal{D}} X$ is defined as the following isomorphism.

$$\begin{array}{ccc} (a,x) \square (a',x') & \xrightarrow{\beta^{\square}} & (a',x') \square (a,x) \\ \| & & \| \\ \big(a \oplus a', X^2_{a,a'}(x,x')\big) & \underset{\cong}{\xrightarrow{(\beta^{\oplus},1)}} & \big(a' \oplus a, X^2_{a',a}(x',x)\big) \end{array} \tag{6.41}$$

This finishes the definition of $\int_{\mathcal{D}} X$. In Lemma 6.42 we will show that $\int_{\mathcal{D}} X$ is a permutative category. ◇

Lemma 6.42. *In the context of Definition 6.37, the quadruple*

$$\Big(\textstyle\int_{\mathcal{D}} X, \square, (0, X^0 *), \beta^{\square}\Big)$$

is a small permutative category.

Proof. By Lemma 6.36 $\int_{\mathcal{D}} X$ is a small category. The functoriality of

$$\textstyle\int_{\mathcal{D}} X \times \int_{\mathcal{D}} X \xrightarrow{\square} \int_{\mathcal{D}} X$$

follows from

- the functoriality of the monoidal product $\oplus : \mathcal{D} \times \mathcal{D} \longrightarrow \mathcal{D}$,
- the functoriality of each component functor of the monoidal constraint X^2 (5.34), and
- the naturality of X^2 (5.35).

The associativity of $\square$ follows from

- the associativity of $\oplus$ in $\mathcal{D}$ and
- the associativity axiom (5.37) for (X, X^2).

The left and right unity of $\square$ follows from

- the unity of $0 \in \mathcal{D}$ and
- the unity axiom (5.36) for (X, X^2, X^0).

Thus $\left(\int_{\mathcal{D}} X, \square, (0, X^0 *)\right)$ is a strict monoidal category.

The braiding $\beta^{\square}$ (6.41) is well defined by the braiding axiom (5.38) for (X, X^2). The naturality of $\beta^{\square}$ follows from

- the naturality of $\beta^{\oplus}$ in $\mathcal{D}$ and
- the braiding axiom (5.38) for (X, X^2).

The symmetry and hexagon axioms (1.12) for $(\int_{\mathcal{D}} X, \beta^{\square})$ follow from

- the same axioms in the permutative category $\mathcal{D}$ and
- the functoriality of each component functor of X^2 (5.34).

Thus $(\int_{\mathcal{D}} X, \beta^{\square})$ is a strict symmetric monoidal category. □

In what follows, we regard the Grothendieck construction $\int_{\mathcal{D}} X$ as a permutative category via Lemma 6.42. Note that, while $\mathcal{D}$ is required to be a small permutative category, *neither* X^0 nor X^2 is required to be an isomorphism.

6.3 The Grothendieck Construction of Additive Natural Transformations

For the rest of this chapter, we assume that

$$(\mathcal{D}, (\oplus, 0, \beta^{\oplus}), (\otimes, 1, \beta^{\otimes}), (\partial^l, \partial^r))$$

is a small *tight* bipermutative category (Definition 2.41) with

- additive structure $\mathcal{D}^{\oplus} = (\mathcal{D}, \oplus, 0, \beta^{\oplus})$ and
- multiplicative structure $\mathcal{D}^{\otimes} = (\mathcal{D}, \otimes, 1, \beta^{\otimes})$.

Recall the Cat-multicategory $\mathcal{D}\mathsf{Cat}$ in Theorem 5.71 with

- additive symmetric monoidal functors $\mathcal{D}^{\oplus} \longrightarrow \mathsf{Cat}$ as objects,
- additive natural transformations as objects in the multimorphism categories, and
- additive modifications as morphisms in the multimorphism categories.

Also recall the Cat-multicategory $\mathsf{PermCat}^{\mathsf{sg}}$ in Theorem 6.31 with

- small permutative categories as objects,
- *strong* multilinear functors as objects in the multimorphism categories, and

- multilinear natural transformations as morphisms in the multimorphism categories.

This section contains the next step in proving that the Grothendieck construction

$$\mathcal{D}\mathsf{Cat} \xrightarrow{\int_{\mathcal{D}}} \mathsf{PermCat}^{\mathsf{sg}}$$

is a pseudo symmetric Cat-multifunctor. The object assignment $X \longmapsto \int_{\mathcal{D}} X$ is given by Lemma 6.42. In this section we show that the Grothendieck construction sends each additive natural transformation between additive symmetric monoidal functors to a strong multilinear functor; see Lemma 6.57.

Arity 0

First we discuss the arity 0 case, which has its own definition and is easier than the positive arity case.

Definition 6.43. Suppose given an additive symmetric monoidal functor (Definition 5.31)

$$(Z, Z^2, Z^0) : \mathcal{D}^{\oplus} \longrightarrow \mathsf{Cat}.$$

Using the definitions (5.62) and (6.23) of arity 0 multimorphism categories in, respectively, $\mathcal{D}\mathsf{Cat}$ and $\mathsf{PermCat}^{\mathsf{sg}}$, we define the functor

$$Z\mathbb{1} = \mathcal{D}\mathsf{Cat}\big(\langle\rangle\,;\,Z\big) \xrightarrow{\int_{\mathcal{D}}} \mathsf{PermCat}^{\mathsf{sg}}\big(\langle\rangle\,;\,\textstyle\int_{\mathcal{D}} Z\big) = \textstyle\int_{\mathcal{D}} Z \tag{6.44}$$

by sending

- each object $x \in Z\mathbb{1}$ to the object

$$(\mathbb{1}, x) \in \textstyle\int_{\mathcal{D}} Z$$

 and

- each morphism $f : x \longrightarrow y$ in $Z\mathbb{1}$ to the morphism

$$(\mathbb{1}, x) \xrightarrow{(1_{\mathbb{1}}, f)} (\mathbb{1}, y) \quad \text{in} \quad \textstyle\int_{\mathcal{D}} Z.$$

These object and morphism assignments define a functor because $Z(1_{\mathbb{1}}) = 1_{Z\mathbb{1}}$ by the functoriality of Z. ◇

As discussed in

- the first paragraph in the proof of Lemma 5.68 and
- the paragraph after (6.29),

proofs involving the arity 0 case require some slight adjustments. With this in mind, in the rest of this chapter we discuss exclusively the positive arity case, using Definition 6.43 to make the necessary adjustments when some inputs have arity 0.

Positive Arity

Next we define the Grothendieck construction on additive natural transformations (Definition 5.39). Recall that an *additive natural transformation* is a natural transformation in $\mathcal{D}^{\otimes}$-Cat (5.5) that satisfies the unity axiom (5.43) and the additivity axiom (5.45). Also recall from Definition 6.4 that an n-*linear functor* is a functor equipped with n linearity constraints (6.6) that satisfy the axioms (6.7) through (6.11).

Definition 6.45. Suppose

$$(Z, Z^2, Z^0) \quad \text{and} \quad \big\langle (X_j, X_j^2, X_j^0) \big\rangle_{j=1}^n : \mathcal{D}^{\oplus} \longrightarrow \mathsf{Cat}$$

are additive symmetric monoidal functors with $n > 0$. Suppose

$$\textstyle\bigotimes_{j=1}^n X_j \xrightarrow{\ \phi\ } Z$$

is an additive natural transformation (5.40). Define the data of an n-linear functor

$$\textstyle\prod_{j=1}^n \big(\int_{\mathcal{D}} X_j \big) \xrightarrow{\ \int_{\mathcal{D}} \phi\ } \int_{\mathcal{D}} Z \tag{6.46}$$

with the following object assignment (6.49), morphism assignment (6.52), and linearity constraints (6.56).

Object Assignment Suppose given an n-tuple of objects

$$\langle (a_j, x_j) \rangle \in \textstyle\prod_{j=1}^n \big(\int_{\mathcal{D}} X_j \big) \tag{6.47}$$

consisting of objects $a_j \in \mathcal{D}$ and $x_j \in X_j a_j$ for $j \in \{1, \ldots, n\}$. First define the following objects.

$$\begin{cases} \langle a \rangle = \langle a_j \rangle_{j=1}^n \in \mathcal{D}^n & a = \bigotimes_{j=1}^n a_j \in \mathcal{D} \\ \langle x \rangle = \langle x_j \rangle_{j=1}^n \in \prod_{j=1}^n X_j a_j & \end{cases} \tag{6.48}$$

Then $\int_{\mathcal{D}} \phi$ sends $\langle (a_j, x_j) \rangle$ to the object

$$\big(\textstyle\int_{\mathcal{D}} \phi \big) \langle (a_j, x_j) \rangle = \big(a \,,\, \phi_{\langle a \rangle} \langle x \rangle \big) \quad \text{in} \quad \textstyle\int_{\mathcal{D}} Z \tag{6.49}$$

with $\phi_{\langle a \rangle}$ the $\langle a \rangle$-component functor of ϕ as in (5.7):

$$\textstyle\prod_{j=1}^n X_j a_j \xrightarrow{\ \phi_{\langle a \rangle}\ } Za.$$

Morphism Assignment Suppose given an n-tuple of morphisms

$$\big\langle (f_j, p_j) : (a_j, x_j) \longrightarrow (b_j, y_j) \big\rangle \in \textstyle\prod_{j=1}^n \big(\int_{\mathcal{D}} X_j \big) \tag{6.50}$$

consisting of, for each $j \in \{1, \ldots, n\}$, morphisms

- $f_j : a_j \longrightarrow b_j$ in $\mathcal{D}$ and
- $p_j : (f_j)_* x_j \longrightarrow y_j$ in $X_j b_j$ with

$$(f_j)_* = X f_j : X a_j \longrightarrow X b_j.$$

In addition to the objects in (6.48), define the following objects and morphisms.

$$\begin{cases} \langle b\rangle = \langle b_j\rangle_{j=1}^n \in \mathcal{D}^n & b = \bigotimes_{j=1}^n b_j \in \mathcal{D} \\ \langle y\rangle = \langle y_j\rangle_{j=1}^n \in \prod_{j=1}^n X_j b_j & f = \bigotimes_{j=1}^n f_j \in \mathcal{D} \\ \langle p\rangle = \langle p_j\rangle_{j=1}^n \in \prod_{j=1}^n X_j b_j & \end{cases} \tag{6.51}$$

Then $\int_{\mathcal{D}}\phi$ sends $\langle (f_j, p_j)\rangle$ to the following morphism in $\int_{\mathcal{D}} Z$.

$$\begin{array}{ccc} (\int_{\mathcal{D}}\phi)\langle (a_j, x_j)\rangle & \xrightarrow{(\int_{\mathcal{D}}\phi)\langle (f_j, p_j)\rangle} & (\int_{\mathcal{D}}\phi)\langle (b_j, y_j)\rangle \\ \| & & \| \\ (a\,,\, \phi_{\langle a\rangle}\langle x\rangle) & \xrightarrow{(f\,,\, \phi_{\langle b\rangle}\langle p\rangle)} & (b\,,\, \phi_{\langle b\rangle}\langle y\rangle) \end{array} \tag{6.52}$$

This is well defined by the naturality diagram (5.9) for ϕ.

Linearity Constraints To define the linearity constraints of $\int_{\mathcal{D}}\phi$, suppose given an index $i \in \{1, \ldots, n\}$ and an object

$$(a_i', x_i') \in \textstyle\int_{\mathcal{D}} X_i$$

with $a_i' \in \mathcal{D}$ and $x_i' \in X_i a_i'$. In addition to the objects in (6.47) and (6.48), define the following objects using the $\circ_i$ notation in (6.2).

$$\begin{cases} a_i'' = a_i \oplus a_i' \in \mathcal{D} & \\ a' = a_1 \otimes \cdots \otimes a_i' \otimes \cdots \otimes a_n \in \mathcal{D} & \langle a'\rangle = \langle a\rangle \circ_i a_i' \in \mathcal{D}^n \\ a'' = a_1 \otimes \cdots \otimes a_i'' \otimes \cdots \otimes a_n \in \mathcal{D} & \langle a''\rangle = \langle a\rangle \circ_i a_i'' \in \mathcal{D}^n \\ x_i'' = (X_i^2)_{a_i, a_i'}(x_i, x_i') \in X_i a_i'' & \\ \langle x'\rangle = \langle x\rangle \circ_i x_i' \in X_1 a_1 \times \cdots \times X_i a_i' \times \cdots \times X_n a_n & \\ \langle x''\rangle = \langle x\rangle \circ_i x_i'' \in X_1 a_1 \times \cdots \times X_i a_i'' \times \cdots \times X_n a_n & \end{cases} \tag{6.53}$$

The above definition of x_i'' uses the monoidal constraint (5.34) of X_i:

$$X_i a_i \times X_i a_i' \xrightarrow{(X_i^2)_{a_i, a_i'}} X_i(a_i \oplus a_i') = X_i a_i''.$$

For the objects $\langle (a_j, x_j)\rangle$ and (a_i', x_i'), the i-th linearity constraint (6.6) of $\int_{\mathcal{D}}\phi$ has the following domain object in $\int_{\mathcal{D}} Z$.

$$\begin{cases} (\int_{\mathcal{D}}\phi)\langle (a_j, x_j)\rangle \,\square\, (\int_{\mathcal{D}}\phi)\big(\langle (a_j, x_j)\rangle \circ_i (a_i', x_i')\big) & \\ = \big(a\,,\, \phi_{\langle a\rangle}\langle x\rangle\big) \,\square\, \big(a'\,,\, \phi_{\langle a'\rangle}\langle x'\rangle\big) & \text{by (6.49)} \\ = \big(a \oplus a'\,,\, Z^2_{a,a'}(\phi_{\langle a\rangle}\langle x\rangle\,,\, \phi_{\langle a'\rangle}\langle x'\rangle)\big) & \text{by (6.39)} \\ = \big(a \oplus a'\,,\, \mathsf{L}_*\phi_{\langle a''\rangle}\langle x''\rangle\big) & \text{by (5.51)} \end{cases} \tag{6.54}$$

In the previous line, L is the unique Laplaza coherence isomorphism in $\mathcal{D}$

$$a'' \xrightarrow[\cong]{\mathsf{L}} a \oplus a'$$

in (5.46) that distributes over the sum $a_i \oplus a_i'$ in a''.

The codomain object in $\int_{\mathcal{D}} Z$ of the same i-th linearity constraint of $\int_{\mathcal{D}} \phi$ is the following.

$$\begin{cases} \big(\int_{\mathcal{D}}\phi\big)\Big(\langle(a_j,x_j)\rangle \circ_i \big((a_i,x_i) \,\square\, (a_i',x_i')\big)\Big) & \\ = \big(\int_{\mathcal{D}}\phi\big)\Big(\langle(a_j,x_j)\rangle \circ_i (a_i'', x_i'')\Big) & \text{by (6.39) and (6.53)} \\ = \big(a'', \phi_{\langle a''\rangle}\langle x''\rangle\big) & \text{by (6.49)} \end{cases} \tag{6.55}$$

Using the last expression in each of (6.54) and (6.55), the *i-th linearity constraint* $(\int_{\mathcal{D}}\phi)_i^2$ of $\int_{\mathcal{D}}\phi$ (6.6) is the isomorphism in $\int_{\mathcal{D}} Z$

$$\big(a \oplus a', \mathsf{L}_*\phi_{\langle a''\rangle}\langle x''\rangle\big) \xrightarrow[\cong]{(\mathsf{L}^{-1},1)} \big(a'', \phi_{\langle a''\rangle}\langle x''\rangle\big) \tag{6.56}$$

with

$$(\mathsf{L}^{-1})_*\mathsf{L}_*\phi_{\langle a''\rangle}\langle x''\rangle = \phi_{\langle a''\rangle}\langle x''\rangle \xrightarrow{1} \phi_{\langle a''\rangle}\langle x''\rangle$$

the identity morphism of $\phi_{\langle a''\rangle}\langle x''\rangle$ in Za''.

This finishes the definition of $\int_{\mathcal{D}}\phi$ in (6.46). ◇

Lemma 6.57. *In the context of Definition 6.45, the data*

$$\prod_{j=1}^n \big(\textstyle\int_{\mathcal{D}} X_j\big) \xrightarrow{\big(\int_{\mathcal{D}}\phi,\ \langle(\int_{\mathcal{D}}\phi)_i^2\rangle_{i=1}^n\big)} \textstyle\int_{\mathcal{D}} Z$$

form a strong n-linear functor.

Proof. By Definition 6.4 we must verify the following statements.

1. $\int_{\mathcal{D}}\phi$, as defined in (6.49) and (6.52), is a functor.
2. For each $i \in \{1,\ldots,n\}$, the i-th linearity constraint $(\int_{\mathcal{D}}\phi)_i^2$ in (6.56) is a natural isomorphism.
3. The axioms (6.7) through (6.11) hold for $\big(\int_{\mathcal{D}}\phi, \langle(\int_{\mathcal{D}}\phi)_i^2\rangle_{i=1}^n\big)$.

We begin with statement 1.

(1) *Functoriality.* First we observe that $\int_{\mathcal{D}}\phi$ preserves identity morphisms. By definition (6.34) an identity morphism in the Grothendieck construction consists of a pair of identity morphisms. In (6.50) if for each $j \in \{1,\ldots,n\}$

$$\begin{aligned} f_j &= 1 : a_j \longrightarrow a_j \in \mathcal{D} \quad \text{and} \\ p_j &= 1 : x_j \longrightarrow x_j \in X_j a_j \end{aligned}$$

then

$$f = \bigotimes_{j=1}^n f_j = 1_a \quad \text{and}$$
$$\phi_{\langle b\rangle}\langle p\rangle = \phi_{\langle a\rangle}\langle 1\rangle = 1$$

by the functoriality of $\otimes$ in $\mathcal{D}$ and the component functor $\phi_{\langle b\rangle} = \phi_{\langle a\rangle}$. Thus $(\int_{\mathcal{D}}\phi)\langle(f_j,p_j)\rangle$ is the identity morphism by definition (6.52).

Similarly, $\int_{\mathcal{D}}\phi$ preserves composition of morphisms (6.35) by

- the functoriality of $\otimes$ in $\mathcal{D}$ and each component functor of ϕ and
- the naturality diagram (5.9) for ϕ with respect to morphisms in $\mathcal{D}$.

Thus $\int_{\mathcal{D}}\phi$ is a functor.

(2) *Naturality.* Next we show that each linearity constraint $(\int_{\mathcal{D}}\phi)_i^2$ in (6.56), which is componentwise an isomorphism, is a natural isomorphism. In addition to the morphisms $\langle(f_j,p_j)\rangle$ in (6.50), suppose given a morphism

$$(a_i', x_i') \xrightarrow{(f_i',p_i')} (b_i', y_i') \quad \text{in} \quad \textstyle\int_{\mathcal{D}} X_i$$

with

- $f_i' : a_i' \longrightarrow b_i'$ a morphism in $\mathcal{D}$ and
- $p_i' : (f_i')_* x_i' \longrightarrow y_i'$ a morphism in Xb_i'.

We extend the notation in (6.53) to

- the objects $\langle(b_j,y_j)\rangle$ and (b_i',y_i') and
- the morphisms $\langle(f_j,p_j)\rangle$ and (f_i',p_i').

Then the naturality of $(\int_{\mathcal{D}}\phi)_i^2$ means the commutativity of the following diagram in $\int_{\mathcal{D}} Z$.

$$\begin{array}{ccc}
(a\oplus a',\, \mathsf{L}_*\phi_{\langle a''\rangle}\langle x''\rangle) & \xrightarrow{(\mathsf{L}^{-1},1)} & (a'',\, \phi_{\langle a''\rangle}\langle x''\rangle) \\
{\scriptstyle (f\oplus f',\, \mathsf{L}_*\phi_{\langle b''\rangle}\langle p''\rangle)}\downarrow & & \downarrow{\scriptstyle (f'',\, \phi_{\langle b''\rangle}\langle p''\rangle)} \\
(b\oplus b',\, \mathsf{L}_*\phi_{\langle b''\rangle}\langle y''\rangle) & \xrightarrow{(\mathsf{L}^{-1},1)} & (b'',\, \phi_{\langle b''\rangle}\langle y''\rangle)
\end{array} \tag{6.58}$$

To check the commutativity of the diagram (6.58), we consider the two arguments separately.

- In the first argument, the morphism equality

  $$\mathsf{L}^{-1}(f\oplus f') = f''\mathsf{L}^{-1} : a\oplus a' \longrightarrow b'' \quad \text{in} \quad \mathcal{D}$$

  follows from the naturality of the Laplaza coherence isomorphism L and the definitions of f, f', and f''.

- The morphism equality in Zb'' in the second argument follows from the following equalities.

$$\begin{aligned}1 \circ \mathsf{L}_*^{-1}\mathsf{L}_*\phi_{\langle b''\rangle}\langle p''\rangle &= \phi_{\langle b''\rangle}\langle p''\rangle \\ &= \phi_{\langle b''\rangle}\langle p''\rangle \circ 1 \\ &= \phi_{\langle b''\rangle}\langle p''\rangle \circ (f''_* 1)\end{aligned}$$

Thus each linearity constraint $(\int_{\mathcal{D}}\phi)_i^2$ is a natural isomorphism.

(3) *Multilinearity.* Next we check the axioms (6.7) through (6.11) for $\int_{\mathcal{D}}\phi$.

Unity (6.7). By the definitions (6.34) and (6.38) of, respectively, an identity morphism and the monoidal unit in the Grothendieck construction, the unity axiom for $\int_{\mathcal{D}}\phi$ states the following equalities in $\int_{\mathcal{D}} Z$.

$$\begin{cases} (\int_{\mathcal{D}}\phi)\big(\langle(a_j, x_j)\rangle \circ_i (0, X_i^0 *)\big) = (0, Z^0 *) \\ (\int_{\mathcal{D}}\phi)\big(\langle(f_j, p_j)\rangle \circ_i (1_0, 1_{X_i^0 *})\big) = (1_0, 1_{Z^0 *}) \end{cases} \tag{6.59}$$

The equalities (6.59) hold by

- the definitions (6.49) and (6.52) of $\int_{\mathcal{D}}\phi$ on, respectively, objects and morphisms,
- the fact that $0 \in \mathcal{D}$ is a strict multiplicative zero (2.31), and
- the unity axiom (5.43) for the additive natural transformation ϕ.

Constraint Unity (6.8). This axiom for $\int_{\mathcal{D}}\phi$ states that, if either

$$\begin{aligned} (a'_i, x'_i) &= (0, X_i^0 *) \in \textstyle\int_{\mathcal{D}} X_i \quad \text{or} \\ (a_j, x_j) &= (0, X_j^0 *) \in \textstyle\int_{\mathcal{D}} X_j \quad \text{for any} \quad j \in \{1, \ldots, n\}, \end{aligned}$$

then the i-th linearity constraint $(\mathsf{L}^{-1}, 1)$ in (6.56) is the identity morphism. This axiom holds because, if either

$$a'_i = 0 \quad \text{or} \quad a_j = 0 \quad \text{for any} \quad j \in \{1, \ldots, n\},$$

then the Laplaza coherence isomorphism $\mathsf{L} : a'' \longrightarrow a \oplus a'$ in (5.46) is the identity morphism by the uniqueness in Theorem 2.90.

Other Constraint Axioms. Similar to the constraint unity axiom, each of

- the constraint associativity axiom (6.9),
- the constraint symmetry axiom (6.10), and
- the constraint 2-by-2 axiom (6.11)

for $\int_{\mathcal{D}}\phi$ holds by the uniqueness in Laplaza's Coherence Theorem 2.90. Indeed, recall that each component of the braiding $\beta^{\square}$ in the Grothendieck construction has the form $(\beta^{\oplus}, 1)$, with

- $\beta^{\oplus}$ the braiding in $\mathcal{D}^{\oplus} = (\mathcal{D}, \oplus, 0, \beta^{\oplus})$ and
- the second argument an identity morphism.

Thus in the diagram for each of the axioms (6.9) through (6.11) for $\int_{\mathcal{D}} \phi$, each composite has second argument given by an identity morphism in some component category of Z. In each case Theorem 2.90 gives the desired morphism equality in the first argument in $\mathcal{D}$. For example, in the constraint symmetry axiom (6.10) for $\int_{\mathcal{D}} \phi$, the first argument is the following diagram in $\mathcal{D}$, with $\otimes$ abbreviated to juxtaposition to save space.

$$\begin{array}{ccc} a \oplus a' & \xrightarrow{\mathsf{L}^{-1}} & a_1 \cdots (a_i \oplus a_i') \cdots a_n \\ {\scriptstyle \beta^{\oplus}} \downarrow & & \downarrow {\scriptstyle 1 \cdots \beta^{\oplus} \cdots 1} \\ a' \oplus a & \xrightarrow{\mathsf{L}^{-1}} & a_1 \cdots (a_i' \oplus a_i) \cdots a_n \end{array}$$

This diagram commutes by Theorem 2.90. □

Remark 6.60 (Unary Case)**.** In Lemma 6.57 with $n = 1$, the pair

$$\left(\int_{\mathcal{D}} \phi, \left(\int_{\mathcal{D}} \phi\right)^2 \right) : \int_{\mathcal{D}} X \longrightarrow \int_{\mathcal{D}} Z$$

is a *strict* symmetric monoidal functor for the following reasons.

- It is a 1-linear functor, which is the same thing as a strictly unital symmetric monoidal functor (Example 6.13).
- The linearity constraint $(\int_{\mathcal{D}} \phi)^2$ is componentwise $(\mathsf{L}^{-1}, 1)$ in (6.56). This is an identity morphism because the Laplaza coherence isomorphism

  $$a \oplus a' \xrightarrow[\cong]{\mathsf{L}} a \oplus a'$$

  is the identity morphism by uniqueness (Theorem 2.90).

For $n > 1$ the linearity constraints $(\int_{\mathcal{D}} \phi)_i^2$ are natural isomorphisms but not identities in general. ◇

6.4 The Grothendieck Construction of Additive Modifications

As in Section 6.3 we continue to assume that $\mathcal{D}$ is a small tight bipermutative category with additive structure $\mathcal{D}^{\oplus}$ and multiplicative structure $\mathcal{D}^{\otimes}$. Recall from Definition 5.52 that an *additive modification* is a modification (5.10) that satisfies the unity axiom (5.54) and the additivity axiom (5.55).

Also recall from Definition 6.16 that an *n-linear natural transformation* is a natural transformation that satisfies the unity axiom (6.18) and the constraint compatibility axiom (6.19).

This section contains the next step in the construction of the pseudo symmetric Cat-multifunctorial Grothendieck construction

$$\mathcal{D}\mathsf{Cat} \xrightarrow{\int_{\mathcal{D}}} \mathsf{PermCat}^{\mathsf{sg}}.$$

The object assignment $X \longmapsto \int_{\mathcal{D}} X$ is given by Lemma 6.42. The object assignment of the multimorphism functors is given in

- Definition 6.43 for arity 0 inputs and
- Definition 6.45 for positive arity inputs.

In this section we show that the Grothendieck construction

- sends additive modifications in $\mathcal{D}\mathsf{Cat}$ to multilinear natural transformations in $\mathsf{PermCat}^{\mathsf{sg}}$ and
- preserves vertical composition.

Since the multimorphism functors in arity 0 are constructed in Definition 6.43, we only need to consider the positive arity case.

- The Grothendieck construction of an additive modification is in Definition 6.61.
- Its naturality and n-linearity are proved in Lemmas 6.64 and 6.67.
- Lemma 6.69 shows that $\int_{\mathcal{D}}$ defines a functor between multimorphism categories.

Definition 6.61. Suppose given

- additive symmetric monoidal functors (Definition 5.31)

$$(Z, Z^2, Z^0) \quad \text{and} \quad \big\langle (X_j, X_j^2, X_j^0) \big\rangle_{j=1}^{n} : \mathcal{D}^{\oplus} \longrightarrow \mathsf{Cat}$$

  with $n > 0$ and

- an additive modification

$$\big\langle (X_j, X_j^2, X_j^0) \big\rangle_{j=1}^{n} \overset{\phi}{\underset{\psi}{\genfrac{}{}{0pt}{}{\longrightarrow}{\longrightarrow}}} (Z, Z^2, Z^0) \qquad \Downarrow \Phi$$

  as in (5.53) with ϕ and ψ additive natural transformations (5.40).

Define the data of an n-linear natural transformation

$$\prod_{j=1}^{n}\left(\int_{\mathcal{D}} X_j\right) \overset{\int_{\mathcal{D}}\phi}{\underset{\int_{\mathcal{D}}\psi}{\Downarrow \int_{\mathcal{D}}\Phi}} \int_{\mathcal{D}} Z \tag{6.62}$$

as follows. Using the notation in (6.47) through (6.49), for each n-tuple of objects

$$\langle (a_j, x_j)\rangle \in \prod_{j=1}^{n}\left(\int_{\mathcal{D}} X_j\right)$$

the component morphism of $\int_{\mathcal{D}}\Phi$ is defined as the following morphism in $\int_{\mathcal{D}} Z$.

$$\begin{array}{ccc} (\int_{\mathcal{D}}\phi)\langle (a_j, x_j)\rangle & \xrightarrow{(\int_{\mathcal{D}}\Phi)_{\langle (a_j, x_j)\rangle}} & (\int_{\mathcal{D}}\psi)\langle (a_j, x_j)\rangle \\ \| & & \| \\ (a\,,\,\phi_{\langle a\rangle}\langle x\rangle) & \xrightarrow{(1_a\,,\,(\Phi_{\langle a\rangle})_{\langle x\rangle})} & (a\,,\,\psi_{\langle a\rangle}\langle x\rangle) \end{array} \tag{6.63}$$

Here

$$\prod_{j=1}^{n} X_j a_j \overset{\phi_{\langle a\rangle}}{\underset{\psi_{\langle a\rangle}}{\Downarrow \Phi_{\langle a\rangle}}} Za$$

is the $\langle a\rangle$-component natural transformation of Φ as in (5.11). This finishes the definition of $\int_{\mathcal{D}}\Phi$ ◇

Lemma 6.64. *In the context of Definition 6.61, $\int_{\mathcal{D}}\Phi$ is a natural transformation.*

Proof. Using the notation in (6.50) through (6.52), suppose given any n-tuple of morphisms

$$\langle (f_j, p_j) : (a_j, x_j) \longrightarrow (b_j, y_j)\rangle \in \prod_{j=1}^{n}\left(\int_{\mathcal{D}} X_j\right).$$

The naturality of $\int_{\mathcal{D}}\Phi$ means the commutativity of the following diagram in $\int_{\mathcal{D}} Z$.

$$\begin{array}{ccc} (a\,,\,\phi_{\langle a\rangle}\langle x\rangle) & \xrightarrow{(1_a\,,\,(\Phi_{\langle a\rangle})_{\langle x\rangle})} & (a\,,\,\psi_{\langle a\rangle}\langle x\rangle) \\ {\scriptstyle (f\,,\,\phi_{\langle b\rangle}\langle p\rangle)}\downarrow & & \downarrow{\scriptstyle (f\,,\,\psi_{\langle b\rangle}\langle p\rangle)} \\ (b\,,\,\phi_{\langle b\rangle}\langle y\rangle) & \xrightarrow{(1_b\,,\,(\Phi_{\langle b\rangle})_{\langle y\rangle})} & (b\,,\,\psi_{\langle b\rangle}\langle y\rangle) \end{array} \tag{6.65}$$

In the first argument in (6.65), each composite is equal to $f : a \longrightarrow b$ in $\mathcal{D}$.

The second argument in (6.65) forms the boundary in the following diagram in Zb.

$$\begin{array}{ccc} f_*\phi_{\langle a\rangle}\langle x\rangle & \xrightarrow{f_*(\Phi_{\langle a\rangle})_{\langle x\rangle}} & f_*\psi_{\langle a\rangle}\langle x\rangle \\ \| & & \| \\ \phi_{\langle b\rangle}\langle (f_j)_* x_j\rangle & \xrightarrow{(\Phi_{\langle b\rangle})_{\langle (f_j)_* x_j\rangle}} & \psi_{\langle b\rangle}\langle (f_j)_* x_j\rangle \\ {\scriptstyle \phi_{\langle b\rangle}\langle p\rangle}\downarrow & & \downarrow{\scriptstyle \psi_{\langle b\rangle}\langle p\rangle} \\ \phi_{\langle b\rangle}\langle y\rangle & \xrightarrow{(\Phi_{\langle b\rangle})_{\langle y\rangle}} & \psi_{\langle b\rangle}\langle y\rangle \end{array} \tag{6.66}$$

The detail of the diagram (6.66) is as follows.

- $f_* = Zf : Za \longrightarrow Zb$.
- $(f_j)_* = X_j f_j : X_j a_j \longrightarrow X_j b_j$.
- The top rectangle commutes by the modification axiom for Φ (5.12).
- The bottom rectangle commutes by the naturality of the component natural transformation $\Phi_{\langle b\rangle}$.

This proves that $\int_{\mathcal{D}} \Phi$ is a natural transformation. □

Lemma 6.67. *In the context of Definition 6.61, $\int_{\mathcal{D}} \Phi$ is an n-linear natural transformation.*

Proof. Lemma 6.64 shows that $\int_{\mathcal{D}} \Phi$ is a natural transformation. It remains to check the n-linearity axioms (6.18) and (6.19).

Unity (6.18). This axiom for $\int_{\mathcal{D}} \Phi$ says that, if

$$(a_i, x_i) = (0, X_i^0 *) \in \textstyle\int_{\mathcal{D}} X_i \quad \text{for some} \quad i \in \{1, \ldots, n\},$$

then

$$\left(\textstyle\int_{\mathcal{D}} \Phi\right)_{\langle (a_j, x_j)\rangle} = \left(1_a \,,\, (\Phi_{\langle a\rangle})_{\langle x\rangle}\right) = \left(1_0 \,,\, 1_{Z^0 *}\right) \in \textstyle\int_{\mathcal{D}} Z.$$

This axiom holds by

- the fact that $0 \in \mathcal{D}$ is the strict multiplicative zero (2.31), which implies

  $$a = \textstyle\bigotimes_{j=1}^n a_j = 0 \in \mathcal{D},$$

  and
- the unity axiom (5.54) for the additive modification Φ.

Constraint Compatibility (6.19). Using the notation in Definition 6.45, the constraint compatibility axiom for $\int_{\mathcal{D}} \Phi$ states the commutativity of the following diagram in $\int_{\mathcal{D}} Z$.

$$\begin{array}{ccc} \left(a \oplus a' \,,\, \mathsf{L}_*\phi_{\langle a''\rangle}\langle x''\rangle\right) & \xrightarrow{(\mathsf{L}^{-1}, 1)} & \left(a'' \,,\, \phi_{\langle a''\rangle}\langle x''\rangle\right) \\ {\scriptstyle \left(1_{a\oplus a'} \,,\, \mathsf{L}_*(\Phi_{\langle a''\rangle})_{\langle x''\rangle}\right)}\downarrow & & \downarrow{\scriptstyle \left(1_{a''} \,,\, (\Phi_{\langle a''\rangle})_{\langle x''\rangle}\right)} \\ \left(a \oplus a' \,,\, \mathsf{L}_*\psi_{\langle a''\rangle}\langle x''\rangle\right) & \xrightarrow{(\mathsf{L}^{-1}, 1)} & \left(a'' \,,\, \psi_{\langle a''\rangle}\langle x''\rangle\right) \end{array} \tag{6.68}$$

The detail of the diagram (6.68) is as follows.

- The top and bottom $(\mathsf{L}^{-1}, 1)$ are the i-th linearity constraints (6.56) of, respectively, $\int_{\mathcal{D}}\phi$ and $\int_{\mathcal{D}}\psi$.
- The left vertical arrow comes from the following computation.

$$\begin{cases} \big(1_a\,,\,(\Phi_{\langle a\rangle})_{\langle x\rangle}\big) \,\square\, \big(1_{a'}\,,\,(\Phi_{\langle a'\rangle})_{\langle x'\rangle}\big) & \\ = \big(1_a \oplus 1_{a'}\,,\, Z^2_{a,a'}\big((\Phi_{\langle a\rangle})_{\langle x\rangle}\,,\,(\Phi_{\langle a'\rangle})_{\langle x'\rangle}\big)\big) & \text{by (6.40)} \\ = \big(1_{a\oplus a'}\,,\, \mathsf{L}_*(\Phi_{\langle a''\rangle})_{\langle x''\rangle}\big) & \text{by (5.57)} \end{cases}$$

- In the first argument in $\mathcal{D}$, each composite is equal to

$$\mathsf{L}^{-1} : a \oplus a' \longrightarrow a''.$$

- In the second argument in Za'', each composite is equal to the morphism

$$\mathsf{L}_*^{-1}\mathsf{L}_*(\Phi_{\langle a''\rangle})_{\langle x''\rangle} = (\Phi_{\langle a''\rangle})_{\langle x''\rangle}.$$

This proves that (6.68) is commutative. □

Lemma 6.69. *In the context of Definitions 6.45 and 6.61, the object and morphism assignments,*

$$\phi \longmapsto \textstyle\int_{\mathcal{D}}\phi \quad \textit{and} \quad \Phi \longmapsto \int_{\mathcal{D}}\Phi,$$

define a functor

$$\mathcal{D}\mathsf{Cat}\big(\langle X\rangle\,;\,Z\big) \xrightarrow{\;\int_{\mathcal{D}}\;} \mathsf{PermCat}^{\mathsf{sg}}\big(\langle \textstyle\int_{\mathcal{D}} X\rangle\,;\,\int_{\mathcal{D}} Z\big) \tag{6.70}$$

with $\langle X\rangle = \langle X_j\rangle_{j=1}^n$ *and* $\langle \int_{\mathcal{D}} X\rangle = \langle \int_{\mathcal{D}} X_j\rangle_{j=1}^n$.

Proof. The object and morphism assignments are well defined by, respectively, Lemmas 6.57 and 6.67. It remains to check that $\int_{\mathcal{D}}$ preserves identity morphisms and composition.

Preservation of Identities. Consider the identity modification

$$1_\phi : \phi \longrightarrow \phi$$

of an additive natural transformation ϕ in $\mathcal{D}\mathsf{Cat}\big(\langle X\rangle\,;\,Z\big)$. Each component natural transformation of 1_ϕ is an identity. By definitions (6.34) and (6.63) each component morphism of $\int_{\mathcal{D}} 1_\phi$ is an identity morphism.

Preservation of Composition. Suppose given vertically composable additive modifications Φ and Ψ in $\mathcal{D}\mathsf{Cat}\big(\langle X\rangle\,;\,Z\big)$ as follows.

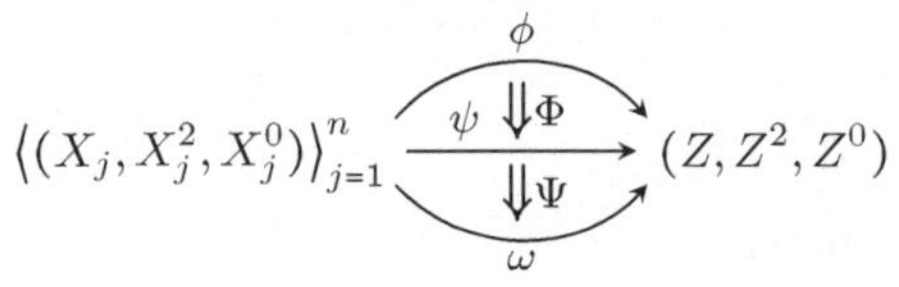

Using the notation in (6.47) through (6.49), suppose given an n-tuple of objects

$$\langle (a_j, x_j) \rangle \in \prod_{j=1}^{n} \big(\textstyle\int_{\mathcal{D}} X_j \big).$$

By definition (6.63) the component morphism of $\int_{\mathcal{D}}(\Psi\Phi)$ at $\langle (a_j, x_j) \rangle$ is the following morphism in $\int_{\mathcal{D}} Z$.

$$(a\,,\,\phi_{\langle a\rangle}\langle x\rangle) \xrightarrow{\big(1_a\,,\,((\Psi\Phi)_{\langle a\rangle})_{\langle x\rangle}\big)} (a\,,\,\omega_{\langle a\rangle}\langle x\rangle) \tag{6.71}$$

By the definitions (1.44) and (6.35) of, respectively, the vertical composite $\Psi\Phi$ and the composition in $\int_{\mathcal{D}} Z$, the morphism (6.71) is equal to the composite

$$(a\,,\,\phi_{\langle a\rangle}\langle x\rangle) \xrightarrow{\big(1_a\,,\,(\Phi_{\langle a\rangle})_{\langle x\rangle}\big)} (a\,,\,\psi_{\langle a\rangle}\langle x\rangle) \xrightarrow{\big(1_a\,,\,(\Psi_{\langle a\rangle})_{\langle x\rangle}\big)} (a\,,\,\omega_{\langle a\rangle}\langle x\rangle). \tag{6.72}$$

By definition (6.63) the composite (6.72) is equal to the component morphism at $\langle (a_j, x_j) \rangle$ of the vertical composite natural transformation

$$\textstyle\int_{\mathcal{D}} \phi \xrightarrow{\int_{\mathcal{D}} \Phi} \int_{\mathcal{D}} \psi \xrightarrow{\int_{\mathcal{D}} \Psi} \int_{\mathcal{D}} \omega.$$

Therefore, $\int_{\mathcal{D}}(\Psi\Phi)$ is equal to $(\int_{\mathcal{D}} \Psi)(\int_{\mathcal{D}} \Phi)$. □

6.5 The Grothendieck Construction Preserves Units and Composition

As in Sections 6.3 and 6.4 we continue to assume that $\mathcal{D}$ is a small tight bipermutative category with

- additive structure $\mathcal{D}^{\oplus} = (\mathcal{D}, \oplus, 0, \beta^{\oplus})$ and
- multiplicative structure $\mathcal{D}^{\otimes} = (\mathcal{D}, \otimes, 1, \beta^{\otimes})$ (Definition 2.41).

In this section we show that the Grothendieck construction $\int_{\mathcal{D}}$ preserves the colored units and multicategorical composition. Recall the following constructions from previous sections.

- The object assignment $X \longmapsto \int_{\mathcal{D}} X$ is given by Lemma 6.42.
- The multimorphism functors in arity 0 are given in Definition 6.43.
- For the multimorphism functors in positive arity, the object and morphism assignments are given in, respectively, Definitions 6.45 and 6.61.
- The functoriality of the multimorphism functors is proved in Lemma 6.69.

Also recall the Cat-multicategories $\mathcal{D}\mathsf{Cat}$ and $\mathsf{PermCat}^{\mathsf{sg}}$ in, respectively, Theorems 5.71 and 6.31.

Lemma 6.73. *The Grothendieck construction*

$$\mathcal{D}\mathsf{Cat} \xrightarrow{\int_{\mathcal{D}}} \mathsf{PermCat}^{\mathsf{sg}}$$

preserves the colored units in the sense of (3.21).

Proof. Suppose given an additive symmetric monoidal functor

$$(Z, Z^2, Z^0) : \mathcal{D}^{\oplus} \longrightarrow \mathsf{Cat}.$$

Consider Definition 6.45 for the identity natural transformation $\phi = 1_Z : Z \longrightarrow Z$.

- By the definitions (6.49) and (6.52) of, respectively, object and morphism assignments,
$$\textstyle\int_{\mathcal{D}} 1_Z : \int_{\mathcal{D}} Z \longrightarrow \int_{\mathcal{D}} Z$$
is the identity functor.
- The only linearity constraint (6.56) of $\int_{\mathcal{D}} 1_Z$ is the identity natural transformation because the unique Laplaza coherence isomorphism
$$a_1 \oplus a_1' \xrightarrow[\cong]{\mathsf{L}} a_1 \oplus a_1' \quad \text{for} \quad a_1, a_1' \in \mathcal{D}$$
is the identity morphism by the uniqueness in Theorem 2.90.

Thus $\int_{\mathcal{D}} 1_Z$ is the identity 1-linear functor of $\int_{\mathcal{D}} Z$. □

Lemma 6.74. *The Grothendieck construction*

$$\mathcal{D}\mathsf{Cat} \xrightarrow{\int_{\mathcal{D}}} \mathsf{PermCat}^{\mathsf{sg}}$$

preserves composition in the sense of (3.22).

Proof. Suppose given additive symmetric monoidal functors

$$(Z, Z^2, Z^0), \big\langle (X_j, X_j^2, X_j^0) \big\rangle_{j=1}^{n}, \big\langle \langle (W_{ji}, W_{ji}^2, W_{ji}^0) \rangle_{i=1}^{\ell_j} \big\rangle_{j=1}^{n} : \mathcal{D}^{\oplus} \longrightarrow \mathsf{Cat} \tag{6.75}$$

with $n > 0$, along with the following notation.

$$\begin{cases} \langle X \rangle = \langle X_j \rangle_{j=1}^{n} & \langle \int_{\mathcal{D}} X \rangle = \langle \int_{\mathcal{D}} X_j \rangle_{j=1}^{n} \\ \langle W_j \rangle = \langle W_{ji} \rangle_{i=1}^{\ell_j} & \langle \int_{\mathcal{D}} W_j \rangle = \langle \int_{\mathcal{D}} W_{ji} \rangle_{i=1}^{\ell_j} \\ \langle W \rangle = \langle \langle W_j \rangle \rangle_{j=1}^{n} & \langle \int_{\mathcal{D}} W \rangle = \langle \langle \int_{\mathcal{D}} W_j \rangle \rangle_{j=1}^{n} \end{cases} \tag{6.76}$$

We must show that the following diagram of functors is commutative, with the left and right γ as in, respectively, Explanation 5.21 and (6.28).

$$\begin{array}{ccc} \mathcal{D}\mathsf{Cat}(\langle X\rangle; Z) \times \prod_{j=1}^{n} \mathcal{D}\mathsf{Cat}(\langle W_j\rangle; X_j) & \xrightarrow{\int_{\mathcal{D}} \times \prod_j \int_{\mathcal{D}}} & \mathsf{PermCat}^{\mathsf{sg}}(\langle \int_{\mathcal{D}} X\rangle; \int_{\mathcal{D}} Z) \times \prod_{j=1}^{n} \mathsf{PermCat}^{\mathsf{sg}}(\langle \int_{\mathcal{D}} W_j\rangle; \int_{\mathcal{D}} X_j) \\ \gamma \downarrow & & \downarrow \gamma \\ \mathcal{D}\mathsf{Cat}(\langle W\rangle; Z) & \xrightarrow{\int_{\mathcal{D}}} & \mathsf{PermCat}^{\mathsf{sg}}(\langle \int_{\mathcal{D}} W\rangle; \int_{\mathcal{D}} Z) \end{array} \tag{6.77}$$

Additive Natural Transformations and Objects. To show that (6.77) is commutative, suppose given additive natural transformations

$$\bigotimes_{j=1}^{n} X_j \xrightarrow{\phi} Z \quad \text{and} \quad \bigotimes_{i=1}^{\ell_j} W_{ji} \xrightarrow{\phi_j} X_j \tag{6.78}$$

for $j \in \{1, \ldots, n\}$ and objects

$$\langle (a_{ji}, w_{ji})\rangle \in \textstyle\prod_{j=1}^{n} \prod_{i=1}^{\ell_j} (\int_{\mathcal{D}} W_{ji}) \tag{6.79}$$

along with the following notation.

$$\begin{cases} a_j = \bigotimes_{i=1}^{\ell_j} a_{ji} \in \mathcal{D} & a = \bigotimes_{j=1}^{n} a_j \in \mathcal{D} \\ \langle a_j\rangle = \langle a_{ji}\rangle_{i=1}^{\ell_j} \in \mathcal{D}^{\ell_j} & \langle a\rangle = \langle a_j\rangle_{j=1}^{n} \in \mathcal{D}^n \\ \langle w_j\rangle = \langle w_{ji}\rangle_{i=1}^{\ell_j} \in \prod_{i=1}^{\ell_j} W_{ji} a_{ji} & \langle w\rangle = \langle \langle w_j\rangle\rangle_{j=1}^{n} \in \prod_{j=1}^{n} \prod_{i=1}^{\ell_j} W_{ji} a_{ji} \end{cases} \tag{6.80}$$

Then the following computation shows that the two composites in the diagram (6.77), when applied to (6.78) and the objects (6.79), are equal.

$$\begin{cases} \big(\int_{\mathcal{D}} \gamma(\phi; \langle \phi_j\rangle)\big)\langle (a_{ji}, w_{ji})\rangle & \\ = \big(a, \gamma(\phi; \langle \phi_j\rangle)_{\langle\langle a_{ji}\rangle_i\rangle_j} \langle w\rangle\big) & \text{by (6.49)} \\ = \big(a, \phi_{\langle a\rangle} \langle (\phi_j)_{\langle a_j\rangle} \langle w_j\rangle\rangle_{j=1}^{n}\big) & \text{by (5.25)} \\ = (\int_{\mathcal{D}} \phi)\big\langle \big(a_j, (\phi_j)_{\langle a_j\rangle}\langle w_j\rangle\big)\big\rangle_{j=1}^{n} & \text{by (6.49)} \\ = (\int_{\mathcal{D}} \phi)\big\langle (\int_{\mathcal{D}} \phi_j)\langle (a_{ji}, w_{ji})\rangle_{i=1}^{\ell_j}\big\rangle_{j=1}^{n} & \text{by (6.49)} \\ = \gamma\big(\int_{\mathcal{D}} \phi; \langle \int_{\mathcal{D}} \phi_j\rangle\big)\langle (a_{ji}, w_{ji})\rangle & \text{by (6.29)} \end{cases} \tag{6.81}$$

Additive Natural Transformations and Morphisms. Suppose given morphisms

$$\langle (f_{ji}, p_{ji})\rangle \in \textstyle\prod_{j=1}^{n} \prod_{i=1}^{\ell_j} (\int_{\mathcal{D}} W_{ji}). \tag{6.82}$$

The computation (6.81), with

- $\langle (f_{ji}, p_{ji})\rangle$ in place of (a_{ji}, w_{ji}) and
- (6.52) in place of (6.49),

shows that the two composites in the diagram (6.77), when applied to (6.78) and the morphisms (6.82), are equal. So far we have shown that the two composites in the diagram (6.77), when applied to (6.78), are equal as functors. We will show that they also have the same linearity constraints after the following paragraph.

Additive Modifications. Suppose given additive modifications

$$(\Phi\,,\langle\Phi_j\rangle)\in\mathcal{D}\mathsf{Cat}(\langle X\rangle\,;Z)\times\textstyle\prod_{j=1}^{n}\mathcal{D}\mathsf{Cat}(\langle W_j\rangle\,;X_j).$$

The computation (6.81), with

- $(\Phi\,,\langle\Phi_j\rangle)$ in place of $(\phi\,,\langle\phi_j\rangle)$,
- (6.63) in place of (6.49), and
- (5.27) in place of (5.25),

shows that the two composites in the diagram (6.77) are equal on additive modifications.

Linearity Constraints. To show that the diagram (6.77) is commutative, it remains to show that the functors

$$\textstyle\int_{\mathcal{D}}\gamma(\phi;\langle\phi_j\rangle)\quad\text{and}\quad\gamma\big(\int_{\mathcal{D}}\phi;\langle\int_{\mathcal{D}}\phi_j\rangle\big):\prod_{j=1}^{n}\prod_{i=1}^{\ell_j}(\int_{\mathcal{D}}W_{ji})\longrightarrow\int_{\mathcal{D}}Z$$

have the same linearity constraints. In addition to the objects in (6.79), suppose given an object

$$(a'_{ji},w'_{ji})\in\textstyle\int_{\mathcal{D}}W_{ji}$$

for some $j\in\{1,\ldots,n\}$ and some $i\in\{1,\ldots,\ell_j\}$, along with the notation in (6.80) and below.

$$\begin{cases}a''_{ji}=a_{ji}\oplus a'_{ji} & k=\ell_1+\cdots+\ell_{j-1}+i\\ a'_j=a_{j1}\otimes\cdots\otimes a'_{ji}\otimes\cdots\otimes a_{j\ell_j} & a'=a_1\otimes\cdots\otimes a'_j\otimes\cdots\otimes a_n\\ a''_j=a_{j1}\otimes\cdots\otimes a''_{ji}\otimes\cdots\otimes a_{j\ell_j} & a''=a_1\otimes\cdots\otimes a''_j\otimes\cdots\otimes a_n\end{cases}$$

With these objects and notation, by definition (6.56) the k-th linearity constraint of $\int_{\mathcal{D}}\gamma(\phi;\langle\phi_j\rangle)$ has the form

$$(\mathsf{L}^{-1},1)\in\textstyle\int_{\mathcal{D}}Z.$$

Here L is the unique Laplaza coherence isomorphism in $\mathcal{D}$ (Theorem 2.90)

$$a''\xrightarrow[\cong]{\mathsf{L}}a\oplus a'\tag{6.83}$$

that distributes over the sum $a_{ji}\oplus a'_{ji}$ in a''.

On the other hand, by definition (6.30) the k-th linearity constraint of $\gamma(\int_{\mathcal{D}}\phi;\langle\int_{\mathcal{D}}\phi_j\rangle)$ is the following composite in $\int_{\mathcal{D}} Z$.

$$\begin{cases} \big((\int_{\mathcal{D}}\phi)\langle 1\circ_j (\int_{\mathcal{D}}\phi_j)_i^2\rangle\big)\circ(\int_{\mathcal{D}}\phi)_j^2 & \\ = \big((\int_{\mathcal{D}}\phi)\langle 1\circ_j(\mathsf{L}_j^{-1},1)\rangle\big)\circ(\mathsf{L}_0^{-1},1) & \text{by (6.56)}\\ = \big(1\otimes\cdots\otimes\mathsf{L}_j^{-1}\otimes\cdots\otimes 1,\phi\langle 1\rangle\big)\circ(\mathsf{L}_0^{-1},1) & \text{by (6.52)}\\ = \big(1\otimes\cdots\otimes\mathsf{L}_j^{-1}\otimes\cdots\otimes 1,1\big)\circ(\mathsf{L}_0^{-1},1) & \text{by functoriality}\\ = \big((1\otimes\cdots\otimes\mathsf{L}_j^{-1}\otimes\cdots\otimes 1)\mathsf{L}_0^{-1},1\big) & \text{by (6.35)}\\ = (\mathsf{L}^{-1},1) & \text{by Theorem 2.90} \end{cases} \tag{6.84}$$

The detail of (6.84) is as follows.

- The first, second, and fourth equalities follow from, respectively, the definitions (6.56), (6.52), and (6.35) as stated.
- L_0 is the unique Laplaza coherence isomorphism in $\mathcal{D}$

$$a_1\otimes\cdots\otimes(a_j\oplus a_j')\otimes\cdots\otimes a_n \xrightarrow[\cong]{\mathsf{L}_0} a\oplus a'$$

  that distributes over the sum $a_j\oplus a_j'$ in the domain.
- L_j is the unique Laplaza coherence isomorphism in $\mathcal{D}$

$$a_j'' \xrightarrow[\cong]{\mathsf{L}_j} a_j\oplus a_j'$$

  that distributes over the sum $a_{ji}\oplus a_{ji}'$ in a_j''. Omitting the $\otimes$ symbols to save space, the composite in $\mathcal{D}$

$$a'' \xrightarrow{1\cdots\mathsf{L}_j\cdots 1} a_1\cdots(a_j\oplus a_j')\cdots a_n \xrightarrow{\mathsf{L}_0} a\oplus a'$$

  is equal to L in (6.83) by the uniqueness in Theorem 2.90. This yields the last equality in (6.84).
- After the second equality, $\langle 1\rangle$ is an n-tuple of identity morphisms. By the functoriality of each component functor of ϕ, $\phi\langle 1\rangle$ is also an identity morphism. This yields the third equality in (6.84).

This proves that

$$\int_{\mathcal{D}}\gamma(\phi;\langle\phi_j\rangle) \quad\text{and}\quad \gamma\big(\int_{\mathcal{D}}\phi;\langle\int_{\mathcal{D}}\phi_j\rangle\big)$$

have the same linearity constraints. □

6.6 Pseudo Symmetry Isomorphisms

For each small tight bipermutative category $\mathcal{D}$, in this section we define the pseudo symmetry isomorphisms for the Grothendieck construction $\int_{\mathcal{D}}$ and prove that they are well-defined natural isomorphisms. The four coherence axioms, (4.3) through (4.6), of a pseudo symmetric Cat-multifunctor are proved in Lemma 6.103. Here is an outline of this section.

- The pseudo symmetry isomorphisms of $\int_{\mathcal{D}}$ are given in Definition 6.85.
- Lemmas 6.92 and 6.95 show that each component of each pseudo symmetry isomorphism is an n-linear natural isomorphism.
- Lemma 6.100 establishes the naturality of the pseudo symmetry isomorphisms.

First we define the natural isomorphisms that will be shown to be a part of the pseudo symmetric Cat-multifunctor structure on the Grothendieck construction.

Definition 6.85. Suppose given additive symmetric monoidal functors Z and $\langle X \rangle$ as in (6.75) and (6.76) and a permutation $\sigma \in \Sigma_n$. Define the data of a natural isomorphism

$$\begin{array}{ccc} \mathcal{D}\mathsf{Cat}(\langle X \rangle; Z) & \xrightarrow{\int_{\mathcal{D}}} & \mathsf{PermCat}^{\mathsf{sg}}(\langle \int_{\mathcal{D}} X \rangle; \int_{\mathcal{D}} Z) \\ \sigma \downarrow & \overset{(\int_{\mathcal{D}})_{\sigma,\langle X \rangle, Z}}{\Longrightarrow} & \downarrow \sigma \\ \mathcal{D}\mathsf{Cat}(\langle X \rangle \sigma; Z) & \xrightarrow[\int_{\mathcal{D}}]{} & \mathsf{PermCat}^{\mathsf{sg}}(\langle \int_{\mathcal{D}} X \rangle \sigma; \int_{\mathcal{D}} Z) \end{array} \tag{6.86}$$

called the *pseudo symmetry isomorphism*, as follows. Suppose given an additive natural transformation (5.40)

$$\bigotimes_{j=1}^n X_j \xrightarrow{\phi} Z$$

with

- image under the σ-action (5.15)

$$\bigotimes_{j=1}^n X_{\sigma(j)} \xrightarrow{\phi^\sigma} Z$$

and

- associated n-linear functor (6.46)

$$\prod_{j=1}^n \left(\int_{\mathcal{D}} X_j\right) \xrightarrow{\int_{\mathcal{D}} \phi} \int_{\mathcal{D}} Z.$$

Suppose given an n-tuple of objects

$$\langle (a_j, x_j) \rangle \in \prod_{j=1}^{n} (\textstyle\int_{\mathcal{D}} X_{\sigma(j)}) \tag{6.87}$$

with each $a_j \in \mathcal{D}$ and $x_j \in X_{\sigma(j)} a_j$. To define the component of $(\int_{\mathcal{D}})_{\sigma,\langle X \rangle,Z}$ at ϕ and $\langle (a_j, x_j) \rangle$, first note that its domain and codomain objects in $\int_{\mathcal{D}} Z$ are given as follows.

$$\begin{cases} (\int_{\mathcal{D}} \phi^{\sigma})\langle (a_j, x_j) \rangle & \\ \quad = \big(a\,,\, \phi^{\sigma}_{\langle a \rangle}\langle x \rangle\big) & \text{by (6.49)} \\ \quad = \big(a\,,\, \sigma_*^{-1} \phi_{\sigma\langle a \rangle} \sigma\langle x \rangle\big) & \text{by (5.15)} \\ (\int_{\mathcal{D}} \phi)^{\sigma}\langle (a_j, x_j) \rangle & \\ \quad = (\int_{\mathcal{D}} \phi)\big\langle \big(a_{\sigma^{-1}(j)}\,,\, x_{\sigma^{-1}(j)}\big) \big\rangle & \text{by (6.25)} \\ \quad = \big(a_{\sigma^{-1}}\,,\, \phi_{\sigma\langle a \rangle} \sigma\langle x \rangle\big) & \text{by (5.16) and (6.49)} \end{cases} \tag{6.88}$$

Using (6.88) we define the component of $(\int_{\mathcal{D}})_{\sigma,\langle X \rangle,Z}$ at ϕ and $\langle (a_j, x_j) \rangle$ as the isomorphism in $\int_{\mathcal{D}} Z$

$$\big(a\,,\, \sigma_*^{-1} \phi_{\sigma\langle a \rangle} \sigma\langle x \rangle\big) \xrightarrow[\cong]{(\sigma,1)} \big(a_{\sigma^{-1}}\,,\, \phi_{\sigma\langle a \rangle} \sigma\langle x \rangle\big). \tag{6.89}$$

This finishes the definition of $(\int_{\mathcal{D}})_{\sigma,\langle X \rangle,Z}$.

As in Section 4.1 we usually abbreviate $(\int_{\mathcal{D}})_{\sigma,\langle X \rangle,Z}$ to $(\int_{\mathcal{D}})_{\sigma}$. Its components at ϕ and $\big(\phi, \langle (a_j, x_j) \rangle\big)$ are denoted by

$$(\textstyle\int_{\mathcal{D}})_{\sigma,\phi} \quad \text{and} \quad (\int_{\mathcal{D}})_{\sigma,\phi,\langle (a_j,x_j) \rangle},$$

respectively. ◇

Explanation 6.90. The two arguments in the isomorphism

$$(\textstyle\int_{\mathcal{D}})_{\sigma,\phi,\langle (a_j,x_j) \rangle} = (\sigma, 1)$$

in (6.89) are given as follows.

- σ is the unique coherence isomorphism in $\mathcal{D}^{\otimes} = (\mathcal{D}, \otimes, 1, \beta^{\otimes})$

$$a = \bigotimes_{j=1}^{n} a_j \xrightarrow[\cong]{\sigma} \bigotimes_{j=1}^{n} a_{\sigma^{-1}(j)} = a_{\sigma^{-1}} \tag{6.91}$$

  that permutes the n factors according to $\sigma \in \Sigma_n$. This is the inverse of the coherence isomorphism in (5.17).

- The second argument 1 is the identity morphism

$$\sigma_* \sigma_*^{-1} \phi_{\sigma\langle a \rangle} \sigma\langle x \rangle \xrightarrow{=} \phi_{\sigma\langle a \rangle} \sigma\langle x \rangle$$

  in $Z a_{\sigma^{-1}}$. ◇

Componentwise Multilinearity

Recall that the n-ary 2-cells in the Cat-multicategory $\mathsf{PermCat}^{\mathsf{sg}}$ are n-linear natural transformations between n-linear functors (Definition 6.16). An n-linear natural transformation is a natural transformation that satisfies the unity axiom (6.18) and the constraint compatibility axiom (6.19). Next we show that each component of the pseudo symmetry isomorphism $(\int_{\mathcal{D}})_\sigma$ is an n-linear natural isomorphism, starting with its naturality.

Lemma 6.92. *In the context of Definition 6.85, the component of $(\int_{\mathcal{D}})_\sigma$ at ϕ*

$$\prod_{j=1}^{n}\left(\textstyle\int_{\mathcal{D}} X_{\sigma(j)}\right) \xrightarrow[\;(\int_{\mathcal{D}}\phi)^\sigma\;]{\;\int_{\mathcal{D}}\phi^\sigma\;} \textstyle\int_{\mathcal{D}} Z \qquad \Downarrow (\int_{\mathcal{D}})_{\sigma,\phi} \tag{6.93}$$

is a natural isomorphism.

Proof. By definition (6.89) each component of $(\int_{\mathcal{D}})_{\sigma,\phi}$ is an isomorphism in $\int_{\mathcal{D}} Z$.

To show that $(\int_{\mathcal{D}})_{\sigma,\phi}$ is a natural transformation, suppose given an n-tuple of morphisms

$$\big\langle (f_j,p_j) : (a_j,x_j) \longrightarrow (b_j,y_j)\big\rangle \in \prod_{j=1}^{n}\left(\textstyle\int_{\mathcal{D}} X_{\sigma(j)}\right)$$

with

- each $f_j : a_j \longrightarrow b_j$ a morphism in $\mathcal{D}$ and
- each $p_j : (f_j)_* x_j \longrightarrow y_j$ a morphism in $X_{\sigma(j)} b_j$, where

$$(f_j)_* = X_{\sigma(j)} f_j : X_{\sigma(j)} a_j \longrightarrow X_{\sigma(j)} b_j .$$

Adapting the notation for $\langle (a_j,x_j)\rangle$ to $\langle (b_j,y_j)\rangle$ and $\langle (f_j,p_j)\rangle$ as in (6.51), we must show that the following naturality diagram for $(\int_{\mathcal{D}})_{\sigma,\phi}$ is commutative in $\int_{\mathcal{D}} Z$.

$$\begin{array}{ccc}
\big(a\,,\, \sigma_*^{-1}\phi_{\sigma\langle a\rangle}\sigma\langle x\rangle\big) & \xrightarrow{(\sigma,1)} & \big(a_{\sigma^{-1}}\,,\, \phi_{\sigma\langle a\rangle}\sigma\langle x\rangle\big) \\
{\scriptstyle (f\,,\,\sigma_*^{-1}\phi_{\sigma\langle b\rangle}\sigma\langle p\rangle)}\downarrow & & \downarrow{\scriptstyle (f_{\sigma^{-1}}\,,\,\phi_{\sigma\langle b\rangle}\sigma\langle p\rangle)} \\
\big(b\,,\, \sigma_*^{-1}\phi_{\sigma\langle b\rangle}\sigma\langle y\rangle\big) & \xrightarrow{(\sigma,1)} & \big(b_{\sigma^{-1}}\,,\, \phi_{\sigma\langle b\rangle}\sigma\langle y\rangle\big)
\end{array} \tag{6.94}$$

The arrows in the diagram (6.94) are given as follows.

- The top and bottom horizontal arrows $(\sigma,1)$ are the components of $(\int_{\mathcal{D}})_{\sigma,\phi}$ in (6.89) at, respectively, $\langle (a_j,x_j)\rangle$ and $\langle (b_j,y_j)\rangle$.
- The left and right vertical arrows are, respectively,

$$\left(\textstyle\int_{\mathcal{D}}\phi^\sigma\right)\langle (f_j,p_j)\rangle \quad\text{and}\quad \left(\textstyle\int_{\mathcal{D}}\phi\right)^\sigma\langle (f_j,p_j)\rangle .$$

The first argument in the diagram (6.94) is the following diagram in $\mathcal{D}^{\otimes} = (\mathcal{D}, \otimes, \beta^{\otimes})$. It commutes by the naturality of the multiplicative braiding $\beta^{\otimes}$.

$$\begin{array}{ccc} a & \xrightarrow{\sigma} & a_{\sigma^{-1}} \\ {\scriptstyle f}\downarrow & & \downarrow{\scriptstyle f_{\sigma^{-1}}} \\ b & \xrightarrow{\sigma} & b_{\sigma^{-1}} \end{array}$$

The second argument in the diagram (6.94) yields two composites in $Zb_{\sigma^{-1}}$. They are equal by the following computation.

$$\begin{aligned} 1 \circ \sigma_* \sigma_*^{-1} \phi_{\sigma\langle b \rangle} \sigma\langle p \rangle &= \phi_{\sigma\langle b \rangle} \sigma\langle p \rangle \\ &= \phi_{\sigma\langle b \rangle} \sigma\langle p \rangle \circ 1 \\ &= \phi_{\sigma\langle b \rangle} \sigma\langle p \rangle \circ (f_{\sigma^{-1}})_* 1 \end{aligned}$$

This shows that the diagram (6.94) is commutative. □

Next we check the multilinearity of $(\int_{\mathcal{D}})_{\sigma,\phi}$.

Lemma 6.95. *In the context of Definition 6.85, the component* $(\int_{\mathcal{D}})_{\sigma,\phi}$ *is an* n*-linear natural isomorphism.*

Proof. Lemma 6.92 shows that $(\int_{\mathcal{D}})_{\sigma,\phi}$ is a natural isomorphism. It remains to check the two axioms in Definition 6.16.

Unity (6.18). The unity axiom for $(\int_{\mathcal{D}})_{\sigma,\phi}$ states that, if for some $i \in \{1, \ldots, n\}$

$$(a_i, x_i) = \left(0\,,\, X^0_{\sigma(i)} *\right) \in \textstyle\int_{\mathcal{D}} X_{\sigma(i)}$$

is the monoidal unit (6.38) in $\int_{\mathcal{D}} X_{\sigma(i)}$, then the component (6.89)

$$(\textstyle\int_{\mathcal{D}})_{\sigma,\phi,\langle (a_j,x_j) \rangle} = (\sigma, 1)$$

is the identity morphism. This axiom holds because, if $a_i = 0$, which is the multiplicative zero in $\mathcal{D}$, then the uniqueness in Theorem 2.90 implies that the coherence isomorphism in $\mathcal{D}$ (6.91)

$$a = 0 \xrightarrow{\sigma} 0 = a_{\sigma^{-1}}$$

is the identity morphism 1_0. Therefore,

$$(\sigma, 1) = (1_0, 1) : (0, Z^0 *) \longrightarrow (0, Z^0 *)$$

is the identity morphism of the monoidal unit in $\int_{\mathcal{D}} Z$ by the unity axiom (5.43) for ϕ.

Constraint Compatibility (6.19). In addition to the objects $\langle (a_j, x_j) \rangle$ in (6.87), suppose given an object

$$(a_i', x_i') \in \textstyle\int_{\mathcal{D}} X_{\sigma(i)}$$

for some $i \in \{1, \ldots, n\}$ and objects $a_i' \in \mathcal{D}$ and $x_i' \in X_{\sigma(i)} a_i'$. We use the notation in

- (5.66) for $a'_{\sigma^{-1}}$ and $a''_{\sigma^{-1}}$ and
- (6.48) and (6.53) for a, a', a'', $\langle a\rangle$, $\langle a'\rangle$, $\langle a''\rangle$, $\langle x\rangle$, $\langle x'\rangle$, and $\langle x''\rangle$ with $X_{\sigma(j)}$ in place of X_j for each $j \in \{1,\dots,n\}$.

With these objects and notation, the constraint compatibility axiom for $(\int_{\mathcal{D}})_{\sigma,\phi}$ states that the following diagram in $\int_{\mathcal{D}} Z$ is commutative.

$$\begin{array}{ccc}
\big(a\,,\,\sigma_*^{-1}\phi_{\sigma\langle a\rangle}\sigma\langle x\rangle\big)\,\square\,\big(a'\,,\,\sigma_*^{-1}\phi_{\sigma\langle a'\rangle}\sigma\langle x'\rangle\big) & \xrightarrow{(\int_{\mathcal{D}}\phi^{\sigma})^2_i} & \big(a''\,,\,\sigma_*^{-1}\phi_{\sigma\langle a''\rangle}\sigma\langle x''\rangle\big) \\
\downarrow{\scriptstyle (\sigma,1)\,\square\,(\sigma,1)} & & \downarrow{\scriptstyle (\sigma,1)} \\
\big(a_{\sigma^{-1}}\,,\,\phi_{\sigma\langle a\rangle}\sigma\langle x\rangle\big)\,\square\,\big(a'_{\sigma^{-1}}\,,\,\phi_{\sigma\langle a'\rangle}\sigma\langle x'\rangle\big) & \xrightarrow{((\int_{\mathcal{D}}\phi)^{\sigma})^2_i} & \big(a''_{\sigma^{-1}}\,,\,\phi_{\sigma\langle a''\rangle}\sigma\langle x''\rangle\big)
\end{array} \tag{6.96}$$

The detail of the diagram (6.96) is as follows.

- By
  - the definitions (6.39) and (6.40) of the monoidal product $\square$ in $\int_{\mathcal{D}} Z$ and
  - the functoriality of each component functor of the monoidal constraint Z^2 (5.34),

  the left vertical morphism in (6.96), $(\sigma,1)\,\square\,(\sigma,1)$, is the following morphism.

$$\begin{array}{c}
\Big(a\oplus a'\,,\,Z^2_{a,a'}\big(\sigma_*^{-1}\phi_{\sigma\langle a\rangle}\sigma\langle x\rangle\,,\,\sigma_*^{-1}\phi_{\sigma\langle a'\rangle}\sigma\langle x'\rangle\big)\Big) \\
\downarrow{\scriptstyle (\sigma\oplus\sigma,1)} \\
\Big(a_{\sigma^{-1}}\oplus a'_{\sigma^{-1}}\,,\,Z^2_{a_{\sigma^{-1}},a'_{\sigma^{-1}}}\big(\phi_{\sigma\langle a\rangle}\sigma\langle x\rangle\,,\,\phi_{\sigma\langle a'\rangle}\sigma\langle x'\rangle\big)\Big)
\end{array} \tag{6.97}$$

- By definition (6.56), the top horizontal morphism in (6.96) is

$$(\textstyle\int_{\mathcal{D}}\phi^{\sigma})^2_i = (\mathsf{L}^{-1},1). \tag{6.98}$$

  Here L is the unique Laplaza coherence isomorphism in $\mathcal{D}$

$$a'' \xrightarrow[\cong]{\mathsf{L}} a\oplus a'$$

  that distributes over the sum $a_i \oplus a'_i$ in the i-th tensor factor in a''.

- By (6.26) and (6.56), the bottom horizontal morphism in (6.96) is

$$\big((\textstyle\int_{\mathcal{D}}\phi)^{\sigma}\big)^2_i = \big(\mathsf{L}^{-1}_{\sigma^{-1}}\,,\,1\big). \tag{6.99}$$

  Here $\mathsf{L}_{\sigma^{-1}}$ is the unique Laplaza coherence isomorphism in $\mathcal{D}$

$$a''_{\sigma^{-1}} \xrightarrow[\cong]{\mathsf{L}_{\sigma^{-1}}} a_{\sigma^{-1}}\oplus a'_{\sigma^{-1}}$$

  that distributes over the sum $a_i \oplus a'_i$ in the $\sigma(i)$-th tensor factor in $a''_{\sigma^{-1}}$.

Using (6.97) through (6.99) the first argument in the diagram (6.96) is the following diagram in $\mathcal{D}$.

$$\begin{array}{ccc} a \oplus a' & \xrightarrow{\mathsf{L}^{-1}} & a'' \\ {\scriptstyle \sigma \oplus \sigma}\downarrow & & \downarrow{\scriptstyle \sigma} \\ a_{\sigma^{-1}} \oplus a'_{\sigma^{-1}} & \xrightarrow{\mathsf{L}^{-1}_{\sigma^{-1}}} & a''_{\sigma^{-1}} \end{array}$$

This diagram commutes by the uniqueness in Theorem 2.90.

Next, each of the two composites in the diagram (6.96) has the second argument given by the identity morphism of $\phi_{\sigma\langle a''\rangle}\sigma\langle x''\rangle$. This is true because, in each of the four morphisms in the diagram (6.96), the second argument is an identity morphism. This shows that $(\int_{\mathcal{D}})_{\sigma,\phi}$ satisfies the constraint compatibility axiom. □

Naturality of the Pseudo Symmetry Isomorphisms

Next we show that the pseudo symmetry isomorphisms for the Grothendieck construction are natural with respect to the morphisms in the multimorphism category $\mathcal{D}\mathsf{Cat}(\langle X\rangle; Z)$.

Lemma 6.100. *In the context of Definition 6.85, the pseudo symmetry isomorphism*

$$\mathcal{D}\mathsf{Cat}(\langle X\rangle; Z) \underset{(\int_{\mathcal{D}}(-))^{\sigma}}{\overset{\int_{\mathcal{D}}(-)^{\sigma}}{\Downarrow (\textstyle\int_{\mathcal{D}})_{\sigma}}} \mathsf{PermCat}^{\mathsf{sg}}(\langle \textstyle\int_{\mathcal{D}} X\rangle\sigma; \int_{\mathcal{D}} Z)$$

is a natural isomorphism.

Proof. Lemma 6.95 shows that each component of $(\int_{\mathcal{D}})_{\sigma}$ is an n-linear natural isomorphism. To show its naturality, suppose given an additive modification Φ as in (5.53)

$$\langle (X_j, X_j^2, X_j^0)\rangle_{j=1}^{n} \underset{\psi}{\overset{\phi}{\Downarrow \Phi}} (Z, Z^2, Z^0).$$

Naturality with respect to Φ means the commutativity of the following diagram of n-linear natural transformations in $\mathsf{PermCat}^{\mathsf{sg}}(\langle \int_{\mathcal{D}} X\rangle\sigma; \int_{\mathcal{D}} Z)$.

$$\begin{array}{ccc} \int_{\mathcal{D}}\phi^{\sigma} & \xrightarrow{(\int_{\mathcal{D}})_{\sigma,\phi}} & (\int_{\mathcal{D}}\phi)^{\sigma} \\ {\scriptstyle \int_{\mathcal{D}}\Phi^{\sigma}}\downarrow & & \downarrow{\scriptstyle (\int_{\mathcal{D}}\Phi)^{\sigma}} \\ \int_{\mathcal{D}}\psi^{\sigma} & \xrightarrow{(\int_{\mathcal{D}})_{\sigma,\psi}} & (\int_{\mathcal{D}}\psi)^{\sigma} \end{array} \tag{6.101}$$

It suffices to check the commutativity of (6.101) when it is evaluated at a general n-tuple of objects as in (6.87)

$$\langle (a_j, x_j) \rangle \in \prod_{j=1}^{n} \left(\int_{\mathcal{D}} X_{\sigma(j)} \right).$$

By definitions (5.18), (6.25), (6.63), and (6.89), the diagram (6.101) evaluated at $\langle (a_j, x_j) \rangle$ is the following diagram in $\int_{\mathcal{D}} Z$.

$$\begin{array}{ccc} \left(a, \sigma_*^{-1}\phi_{\sigma\langle a\rangle}\sigma\langle x\rangle\right) & \xrightarrow{(\sigma,1)} & \left(a_{\sigma^{-1}}, \phi_{\sigma\langle a\rangle}\sigma\langle x\rangle\right) \\ {\scriptstyle (1_a,\, \sigma_*^{-1}(\Phi_{\sigma\langle a\rangle})_{\sigma\langle x\rangle})}\downarrow & & \downarrow{\scriptstyle (1_{a_{\sigma^{-1}}},\, (\Phi_{\sigma\langle a\rangle})_{\sigma\langle x\rangle})} \\ \left(a, \sigma_*^{-1}\psi_{\sigma\langle a\rangle}\sigma\langle x\rangle\right) & \xrightarrow{(\sigma,1)} & \left(a_{\sigma^{-1}}, \psi_{\sigma\langle a\rangle}\sigma\langle x\rangle\right) \end{array} \tag{6.102}$$

The detail of the diagram (6.102) is as follows.

- The first argument of each of the two composites is the coherence isomorphism

$$a \xrightarrow[\cong]{\sigma} a_{\sigma^{-1}} \quad \text{in} \quad \mathcal{D}^{\otimes}.$$

- The second argument in the diagram (6.102) yields two composites in $Z a_{\sigma^{-1}}$. They are equal by the following computation.

$$\begin{aligned} 1 \circ \sigma_* \sigma_*^{-1}(\Phi_{\sigma\langle a\rangle})_{\sigma\langle x\rangle} &= (\Phi_{\sigma\langle a\rangle})_{\sigma\langle x\rangle} \\ &= (\Phi_{\sigma\langle a\rangle})_{\sigma\langle x\rangle} \circ 1 \\ &= (\Phi_{\sigma\langle a\rangle})_{\sigma\langle x\rangle} \circ (1_{a_{\sigma^{-1}}})_* 1 \end{aligned}$$

This shows that the diagram (6.102) is commutative. □

6.7 Pseudo Symmetric Cat-Multifunctoriality

As in Sections 6.3 through 6.6 we continue to assume that $\mathcal{D}$ is a small tight bipermutative category with

- additive structure $\mathcal{D}^{\oplus} = (\mathcal{D}, \oplus, 0, \beta^{\oplus})$ and
- multiplicative structure $\mathcal{D}^{\otimes} = (\mathcal{D}, \otimes, 1, \beta^{\otimes})$ (Definition 2.41).

In Lemmas 6.73 and 6.74 we showed that the Grothendieck construction $\int_{\mathcal{D}}$ strictly preserves the colored units and multicategorical composition. In Section 6.6 we constructed the pseudo symmetry isomorphisms for the Grothendieck construction and proved that they are natural isomorphisms. To show that $\int_{\mathcal{D}}$ is a pseudo symmetric Cat-multifunctor, next we check that its pseudo symmetry isomorphisms satisfy the four axioms in Definition 4.1.

Lemma 6.103. *The pseudo symmetry isomorphisms* $(\int_{\mathcal{D}})_\sigma$ *in* (6.86) *satisfy the axioms* (4.3) *through* (4.6).

Proof. Suppose given additive symmetric monoidal functors

$$Z\,,\, \langle X_j\rangle_{j=1}^n\,,\, \big\langle \langle W_{ji}\rangle_{i=1}^{\ell_j}\big\rangle_{j=1}^n : \mathcal{D}^{\oplus} \longrightarrow \mathsf{Cat}$$

as in (6.75) and (6.76).

The Unit Permutation Axiom (4.3).
Suppose $\mathrm{id}_n \in \Sigma_n$ is the identity permutation. By definition (6.89) there is an equality of morphisms in $\int_{\mathcal{D}} Z$

$$(\textstyle\int_{\mathcal{D}})_{\mathrm{id}_n,\phi,\langle (a_j,x_j)\rangle} = (1_a, 1) : \big(a\,,\, \phi_{\langle a\rangle}\langle x\rangle\big) \longrightarrow \big(a\,,\, \phi_{\langle a\rangle}\langle x\rangle\big)$$

for

- each additive natural transformation $\phi \in \mathcal{D}\mathsf{Cat}\big(\langle X\rangle\,;\, Z\big)$ as in (5.40) and
- each n-tuple of objects as in (6.87)

$$\langle (a_j,x_j)\rangle \in \textstyle\prod_{j=1}^n (\int_{\mathcal{D}} X_j).$$

Since $(1_a, 1)$ is an identity morphism (6.34), this shows that $(\int_{\mathcal{D}})_{\mathrm{id}_n}$ is the identity natural transformation of $\int_{\mathcal{D}}$.

The Product Permutation Axiom (4.4).
Suppose given

- permutations $\sigma, \tau \in \Sigma_n$,
- an additive natural transformation $\phi \in \mathcal{D}\mathsf{Cat}\big(\langle X\rangle\,;\, Z\big)$, and
- an n-tuple of objects

$$\langle (a_j,x_j)\rangle \in \textstyle\prod_{j=1}^n (\int_{\mathcal{D}} X_{\sigma\tau(j)})$$

with each $a_j \in \mathcal{D}$ and $x_j \in X_{\sigma\tau(j)} a_j$.

By definition (6.25) and (6.89), for the above data the left pasting diagram in the product permutation axiom (4.4) is the composite in $\int_{\mathcal{D}} Z$

$$(\sigma, 1) \circ (\tau, 1) = (\sigma\tau, 1).$$

The right-hand side above is the component of $(\int_{\mathcal{D}})_{\sigma\tau}$ at ϕ and $\langle (a_j,x_j)\rangle$. This proves the product permutation axiom.

The Top Equivariance Axiom (4.5).
Suppose given

- a permutation $\sigma \in \Sigma_n$,

- additive natural transformations

$$\bigotimes_{j=1}^{n} X_j \xrightarrow{\phi} Z \quad \text{and} \quad \bigotimes_{i=1}^{\ell_j} W_{ji} \xrightarrow{\phi_j} X_j \tag{6.104}$$

for $j \in \{1, \ldots, n\}$ and $i \in \{1, \ldots, \ell_j\}$, and

- an $(\ell_{\sigma(1)} + \cdots + \ell_{\sigma(n)})$-tuple of objects

$$\langle a, w\rangle = \langle\langle(a_{ji}, w_{ji})\rangle_i\rangle_j \in \prod_{j=1}^{n}\prod_{i=1}^{\ell_{\sigma(j)}} \left(\textstyle\int_{\mathcal{D}} W_{\sigma(j),i}\right)$$

with each $a_{ji} \in \mathcal{D}$ and $w_{ji} \in W_{\sigma(j),i}\, a_{ji}$.

We use the notation below, where $\overline{\sigma}$ is the block permutation induced by σ that permutes n blocks of lengths $\{\ell_{\sigma(j)}\}_{j=1}^{n}$.

$$\begin{cases} \langle\phi_j\rangle = (\phi_1, \ldots, \phi_n) & g = \gamma(\phi; \langle\phi_j\rangle) \qquad \phi^{\sigma(j)} = \int_{\mathcal{D}} \phi_{\sigma(j)} \\ \ell = \ell_{\sigma(1)} + \cdots + \ell_{\sigma(n)} & \overline{\sigma} = \sigma\langle \ell_{\sigma(1)}, \ldots, \ell_{\sigma(n)}\rangle \in \Sigma_\ell \\ \langle a_j\rangle = \langle a_{ji}\rangle_{i=1}^{\ell_{\sigma(j)}} \in \mathcal{D}^{\ell_{\sigma(j)}} & \langle w_j\rangle = \langle w_{ji}\rangle_{i=1}^{\ell_{\sigma(j)}} \in \prod_{i=1}^{\ell_{\sigma(j)}} W_{\sigma(j),i}\, a_{ji} \\ \langle a\rangle = \langle\langle a_j\rangle\rangle_{j=1}^{n} \in \mathcal{D}^{\ell} & \langle w\rangle = \langle\langle w_j\rangle\rangle_{j=1}^{n} \in \prod_{j=1}^{n}\prod_{i=1}^{\ell_{\sigma(j)}} W_{\sigma(j),i}\, a_{ji} \\ a_j = \bigotimes_{i=1}^{\ell_{\sigma(j)}} a_{ji} \in \mathcal{D} & a = \bigotimes_{j=1}^{n} a_j \in \mathcal{D} \qquad a_{\overline{\sigma}^{-1}} = \bigotimes_{j=1}^{n} a_{\sigma^{-1}(j)} \in \mathcal{D} \end{cases} \tag{6.105}$$

By definition (6.89), for the above data the left pasting diagram in (4.5) yields the following morphism in $\int_{\mathcal{D}} Z$.

$$\left(a\,,\, \overline{\sigma}_*^{-1} g_{\overline{\sigma}\langle a\rangle}\overline{\sigma}\langle w\rangle\right) \xrightarrow[= (\overline{\sigma}, 1)]{(\int_{\mathcal{D}})_{\overline{\sigma}, g, \langle a,w\rangle}} \left(a_{\overline{\sigma}^{-1}}\,,\, g_{\overline{\sigma}\langle a\rangle}\overline{\sigma}\langle w\rangle\right) \tag{6.106}$$

The right pasting diagram in (4.5) yields the following morphism in $\int_{\mathcal{D}} Z$, with the same domain, respectively, codomain, as the arrow in (6.106).

$$\begin{cases} \gamma\left((\textstyle\int_{\mathcal{D}})_{\sigma,\phi}\,;\, \langle 1_{\phi^{\sigma(j)}}\rangle_{j=1}^{n}\right)_{\langle a,w\rangle} & \\ = \left(\textstyle\int_{\mathcal{D}}\right)_{\sigma,\phi,\left\langle \phi^{\sigma(j)}\langle(a_{ji},w_{ji})\rangle_{i=1}^{\ell_{\sigma(j)}}\right\rangle_{j=1}^{n}} & \text{by (6.29)} \\ = \left(\textstyle\int_{\mathcal{D}}\right)_{\sigma,\phi,\left\langle (a_j, (\phi_{\sigma(j)})_{\langle a_j\rangle}\langle w_j\rangle)\right\rangle_{j=1}^{n}} & \text{by (6.49)} \\ = (\sigma, 1) & \text{by (6.89)} \end{cases} \tag{6.107}$$

The morphisms $(\overline{\sigma}, 1)$ and $(\sigma, 1)$ in, respectively, (6.106) and (6.107), are equal because the following diagram in $\mathcal{D}$ is commutative by the definition of the block permutation $\overline{\sigma}$.

$$\begin{array}{ccc} \bigotimes_{j=1}^{n}\bigotimes_{i=1}^{\ell_{\sigma(j)}} a_{ji} & \xrightarrow{\overline{\sigma}} & \bigotimes_{j=1}^{n}\bigotimes_{i=1}^{\ell_j} a_{\sigma^{-1}(j),i} \\ \| & & \| \\ \bigotimes_{j=1}^{n} a_j & \xrightarrow{\sigma} & \bigotimes_{j=1}^{n} a_{\sigma^{-1}(j)} \end{array}$$

This proves the top equivariance axiom.

The Bottom Equivariance Axiom (4.6).

Suppose given

- permutations $\tau_j \in \Sigma_{\ell_j}$ for $j \in \{1, \ldots, n\}$,
- additive natural transformations ϕ and ϕ_j as in (6.104), and
- an $(\ell_1 + \cdots + \ell_n)$-tuple of objects

$$\langle a, w \rangle = \langle \langle (a_{ji}, w_{ji}) \rangle_i \rangle_j \in \textstyle\prod_{j=1}^n \prod_{i=1}^{\ell_j} (\int_{\mathcal{D}} W_{j, \tau_j(i)})$$

with each $a_{ji} \in \mathcal{D}$ and $w_{ji} \in W_{j,\tau_j(i)}\, a_{ji}$.

We use the following notation, along with $\langle \phi_j \rangle$ and g in (6.105).

$$\begin{cases} \overline{\phi} = \int_{\mathcal{D}} \phi & \ell = \ell_1 + \cdots + \ell_n \\ \tau^{\times} = \tau_1 \times \cdots \times \tau_n \in \Sigma_\ell & \tau^{-\times} = (\tau^{\times})^{-1} \\ \langle a_j \rangle = \langle a_{ji} \rangle_{i=1}^{\ell_j} \in \mathcal{D}^{\ell_j} & \langle w_j \rangle = \langle w_{ji} \rangle_{i=1}^{\ell_j} \in \prod_{i=1}^{\ell_j} W_{j,\tau_j(i)}\, a_{ji} \\ \langle a \rangle = \langle \langle a_j \rangle \rangle_{j=1}^n \in \mathcal{D}^\ell & \langle w \rangle = \langle \langle w_j \rangle \rangle_{j=1}^n \in \prod_{j=1}^n \prod_{i=1}^{\ell_j} W_{j,\tau_j(i)}\, a_{ji} \\ a_j = \bigotimes_{i=1}^{\ell_j} a_{ji} \in \mathcal{D} & a = \bigotimes_{j=1}^n a_j \in \mathcal{D} \\ (a_j)_{\tau_j^{-1}} = \bigotimes_{i=1}^{\ell_j} a_{j,\tau_j^{-1}(i)} & a_{\tau^{-\times}} = \bigotimes_{j=1}^n (a_j)_{\tau_j^{-1}} \end{cases}$$

By definition (6.89), for the above data the left pasting diagram in (4.6) yields the following morphism in $\int_{\mathcal{D}} Z$.

$$\big(a\,,\, \tau_*^{-\times} g_{\tau^\times \langle a \rangle} \tau^\times \langle w \rangle\big) \xrightarrow[= (\tau^\times, 1)]{(\int_{\mathcal{D}})_{\tau^\times, g, \langle a, w \rangle}} \big(a_{\tau^{-\times}}\,,\, g_{\tau^\times \langle a \rangle} \tau^\times \langle w \rangle\big) \tag{6.108}$$

The right pasting diagram in (4.6) yields the following morphism in $\int_{\mathcal{D}} Z$, with the same domain, respectively, codomain, as the arrow in (6.108).

$$\begin{cases} \gamma\big(1_{\overline{\phi}}; \langle (\int_{\mathcal{D}})_{\tau_j, \phi_j} \rangle_{j=1}^n\big)_{\langle a, w \rangle} & \\ = \overline{\phi} \big\langle (\int_{\mathcal{D}})_{\tau_j, \phi_j, \langle (a_{ji}, w_{ji}) \rangle_{i=1}^{\ell_j}} \big\rangle_{j=1}^n & \text{by (6.29)} \\ = \overline{\phi} \big\langle (\tau_j, 1) \big\rangle_{j=1}^n & \text{by (6.89)} \\ = \big(\bigotimes_{j=1}^n \tau_j\,,\, \phi\langle 1 \rangle \big) & \text{by (6.52)} \\ = \big(\bigotimes_{j=1}^n \tau_j\,,\, 1 \big) & \text{by functoriality} \end{cases} \tag{6.109}$$

In the middle line of (6.109), $(\tau_j, 1)$ is the following morphism in $\int_{\mathcal{D}} X_j$.

$$\big(a_j\,,\, (\tau_j^{-1})_* (\phi_j)_{\tau_j \langle a_j \rangle} \tau_j \langle w_j \rangle\big) \xrightarrow{(\tau_j, 1)} \big((a_j)_{\tau_j^{-1}}\,,\, (\phi_j)_{\tau_j \langle a_j \rangle} \tau_j \langle w_j \rangle\big)$$

In the fourth line of (6.109), $\phi\langle 1 \rangle$ is the morphism in $Z a_{\tau^{-\times}}$

$$\phi \langle 1 \rangle = \phi_{\langle (a_j)_{\tau_j^{-1}} \rangle_{j=1}^n} (\overbrace{1, \ldots, 1}^{n}).$$

This is the identity morphism by the functoriality of each component functor of ϕ. This yields the last equality in (6.109).

The morphisms $(\tau^\times, 1)$ and $\big(\bigotimes_{j=1}^n \tau_j\,, 1\big)$ in, respectively, (6.108) and (6.109), are equal because the following diagram in $\mathcal{D}$ is commutative by the definition of the permutation $\tau^\times$.

$$\begin{array}{ccc}
\bigotimes_{j=1}^n \bigotimes_{i=1}^{\ell_j} a_{ji} & \xrightarrow{\tau^\times} & \bigotimes_{j=1}^n \bigotimes_{i=1}^{\ell_j} a_{j,\tau_j^{-1}(i)} \\
\| & & \| \\
\bigotimes_{j=1}^n a_j & \xrightarrow{\bigotimes_j \tau_j} & \bigotimes_{j=1}^n (a_j)_{\tau_j^{-1}}
\end{array}$$

This proves the bottom equivariance axiom. □

Next is the main result of this chapter. It relates the Cat-multicategories $\mathcal{D}$Cat and PermCat^{sg} in, respectively, Theorems 5.71 and 6.31, via the Grothendieck construction.

Theorem 6.110. *Suppose $\mathcal{D}$ is a small tight bipermutative category.*

1. *The Grothendieck construction*

$$\mathcal{D}\mathsf{Cat} \xrightarrow{\int_{\mathcal{D}}} \mathsf{PermCat}^{\mathsf{sg}} \tag{6.111}$$

   is a pseudo symmetric Cat-*multifunctor.*

2. *If the multiplicative braiding in $\mathcal{D}$ is the identity natural transformation, then $\int_{\mathcal{D}}$ is a* Cat-*multifunctor.*

3. *If $\int_{\mathcal{D}}$ is a* Cat-*multifunctor, then*

$$x \otimes y = y \otimes x \tag{6.112}$$

   for all objects $x, y \in \mathcal{D}$.

Proof. For the Grothendieck construction $\int_{\mathcal{D}}$ we established the following.

- The object assignment $X \longmapsto \int_{\mathcal{D}} X$ is given in Lemma 6.42.
- The multimorphism functors

$$\mathcal{D}\mathsf{Cat}(\langle X\rangle; Z) \xrightarrow{\int_{\mathcal{D}}} \mathsf{PermCat}^{\mathsf{sg}}(\langle \textstyle\int_{\mathcal{D}} X\rangle; \int_{\mathcal{D}} Z)$$

  are given in Definition 6.43 and Lemma 6.69 for, respectively, arity 0 and positive arity.
- It preserves the colored units by Lemma 6.73.
- It preserves the composition by Lemma 6.74.
- Its pseudo symmetry isomorphisms $(\int_{\mathcal{D}})_\sigma$ are given in Definition 6.85.

- These pseudo symmetry isomorphisms are well-defined natural isomorphisms by Lemmas 6.92, 6.95, and 6.100.

- The four axioms in Definition 4.1 for the pseudo symmetry isomorphisms are proved in Lemma 6.103.

This shows that $\int_{\mathcal{D}}$ is a pseudo symmetric Cat-multifunctor, proving assertion 1.

For assertion 2, suppose that the multiplicative braiding $\beta^{\otimes}$ in $\mathcal{D}$ is the identity. Since $\int_{\mathcal{D}}$ preserves the colored units and composition (Lemmas 6.73 and 6.74), we must show that $\int_{\mathcal{D}}$ preserves the symmetric group action. In other words, we want to show that the boundary of the diagram (6.86) is commutative. Since $\beta^{\otimes} = 1$, the two objects in (6.88) are equal. So

$$\int_{\mathcal{D}} \phi^{\sigma} = \left(\int_{\mathcal{D}} \phi\right)^{\sigma}$$

for each additive natural transformation ϕ in $\mathcal{D}\mathsf{Cat}(\langle X \rangle; Z)$. This means that the boundary diagram (6.86) is commutative on objects.

To prove the commutativity of the boundary diagram (6.86) on morphisms, suppose $\Phi : \phi \longrightarrow \psi$ is an additive modification in $\mathcal{D}\mathsf{Cat}(\langle X \rangle; Z)$ and $\langle (a_j, x_j) \rangle$ is an n-tuple of objects as in (6.87). Then there are equalities as follows.

$$\begin{cases} \left(\int_{\mathcal{D}} \Phi\right)^{\sigma}_{\langle (a_j, x_j) \rangle} & \\ = \left(\int_{\mathcal{D}} \Phi\right)_{\langle a_{\sigma^{-1}(j)}, x_{\sigma^{-1}(j)} \rangle} & \text{by (6.25)} \\ = \left(1_{a_{\sigma^{-1}}}, \left(\Phi_{\sigma\langle a \rangle}\right)_{\sigma\langle x \rangle}\right) & \text{by (6.63)} \\ = \left(1_a, Z(\mathrm{id}_n)\left(\Phi_{\sigma\langle a \rangle}\right)_{\sigma\langle x \rangle}\right) & \text{by } \beta^{\otimes} = 1 \text{ and functoriality of } Z \\ = \left(1_a, Z(\sigma^{-1})\left(\Phi_{\sigma\langle a \rangle}\right)_{\sigma\langle x \rangle}\right) & \text{by } \beta^{\otimes} = 1 \\ = \left(1_a, \left(\Phi^{\sigma}_{\langle a \rangle}\right)_{\langle x \rangle}\right) & \text{by (5.18)} \\ = \left(\int_{\mathcal{D}} \Phi^{\sigma}\right)_{\langle (a_j, x_j) \rangle} & \text{by (6.63)} \end{cases} \tag{6.113}$$

This shows that the boundary diagram (6.86) is commutative on morphisms. Thus $\int_{\mathcal{D}}$ is a Cat-multifunctor, proving assertion 2.

For assertion 3, suppose that $\int_{\mathcal{D}}$ is a Cat-multifunctor. So $\int_{\mathcal{D}}$ preserves the symmetric group action, which means that the boundary of the diagram (6.86) is commutative. The first component of the computation (6.88), which uses the notation in (5.17), shows that

$$\bigotimes_{j=1}^{n} a_j = a = a_{\sigma^{-1}} = \bigotimes_{j=1}^{n} a_{\sigma^{-1}(j)}.$$

This implies the condition (6.112). □

Example 6.114 (Obstruction to Cat-Multifunctoriality)**.** For the small tight bipermutative category $\mathcal{A}$ in Example 2.63, the condition (6.112) is *not* satisfied by the definition (2.68) of $\otimes$ in $\mathcal{A}$. By Theorem 6.110 (1) and (3) the

Grothendieck construction $\int_{\mathcal{A}}$ is a pseudo symmetric Cat-multifunctor but *not* a Cat-multifunctor. ◇

Explanation 6.115 (Further Criteria for Cat-Multifunctoriality)**.** Under the hypothesis of Theorem 6.110 (3)—that $\int_{\mathcal{D}}$ is a Cat-multifunctor—something beyond the condition (6.112) is true. Since

$$\textstyle\int_{\mathcal{D}} \phi^{\sigma} = \left(\int_{\mathcal{D}} \phi\right)^{\sigma},$$

the second component of the computation (6.88) shows that the diagram

$$\begin{array}{ccc} \prod_{j=1}^{n} X_j a_j & \xrightarrow{\phi_{\langle a \rangle}} & Za \\ {\scriptstyle \phi_{\langle a \rangle}}\big\downarrow & & \big\| \\ Za & \xrightarrow{Z(\sigma)} & Za_{\sigma^{-1}} \end{array}$$

is commutative on objects for any

- permutation $\sigma \in \Sigma_n$,
- object $\langle a \rangle = \langle a_j \rangle_{j=1}^{n} \in \mathcal{D}^n$,
- additive symmetric monoidal functors Z and $\langle X \rangle = \langle X_j \rangle_{j=1}^{n}$ in $\mathcal{D}$Cat, and
- additive natural transformation $\phi \in \mathcal{D}\mathsf{Cat}(\langle X \rangle; Z)$.

Moreover, for any additive modification $\Phi : \phi \longrightarrow \psi$ in $\mathcal{D}\mathsf{Cat}(\langle X \rangle; Z)$, the second component of the computation (6.113) yields the morphism equality

$$\left(\Phi_{\langle a \rangle}\right)_{\langle x \rangle} = Z(\sigma)\left(\Phi_{\langle a \rangle}\right)_{\langle x \rangle}$$

in Za for each object $\langle x \rangle \in \prod_{j=1}^{n} X_j a_j$.

In the conclusion of Theorem 6.110 (3), we only include the condition (6.112) because it is easy to state and sufficient for our main application, which is the case $\mathcal{A}$ in Example 6.114. ◇

7

Permutative Opfibrations from Bipermutative-Indexed Categories

Context

Suppose $(\mathcal{D}, \oplus, \otimes)$ is a small tight bipermutative category (Definition 2.41). We proved in Theorem 6.110 that there is a pseudo symmetric Cat-multifunctor (Definition 4.1)

$$\mathcal{D}\mathsf{Cat} \xrightarrow{\int_{\mathcal{D}}} \mathsf{PermCat}^{\mathsf{sg}} \tag{7.1}$$

with object assignment given by the Grothendieck construction

$$(X, X^2, X^0) \longmapsto \left(\textstyle\int_{\mathcal{D}} X, \square, (0, X^0 *), \beta^{\square}\right).$$

Moreover, $\int_{\mathcal{D}}$ is a Cat-multifunctor if the multiplicative braiding $\beta^{\otimes}$ in $\mathcal{D}$ is the identity natural transformation. The weaker converse says that, if $\int_{\mathcal{D}}$ is a Cat-multifunctor, then $x \otimes y = y \otimes x$ for all objects $x, y \in \mathcal{D}$.

The Grothendieck construction $\int_{\mathcal{D}} X$ has a canonical first-factor projection functor

$$\textstyle\int_{\mathcal{D}} X \xrightarrow{U_X} \mathcal{D}.$$

This functor U_X has some nice lifting properties that make it into a permutative opfibration over $\mathcal{D}$. As the natural codomain of the Grothendieck construction, we would like to consider not just small permutative categories but small permutative opfibrations over $\mathcal{D}$.

Purpose

The main purpose of this chapter is to lift the Grothendieck construction (7.1) to a non-symmetric Cat-multifunctor

$$\mathcal{D}\mathsf{Cat} \xrightarrow{\int_{\mathcal{D}}} \mathsf{PFib}^{\mathcal{D}}. \tag{7.2}$$

- The codomain $\mathsf{PFib}^{\mathcal{D}}$ is a non-symmetric Cat-multicategory with small permutative opfibrations over $\mathcal{D}$ as objects.

DOI: 10.1201/9781003451235-7

- The objects in the multimorphism categories in $\mathsf{PFib}^{\mathcal{D}}$ are called *opcartesian strong n-linear functors*. These are strong n-linear functors with extra properties that generalize Cartesian functors between Grothendieck fibrations.

- The morphisms in the multimorphism categories in $\mathsf{PFib}^{\mathcal{D}}$ are called *opcartesian n-linear transformations*. These are n-linear natural transformations with an extra property that generalize vertical natural transformations between Cartesian functors.

- The non-symmetric Cat-multicategory structure on $\mathsf{PFib}^{\mathcal{D}}$ is inherited from $\mathsf{PermCat}^{\mathsf{sg}}$.

Permutative opfibrations over $\mathcal{D}$ are defined using only the additive structure $\mathcal{D}^{\oplus}$. On the other hand, opcartesian n-linear functors require the additive structure $\mathcal{D}^{\oplus}$, the multiplicative structure $\mathcal{D}^{\otimes}$, and Laplaza's Coherence Theorem 2.90.

The Grothendieck constructions (7.1) and (7.2) fit together in the following commutative diagram of non-symmetric Cat-multifunctors.

$$\begin{array}{ccc} & \overset{\int_{\mathcal{D}}}{\nearrow} & \mathsf{PFib}^{\mathcal{D}} \\ & & \downarrow \mathsf{U} \\ \mathcal{D}\mathsf{Cat} & \xrightarrow{\int_{\mathcal{D}}} & \mathsf{PermCat}^{\mathsf{sg}} \end{array}$$

The non-symmetric Cat-multifunctor U sends each small permutative opfibration over $\mathcal{D}$ to its domain small permutative category. Between multimorphism categories, U is given by subcategory inclusions. Moreover, if the multiplicative braiding $\beta^{\otimes}$ in $\mathcal{D}$ is the identity natural transformation, then the following two statements hold:

- The symmetric group action in $\mathsf{PermCat}^{\mathsf{sg}}$ restricts to $\mathsf{PFib}^{\mathcal{D}}$ to make it into a Cat-multicategory.

- Both U and $\int_{\mathcal{D}} : \mathcal{D}\mathsf{Cat} \longrightarrow \mathsf{PFib}^{\mathcal{D}}$ are Cat-multifunctors.

A weaker converse is also true: if $\mathsf{PFib}^{\mathcal{D}}$ inherits a well-defined symmetric group action from $\mathsf{PermCat}^{\mathsf{sg}}$, then $\otimes = \otimes^{\mathsf{op}}$ in $\mathcal{D}$. The very restrictive condition $\otimes = \otimes^{\mathsf{op}}$ fails in most bipermutative categories of interest, including $\mathsf{Fin}^{\mathsf{sk}}$, Σ, $\mathcal{F}$, and $\mathcal{A}$ in Section 2.3. Therefore, in each of those cases, $\mathsf{PFib}^{\mathcal{D}}$ is genuinely a non-symmetric Cat-multicategory and *not* a Cat-multicategory.

In Chapter 8 we show that the non-symmetric Cat-multifunctor $\int_{\mathcal{D}}$ in (7.2) is actually a non-symmetric Cat-*multiequivalence*. This means that $\int_{\mathcal{D}}$ is

- essentially surjective on objects between the underlying 1-ary categories and

- an isomorphism between multimorphism categories.

Furthermore, $\int_{\mathcal{D}}$ is a Cat-multiequivalence if $\beta^{\otimes}$ in $\mathcal{D}$ is the identity.

Summary

Below is a summary of the non-symmetric Cat-multicategory $\mathsf{PFib}^{\mathcal{D}}$ and the lifted Grothendieck construction $\int_{\mathcal{D}}$.

- $\mathsf{PFib}^{\mathcal{D}}$ (Theorem 7.51 and Corollary 7.52).
  - Its objects are small permutative opfibrations over $\mathcal{D}$ (Definition 7.6).
  - Its n-ary 1-cells are opcartesian strong n-linear functors (Definition 7.11).
  - Its n-ary 2-cells are opcartesian n-linear transformations (Definition 7.27).
  - Its units are identity opcartesian 1-linear functors (Example 7.23).
  - Its composition is given in Lemmas 7.37 and 7.42.
  - Its behavior relative to symmetry is given in Lemmas 7.44 and 7.48.
  - It is *not* induced by a symmetric monoidal structure (Lemma 7.44).
- $\int_{\mathcal{D}}$ (Theorem 7.64).
  - Its object assignment is given in Lemma 7.56.
  - Its 1-cell and 2-cell assignments are given in Lemmas 7.61 and 7.62.
  - It preserves units by Lemma 6.73 and Definition 7.34.
  - It preserves composition by Lemmas 6.74, 7.61, and 7.62.

Organization

In Section 7.1 we first review the definition of a split opfibration. A functor $P : \mathsf{E} \longrightarrow \mathsf{B}$ is an opfibration if and only if its opposite functor P^{op} is a Grothendieck fibration. A split opfibration is an opfibration with a cleavage that satisfies the unitarity and multiplicativity conditions. For a permutative category E, a *permutative opfibration* $P : \mathsf{E} \longrightarrow \mathcal{D}^{\oplus}$ is a split opfibration that satisfies the following three conditions:

- P is a strict symmetric monoidal functor.
- Chosen opcartesian lifts are closed under the monoidal product in E.
- Each component of the braiding in E is the chosen opcartesian lift of the corresponding component of the braiding in $\mathcal{D}$.

Small permutative opfibrations over $\mathcal{D}$ are the objects in $\mathsf{PFib}^{\mathcal{D}}$.

In Section 7.2 we define opcartesian strong n-linear functors and opcartesian n-linear transformations. They are, respectively, the n-ary 1-cells and n-ary 2-cells in $\mathsf{PFib}^{\mathcal{D}}$. By definition an opcartesian n-linear functor is an n-linear functor with extra properties but *no* extra structure, and the same remark applies to opcartesian n-linear transformations. Each multimorphism category in $\mathsf{PFib}^{\mathcal{D}}$ is defined as a subcategory of the corresponding multimorphism category in $\mathsf{PermCat}^{\mathsf{sg}}$.

In Section 7.3 we define the (non-symmetric) Cat-multicategory structure on $\mathsf{PFib}^{\mathcal{D}}$ as the one inherited from $\mathsf{PermCat}^{\mathsf{sg}}$; see Theorem 7.51 and Corollary 7.52. First we check that the composition in $\mathsf{PFib}^{\mathcal{D}}$ is well defined. Then we check that, if $\beta^{\otimes} = 1$ in $\mathcal{D}$, then the symmetric group action in $\mathsf{PFib}^{\mathcal{D}}$ is well defined. The weaker converse says that if the symmetric group action in $\mathsf{PFib}^{\mathcal{D}}$ is well defined, then $\otimes = \otimes^{\mathsf{op}}$ in $\mathcal{D}$.

In Section 7.4 we construct the lifted Grothendieck construction in (7.2); see Theorem 7.64. We already know that

- $\int_{\mathcal{D}}$ in (7.1) is a pseudo symmetric Cat-multifunctor and
- each multimorphism category in $\mathsf{PFib}^{\mathcal{D}}$ is a subcategory of the corresponding multimorphism category in $\mathsf{PermCat}^{\mathsf{sg}}$.

Thus most of the work in lifting $\int_{\mathcal{D}}$ to $\mathsf{PFib}^{\mathcal{D}}$ is to show that $\int_{\mathcal{D}}$ lands in the multimorphism categories in $\mathsf{PFib}^{\mathcal{D}}$.

We remind the reader of our left normalized bracketing Convention 1.24 for iterated monoidal product.

7.1 Permutative Opfibrations

To define permutative opfibrations, we first define opfibrations. A general reference for opfibrations is [JY21, Ch. 9], which discusses fibrations. Results there apply here as well by considering the opposite functor between the opposite categories. Permutative opfibrations are given in Definition 7.6. Small permutative opfibrations are the objects in a non-symmetric Cat-multicategory $\mathsf{PFib}^{\mathcal{D}}$ that we will discuss in subsequent sections.

Definition 7.3. Suppose $P : \mathsf{E} \longrightarrow \mathsf{B}$ is a functor between categories. Define the following.

1. A *fore-lift* is a pair $\langle y\,; f\rangle$ consisting of
   - an object y in E and
   - a morphism $f : Py \longrightarrow a$ in B for some object a.
2. A *lift* of a fore-lift $\langle y\,; f\rangle$ is a morphism $\overline{f} : y \longrightarrow y_f$ in E such that

   $$Py_f = a \quad \text{and} \quad P\overline{f} = f \quad \text{in} \quad B,$$

   which we depict as follows.

   $$\begin{array}{ccc} y & & Py \\ {\scriptstyle \overline{f}}\downarrow & \overset{P}{\longmapsto} & \downarrow{\scriptstyle f} \\ y_f & & a \end{array}$$

   If y is clear from the context, then we also say that $\overline{f}$ is a *lift of f*.

3. A *fore-raise* is a triple $\langle g, h; f\rangle$ consisting of

   - two morphisms $g : y \longrightarrow x$ and $h : y \longrightarrow z$ in E with the same domain and
   - a morphism $f : Px \longrightarrow Pz$ in B such that

   $$f(Pg) = Ph.$$

4. A *raise* of a fore-raise $\langle g, h; f\rangle$ is a morphism $\widetilde{f} : x \longrightarrow z$ in E such that

   $$\widetilde{f}g = h \in \mathsf{E} \quad \text{and} \quad P\widetilde{f} = f \in \mathsf{B},$$

   which we depict as follows.

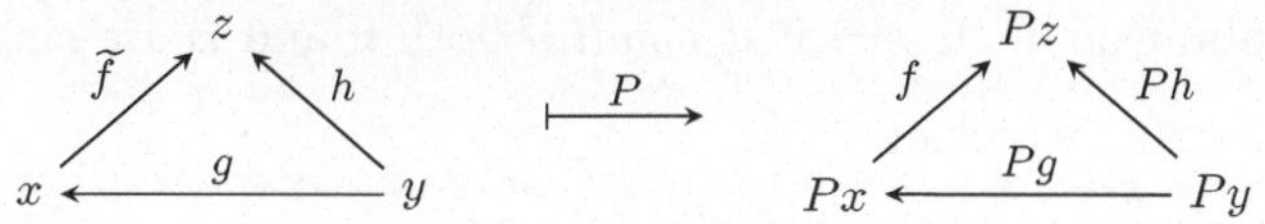

5. A morphism $g : y \longrightarrow x$ in E is called an *opcartesian morphism* if each fore-raise of the form $\langle g, h; f\rangle$ has a *unique* raise.

6. A lift of a fore-lift $\langle y\,; f\rangle$ is called an *opcartesian lift* if it is also an opcartesian morphism.

7. $P : \mathsf{E} \longrightarrow \mathsf{B}$ is called an *opfibration* over B if each fore-lift admits an opcartesian lift. In this case, P, E, and B are called, respectively, the *projection functor*, the *total category*, and the *base category*. A *cleavage* is a choice of an opcartesian lift $\overline{f}$ for each fore-lift $\langle y\,; f\rangle$, in which case $\overline{f}$ is called the *chosen opcartesian lift* of $\langle y\,; f\rangle$. An opfibration equipped with a cleavage is called a *cloven opfibration.*

8. A *split opfibration* is a cloven opfibration $P : \mathsf{E} \longrightarrow \mathsf{B}$ such that the following two axioms hold.

   Unitarity For each object $y \in \mathsf{E}$, the identity morphism 1_y is the chosen opcartesian lift of $\langle y\,; 1_{Py}\rangle$.

   Multiplicativity Suppose given

   - an object y in E and
   - composable morphisms $f : Py \longrightarrow a$ and $g : a \longrightarrow a'$ in B.

   Suppose

   - $\overline{f} : y \longrightarrow y_f$ is the chosen opcartesian lift of $\langle y\,; f\rangle$ and
   - $\overline{g} : y_f \longrightarrow (y_f)_g$ is the chosen opcartesian lift of $\langle y_f\,; g\rangle$, as depicted below.

$$\begin{array}{ccc} y & & Py \\ {\scriptstyle \overline{f}}\big\downarrow & & \big\downarrow{\scriptstyle f} \\ y_f & \overset{P}{\longmapsto} & a \\ {\scriptstyle \overline{g}}\big\downarrow & & \big\downarrow{\scriptstyle g} \\ (y_f)_g & & a' \end{array}$$

Then the composite

$$y \xrightarrow{\overline{g}\,\overline{f}} (y_f)_g \quad \text{in} \quad \mathsf{E}$$

is the chosen opcartesian lift of the fore-lift $\langle y\,; gf\rangle$.

9. An opfibration $P : \mathsf{E} \longrightarrow \mathsf{B}$ is *small* if both E and B are small categories.

This finishes the definition. $\diamond$

Explanation 7.4. Consider Definition 7.3.

1. A functor $P : \mathsf{E} \longrightarrow \mathsf{B}$ is an opfibration if and only if its opposite functor $P^{\mathsf{op}} : \mathsf{E}^{\mathsf{op}} \longrightarrow \mathsf{B}^{\mathsf{op}}$ is a fibration in the sense of [JY21, 9.1.1]. In the literature an opfibration is sometimes called a *cofibred category.*

2. The data $\langle y\,; f\rangle$ and $\langle g, h; f\rangle$ for, respective, a fore-lift and a fore-raise are often not given names in the literature. In the setting of fibrations in [JY21, Ch. 9], the corresponding concepts are called, respectively, a *pre-lift* and a *pre-raise.* A lift is sometimes called a *transport morphism* in the literature.

3. A cloven opfibration can always be arranged to satisfy the unitarity axiom. However, the multiplicativity axiom is not automatic. By the Grothendieck Fibration Theorem [JY21, 9.5.6], for each small category B, there is an explicitly-constructed 2-monad on Cat/B whose pseudo, respectively, strict, algebras are in canonical bijections with cloven, respectively, split, opfibrations over B.

4. An opfibration is a functor with additional properties, namely, the existence of an opcartesian lift for each fore-lift, but no extra structure. On the other hand, a *split* opfibration is an opfibration with extra structure, namely, the *chosen* opcartesian lift of each fore-lift. As a consequence, notions that are built upon split opfibrations usually have appropriate axioms regarding the chosen opcartesian lifts. Examples include Definitions 7.6 and 7.11. $\diamond$

Example 7.5. Consider Definition 7.3.

1. For a functor $P : \mathsf{E} \longrightarrow \mathsf{B}$, each isomorphism in E is an opcartesian morphism. In particular, each identity morphism in E is an opcartesian morphism.

2. Each identity functor $1_{\mathsf{B}} : \mathsf{B} \longrightarrow \mathsf{B}$ is a split opfibration. For each fore-lift $\langle y\,; f\rangle$, the chosen opcartesian lift is f. ◇

Next we consider the situation when each of P, E, and B is equipped with a permutative structure (Definitions 1.11 and 1.17). Eventually, we will assume that the base category B is a small tight bipermutative category, but for the next definition it only needs a permutative structure.

Definition 7.6. Suppose $(\mathsf{E}, \oplus, e, \beta^{\oplus})$ and $(\mathcal{D}, \oplus, e, \beta^{\oplus})$ are permutative categories. A *permutative opfibration* over $\mathcal{D}$

$$\mathsf{E} \xrightarrow{P} \mathcal{D}$$

is a split opfibration (Definition 7.3 (8)) that satisfies the following three axioms.

1. P is a strict symmetric monoidal functor. This is called the *strict symmetry axiom.*

2. Suppose given fore-lifts

   - $\langle x\,; f\rangle$ with chosen opcartesian lift $\overline{f}$ and
   - $\langle y\,; g\rangle$ with chosen opcartesian lift $\overline{g}$.

   Then $\overline{f} \oplus \overline{g}$ is the chosen opcartesian lift of the fore-lift $\langle x \oplus y\,; f \oplus g\rangle$. This is called the *monoidal product axiom.*

3. For any two objects $x, y \in \mathsf{E}$, the braiding

   $$x \oplus y \xrightarrow{\beta^{\oplus}_{x,y}} y \oplus x \quad \text{in} \quad \mathsf{E}$$

   is the chosen opcartesian lift of the fore-lift

   $$\big\langle x \oplus y\,; \beta^{\oplus}_{Px,Py} : Px \oplus Py \longrightarrow Py \oplus Px\big\rangle.$$

   This is called the *braiding axiom.*

This finishes the definition of a permutative opfibration. Moreover, a permutative opfibration as above is *small* if both E and $\mathcal{D}$ are small categories. ◇

Explanation 7.7. Consider Definition 7.6.

1. A permutative opfibration is a split opfibration with additional properties, namely, the axioms 1 through 3, but *no* extra structure.

2. The strict symmetry axiom 1 means that the unit constraint P^0 and the monoidal constraint P^2 are identities (Definition 1.17). They imply the following equalities for all objects $x, y \in \mathsf{E}$ and morphisms $p, q \in \mathsf{E}$.

   $$\begin{cases} Pe = e \\ P(x \oplus y) = Px \oplus Py \\ P(p \oplus q) = Pp \oplus Pq \\ P\beta^{\oplus}_{x,y} = \beta^{\oplus}_{Px,Py} \end{cases} \tag{7.8}$$

The last three equalities in (7.8) ensure that axioms 2 and 3 are well defined. Conversely, if the functor P satisfies the equalities in (7.8), then we can define the unit constraint P^0 and the monoidal constraint P^2 to be identities, and $(P, 1, 1)$ is a strict symmetric monoidal functor.

3. The braiding axiom 3 subsumes the unitarity condition of a split opfibration. Indeed, suppose x is the monoidal unit e in E. Then

$$\begin{aligned} \beta^{\oplus}_{e,y} &= 1_y \in \mathsf{E} \quad \text{and} \\ \beta^{\oplus}_{Pe,Py} = \beta^{\oplus}_{e,Py} &= 1_{Py} \in \mathcal{D} \end{aligned}$$

by the first equality in (7.8) and the right unit diagram (1.10) in E and $\mathcal{D}$. Thus in this case the braiding axiom says that 1_y is the chosen opcartesian lift of $\langle y\,; 1_{Py}\rangle$, which is the unitarity condition.

See Question A.3 for a potential 2-monadic description of permutative opfibrations that is analogous to the Grothendieck Fibration Theorem mentioned in Explanation 7.4 (3) ◇

Example 7.9 (Identity Permutative Opfibrations)**.** For each permutative category $\mathcal{D}$, the identity functor $1_{\mathcal{D}} : \mathcal{D} \longrightarrow \mathcal{D}$ is a permutative opfibration. In fact, the identity functor is a split opfibration (Example 7.5 (2)) and satisfies the three axioms in Definition 7.6. ◇

7.2 Opcartesian Multilinear Functors and Transformations

Throughout the rest of this chapter, we assume that

$$\big(\mathcal{D}, (\oplus, 0, \beta^{\oplus}), (\otimes, 1, \beta^{\otimes}), (\partial^l, \partial^r)\big) \tag{7.10}$$

is a small tight bipermutative category (Definition 2.41) with

- additive structure $\mathcal{D}^{\oplus} = (\mathcal{D}, \oplus, 0, \beta^{\oplus})$ and
- multiplicative structure $\mathcal{D}^{\otimes} = (\mathcal{D}, \otimes, 1, \beta^{\otimes})$.

We use $\mathcal{D}$ as our base category. As the next step in constructing the non-symmetric Cat-multicategory $\mathsf{PFib}^{\mathcal{D}}$ with small permutative opfibrations over $\mathcal{D}$ (Definition 7.6) as objects, in this section we define its multimorphism categories.

- The n-ary 1-cells (Definition 7.11) are *opcartesian strong n-linear functors.* They are strong n-linear functors that are also compatible with the given permutative opfibrations.

- The n-ary 2-cells (Definition 7.27) are *opcartesian n-linear transformations.* They are n-linear natural transformations that are also compatible with the given permutative opfibrations.
- The multimorphism categories of $\mathsf{PFib}^{\mathcal{D}}$ are in Definition 7.34. They are subcategories of corresponding multimorphism categories of the Cat-multicategory $\mathsf{PermCat}^{\mathsf{sg}}$ (Theorem 6.31). We check that the multimorphism categories of $\mathsf{PFib}^{\mathcal{D}}$ are well defined in Lemma 7.35.

The rest of the non-symmetric Cat-multicategory structure of $\mathsf{PFib}^{\mathcal{D}}$ is discussed in Section 7.3.

Opcartesian Multilinear Functors

For fibrations

$$P : \mathsf{E} \longrightarrow \mathsf{B} \quad \text{and} \quad P' : \mathsf{E}' \longrightarrow \mathsf{B},$$

a *Cartesian functor* $F : P \longrightarrow P'$ [JY21, 9.1.14 (1)] is a functor $F : \mathsf{E} \longrightarrow \mathsf{E}'$ that

- satisfies $P = P'F$ and
- sends Cartesian morphisms in E to Cartesian morphisms in E'.

Recall that an *n-linear functor* is a functor F equipped with n linearity constraints $\langle F_j^2 \rangle_{j=1}^n$ and is subject to five axioms (Definition 6.4). The multilinear analogues of Cartesian functors for permutative opfibrations with base category $\mathcal{D}$ are multilinear functors that are compatible with the split opfibration structure as in the next definition. Recall the notation $\circ_i$ in (6.2). Unless otherwise specified, in a permutative category we denote its monoidal product, monoidal unit, and braiding by, respectively, $\oplus$, e, and $\beta^{\oplus}$.

Definition 7.11. Suppose given permutative opfibrations (Definition 7.6)

$$\mathsf{C}_k \xrightarrow{P_k} \mathcal{D} \quad \text{for} \quad k \in \{0, 1, \ldots, n\}$$

with $\langle P \rangle = \langle P_j \rangle_{j=1}^n$ and $n \geq 0$. An *opcartesian n-linear functor*

$$\langle P \rangle \xrightarrow{\left(F, \langle F_j^2 \rangle_{j=1}^n\right)} P_0 \tag{7.12}$$

is an n-linear functor (6.5)

$$\textstyle\prod_{j=1}^n \mathsf{C}_j \xrightarrow{\left(F, \langle F_j^2 \rangle_{j=1}^n\right)} \mathsf{C}_0$$

that satisfies the axioms 1 through 4 below.

1. If $\langle f_j \rangle \in \prod_{j=1}^n \mathsf{C}_j$ is an n-tuple of opcartesian morphisms, then $F\langle f_j \rangle \in \mathsf{C}_0$ is an opcartesian morphism.

2. The following diagram of functors is commutative, where $\otimes$ denotes any one of the equal iterates of $\otimes : \mathcal{D}^2 \longrightarrow \mathcal{D}$. The empty tensor product in $\mathcal{D}$ is interpreted as the multiplicative unit $1 \in \mathcal{D}$.

$$\begin{array}{ccc} \prod_{j=1}^n \mathsf{C}_j & \xrightarrow{F} & \mathsf{C}_0 \\ {\scriptstyle \prod_j P_j}\downarrow & & \downarrow{\scriptstyle P_0} \\ \mathcal{D}^n & \xrightarrow{\otimes} & \mathcal{D} \end{array} \tag{7.13}$$

We call (7.13) the *projection axiom.*

3. Suppose that $\overline{f}_j \in \mathsf{C}_j$ is the chosen opcartesian lift of a fore-lift $\langle x_j\,; f_j\rangle$ for $j \in \{1,\ldots,n\}$. Then $F\langle\overline{f}_j\rangle \in \mathsf{C}_0$ is the chosen opcartesian lift of the fore-lift

$$\big\langle F\langle x_j\rangle\,; \otimes_{j=1}^n f_j\big\rangle.$$

4. Suppose given objects $\langle x_j\rangle \in \prod_{j=1}^n \mathsf{C}_j$ and $x_i' \in \mathsf{C}_i$ for some $i \in \{1,\ldots,n\}$, along with the following notation.

$$\begin{cases} a = P_1x_1 \otimes \cdots \otimes P_ix_i \otimes \cdots \otimes P_nx_n \\ a' = P_1x_1 \otimes \cdots \otimes P_ix_i' \otimes \cdots \otimes P_nx_n \\ a'' = P_1x_1 \otimes \cdots \otimes (P_ix_i \oplus P_ix_i') \otimes \cdots \otimes P_nx_n \end{cases} \tag{7.14}$$

Denote by

$$a'' \xrightarrow[\cong]{\mathsf{L}} a \oplus a' \quad \text{in} \quad \mathcal{D} \tag{7.15}$$

the unique Laplaza coherence isomorphism (Theorem 2.90) that distributes over the sum $P_ix_i \oplus P_ix_i'$ in the i-th tensor factor in a''. Then the component of the i-th linearity constraint

$$F\langle x_j\rangle \oplus F\langle x_j \circ_i x_i'\rangle \xrightarrow{F_i^2} F\langle x_j \circ_i (x_i \oplus x_i')\rangle \tag{7.16}$$

is the chosen opcartesian lift of the fore-lift

$$\big\langle F\langle x_j\rangle \oplus F\langle x_j \circ_i x_i'\rangle\,; \mathsf{L}^{-1}\big\rangle \quad \text{in} \quad \mathsf{C}_0. \tag{7.17}$$

We call this the *constraint lift axiom.*

This finishes the definition of an opcartesian n-linear functor.

Moreover, we define the following.

- An *opcartesian strong n-linear functor* is an opcartesian n-linear functor $(F, \langle F_j^2\rangle)$ in which each linearity constraint F_j^2 is a natural isomorphism.
- An *opcartesian (strong) multilinear functor* is an opcartesian (strong) n-linear functor for some $n \geq 0$.

- If the projection functors $\langle P_k \rangle_{k=0}^n$ are unambiguously determined by the context, then we sometimes denote an opcartesian n-linear functor $\langle P \rangle \longrightarrow P_0$ by $\langle \mathsf{C} \rangle \longrightarrow \mathsf{C}_0$, where $\langle \mathsf{C} \rangle = \langle \mathsf{C}_j \rangle_{j=1}^n$. $\diamond$

Explanation 7.18. Consider Definition 7.11.

1. An opcartesian n-linear functor is an n-linear functor with the extra properties 1 through 4 but *no* extra structure.

2. By definition a 0-linear functor is a functor $F : \mathbf{1} \longrightarrow \mathsf{C}_0$, which is determined by the object $F* \in \mathsf{C}_0$. In this case, axioms 1, 3, and 4 are vacuous, while the projection axiom 2 is the equality

$$P_0(F*) = 1 \quad \text{in} \quad \mathcal{D}. \tag{7.19}$$

   In other words, an *opcartesian 0-linear functor* is a choice of an object in the P_0-preimage of $1 \in \mathcal{D}$.

3. Axioms 1 and 2 are n-variable generalizations of the two axioms of a Cartesian functor, which we recalled before Definition 7.11. Axiom 1 actually follows from axiom 3; see Lemma 7.21.

4. Axiom 3 says that the functor F preserves chosen opcartesian lifts. It is well defined because

$$P_0 F\langle \overline{f}_j \rangle = \otimes_{j=1}^n P_j \overline{f}_j = \otimes_{j=1}^n f_j.$$

   The first equality follows from the projection axiom (7.13). The second equality follows from the assumption that $\overline{f}_j$ is the chosen opcartesian lift of $\langle x_j \,;\, f_j \rangle$.

5. A part of the constraint lift axiom 4 is the morphism equality

$$P_0 F_i^2 = \mathsf{L}^{-1} : a \oplus a' \longrightarrow a'' \quad \text{in} \quad \mathcal{D} \tag{7.20}$$

   with L in (7.15) and F_i^2 in (7.16). For the domain, respectively, codomain, in (7.20) we use the following equalities.

$$\begin{cases} P_0\big(F\langle x_j \rangle \oplus F\langle x_j \circ_i x_i' \rangle\big) & \\ \quad = P_0F\langle x_j \rangle \oplus P_0F\langle x_j \circ_i x_i' \rangle & \text{by (7.8)} \\ \quad = a \oplus a' & \text{by (7.13)} \\ P_0F\langle x_j \circ_i (x_i \oplus x_i') \rangle & \\ \quad = P_1x_1 \otimes \cdots \otimes P_i(x_i \oplus x_i') \otimes \cdots \otimes P_nx_n & \text{by (7.13)} \\ \quad = P_1x_1 \otimes \cdots \otimes (P_ix_i \oplus P_ix_i') \otimes \cdots \otimes P_nx_n & \text{by (7.8)} \\ \quad = a'' & \end{cases}$$

   The Laplaza coherence isomorphism L in (7.15) has appeared several times before, namely

- in (5.46) in the additivity axiom of an additive natural transformation,
- in (5.55) in the additivity axiom of an additive modification, and
- in (6.54) and (6.56) in the linearity constraints of $\int_{\mathcal{D}} \phi$ for an additive natural transformation ϕ. $\diamond$

While Definition 7.11 (1) is the natural n-variable variant of the preservation of Cartesian morphisms by Cartesian functors, the following observation says that it is actually redundant.

Lemma 7.21. *In the context of Definition 7.11, axiom 1 follows from axiom 3.*

Proof. Suppose $f_j : x_j \longrightarrow y_j$ in C_j is an opcartesian morphism for each $j \in \{1, \ldots, n\}$. We want to show that $F\langle f_j \rangle$ is an opcartesian morphism as a consequence of axiom 3. We use the following facts from [JY21, 9.1.4 (1), (4), and (5)] for a functor $P : \mathsf{E} \longrightarrow \mathsf{B}$:

- Each isomorphism in E is an opcartesian morphism.
- Opcartesian morphisms are closed under composition.
- If both $\overline{f} : y \longrightarrow z$ and $\overline{f}' : y \longrightarrow z'$ are opcartesian lifts of a fore-lift $\langle y ; f \rangle$, then there exists a unique isomorphism $g : z \xrightarrow{\cong} z'$ in E such that

$$g\overline{f} = \overline{f}' \quad \text{and} \quad Pg = 1_{Pz}.$$

Consider the fore-lift $\langle x_j ; P_j f_j \rangle$ with respect to $P_j : \mathsf{C}_j \longrightarrow \mathcal{D}$.

- Since f_j is an opcartesian morphism, it is an opcartesian lift of $\langle x_j ; P_j f_j \rangle$.
- Since P_j is a split opfibration, $\langle x_j ; P_j f_j \rangle$ has a *chosen* opcartesian lift $\overline{f}_j : x_j \longrightarrow w_j$.

Therefore, there is a unique isomorphism $g_j : w_j \xrightarrow{\cong} y_j$ in C_j such that

$$g_j \overline{f}_j = f_j \quad \text{and} \quad P_j g_j = 1_{P_j y_j}.$$

The functoriality of F implies

$$F\langle f_j \rangle = F\langle g_j \overline{f}_j \rangle = F\langle g_j \rangle \circ F\langle \overline{f}_j \rangle \quad \text{in} \quad \mathsf{C}_0. \tag{7.22}$$

Now we observe the following.

- By axiom 3 $F\langle \overline{f}_j \rangle$ is the chosen opcartesian lift of $\big\langle F\langle x_j \rangle ; \otimes_j P_j f_j \big\rangle$.
- Since each g_j is an isomorphism, so is $F\langle g_j \rangle$ by the functoriality of F. Thus $F\langle g_j \rangle$ is an opcartesian morphism.

The factorization (7.22) shows that $F\langle f_j \rangle$ is the composite of two opcartesian morphisms, so it is also an opcartesian morphism. □

In view of Lemma 7.21, to check that an n-linear functor is opcartesian, it is unnecessary to check axiom 1. We include axiom 1 in Definition 7.11 solely to preserve the symmetry with the definition of a Cartesian functor.

Example 7.23 (Identity Opcartesian 1-Linear Functors)**.** Suppose $P : \mathsf{C} \longrightarrow \mathcal{D}$ is a permutative opfibration (Definition 7.6). Then the identity 1-linear functor

$$\mathsf{C} \xrightarrow{1_{\mathsf{C}}} \mathsf{C}$$

in Example 6.12 is an opcartesian 1-linear functor. Indeed, axioms 1 through 3 in Definition 7.11 follow from the fact that 1_{C} is the identity functor on C.

For the constraint lift axiom 4, the Laplaza coherence isomorphism (7.15)

$$Px \oplus Px' \xrightarrow[\cong]{\mathsf{L}} Px \oplus Px' \quad \text{for} \quad x, x' \in \mathsf{C}$$

is the identity morphism by uniqueness. So L^{-1} is also the identity morphism. The only linearity constraint of 1_{C} is the identity natural transformation by definition. The chosen opcartesian lift of the fore-lift (7.17)

$$\big\langle x \oplus x' \,; 1_{Px \oplus Px'} \big\rangle = \big\langle x \oplus x' \,; 1_{P(x \oplus x')} \big\rangle$$

is the identity morphism $1_{x \oplus x'}$ by the unitarity condition of a split opfibration (Definition 7.3 (8)), including P. ◇

Example 7.24 (Opcartesian 1-Linear Functors)**.** Generalizing Example 7.23, for permutative opfibrations

$$\mathsf{C}_1 \xrightarrow{P_1} \mathcal{D} \quad \text{and} \quad \mathsf{C}_0 \xrightarrow{P_0} \mathcal{D},$$

an opcartesian 1-linear functor

$$P_1 \xrightarrow{F} P_0$$

is precisely a *strict* symmetric monoidal functor

$$\mathsf{C}_1 \xrightarrow{F} \mathsf{C}_0$$

that

- satisfies $P_1 = P_0 F$ and
- preserves chosen opcartesian lifts.

Indeed, the conditions in the previous two items are axioms 2 and 3 in Definition 7.11. Axiom 1 holds by Lemma 7.21. A 1-linear functor (Definition 6.4)

$$\mathsf{C}_1 \xrightarrow{(F, F^2)} \mathsf{C}_0$$

is precisely a strictly unital symmetric monoidal functor. Moreover, since $n = 1$, the Laplaza coherence isomorphism L in (7.15) is the identity morphism. Then

- axiom 4 in Definition 7.11 and
- the unitarity condition (Definition 7.3 (8)) of the split opfibration P_0

imply that the only linearity constraint F^2 is the identity. Thus F is, in fact, a strict symmetric monoidal functor. ◇

Example 7.25 (Bipermutative Categories)**.** Since $(\mathcal{D}, \oplus, \otimes)$ is a small tight bipermutative category (Definition 2.41), the identity functor

$$\mathcal{D}^{\oplus} \xrightarrow{1_{\mathcal{D}^{\oplus}}} \mathcal{D}^{\oplus}$$

is a permutative opfibration (Example 7.9). For each $n \geq 0$ there is an opcartesian strong n-linear functor

$$\langle 1_{\mathcal{D}^{\oplus}} \rangle_{j=1}^{n} \xrightarrow{\left(\otimes\, , \langle \mathsf{L}_j^{-1} \rangle_{j=1}^{n}\right)} 1_{\mathcal{D}^{\oplus}}. \tag{7.26}$$

Indeed, by Example 6.14 there is a strong n-linear functor

$$\prod_{j=1}^{n} \mathcal{D}^{\oplus} \xrightarrow{\left(\otimes\, , \langle \mathsf{L}_j^{-1} \rangle_{j=1}^{n}\right)} \mathcal{D}^{\oplus}.$$

Its j-th linearity constraint is (6.15)

$$\otimes_j^2 = \mathsf{L}_j^{-1}$$

where L_j is the unique Laplaza coherence isomorphism (Example 2.91) that distributes the sum in the j-th tensor factor.

- Axioms 1 through 3 in Definition 7.11 hold because each $1_{\mathcal{D}^{\oplus}}$ is the identity functor.
- The constraint lift axiom 4 holds because $\otimes_i^2 = \mathsf{L}_i^{-1}$. ◇

Opcartesian Multilinear Transformations

For

- fibrations $P : \mathsf{E} \longrightarrow \mathsf{B}$ and $P' : \mathsf{E}' \longrightarrow \mathsf{B}$ and
- Cartesian functors $F, G : P \longrightarrow P'$,

a *vertical natural transformation* $\theta : F \longrightarrow G$ [JY21, 9.1.14 (2)] is a natural transformation that satisfies

$$1_{P'} * \theta = 1_P.$$

Recall that an *n-linear natural transformation* is a natural transformation that satisfies the unity axiom and the constraint compatibility axiom (Definition 6.16). The multilinear analogues of vertical natural transformations are compatible with the functors $\langle P_k \rangle_{k=0}^{n}$ in the following sense.

Definition 7.27. Suppose given permutative opfibrations (Definition 7.6)

$$\mathsf{C}_k \xrightarrow{P_k} \mathcal{D} \quad \text{for} \quad k \in \{0, 1, \ldots, n\}$$

with $\langle P \rangle = \langle P_j \rangle_{j=1}^n$ and $n \geq 0$. Suppose given two opcartesian n-linear functors (Definition 7.11)

$$\langle P \rangle \overset{\left(F, \langle F_j^2 \rangle_{j=1}^n\right)}{\underset{\left(G, \langle G_j^2 \rangle_{j=1}^n\right)}{\rightrightarrows}} P_0.$$

An *opcartesian n-linear transformation*

$$\left(F, \langle F_j^2 \rangle_{j=1}^n\right) \xrightarrow{\theta} \left(G, \langle G_j^2 \rangle_{j=1}^n\right) \tag{7.28}$$

is an n-linear natural transformation (6.17)

$$\prod_{j=1}^n \mathsf{C}_j \overset{(F, \langle F_j^2 \rangle)}{\underset{(G, \langle G_j^2 \rangle)}{\Downarrow \theta}} \mathsf{C}_0$$

such that the following two whiskered natural transformations are equal.

$$\begin{array}{ccc} \prod_{j=1}^n \mathsf{C}_j & \overset{F}{\underset{G}{\Downarrow \theta}} & \mathsf{C}_0 \\ {\scriptstyle \prod_j P_j} \downarrow & & \downarrow {\scriptstyle P_0} \\ \mathcal{D}^n & \xrightarrow{\otimes} & \mathcal{D} \end{array} \tag{7.29}$$

We call (7.29) the *transformation projection axiom.* An *opcartesian multilinear transformation* is an opcartesian n-linear transformation for some $n \geq 0$. ◇

Explanation 7.30. Consider Definition 7.27.

1. An opcartesian n-linear transformation is an n-linear natural transformation with the extra property (7.29) but *no* extra structure.

2. By definition a 0-linear natural transformation $\theta : F \longrightarrow G$ is a morphism

$$F* \xrightarrow{\theta_*} G* \quad \text{in} \quad \mathsf{C}_0. \tag{7.31}$$

Using (7.19), in this case the transformation projection axiom (7.29) is the morphism equality

$$P_0 \theta_* = 1_1 : 1 \longrightarrow 1 \quad \text{in} \quad \mathcal{D}. \tag{7.32}$$

In other words, an *opcartesian 0-linear transformation* is a morphism as in (7.31) that is in the P_0-preimage of $1 \in \mathcal{D}$.

3. For $n \geq 1$ the three composite functors in the transformation projection axiom (7.29) are equal by the projection axiom (7.13) for F and G. The axiom (7.29) means that, for each n-tuple of objects $\langle x_j \rangle \in \prod_{j=1}^n \mathsf{C}_j$, there is a morphism equality

$$P_0 \theta_{\langle x_j \rangle} = 1_{\otimes_{j=1}^n P_j x_j} : P_0 F\langle x_j \rangle \longrightarrow P_0 G\langle x_j \rangle \tag{7.33}$$

in $\mathcal{D}$. ◇

Multimorphism Categories

For small permutative categories C_0 and $\langle \mathsf{C} \rangle = \langle \mathsf{C}_j \rangle_{j=1}^n$, recall from (6.22) and (6.23) the category

$$\mathsf{PermCat}^{\mathsf{sg}}\big(\langle \mathsf{C} \rangle\, ;\, \mathsf{C}_0\big) = \mathsf{PermCat}^{\mathsf{sg}}\big(\mathsf{C}_1, \ldots, \mathsf{C}_n\, ;\, \mathsf{C}_0\big)$$

with

- *strong* n-linear functors $\prod_{j=1}^n \mathsf{C}_j \longrightarrow \mathsf{C}_0$ as objects,
- n-linear natural transformations as morphisms,
- identity natural transformations as identity morphisms, and
- vertical composition of natural transformations as composition.

In the arity 0 case, $\mathsf{PermCat}^{\mathsf{sg}}(\langle\rangle\, ;\, \mathsf{C}_0)$ is the category C_0.

Definition 7.34. Suppose given small permutative opfibrations (Definition 7.6)

$$\mathsf{C}_k \xrightarrow{P_k} \mathcal{D} \quad \text{for} \quad k \in \{0, 1, \ldots, n\}$$

with $\langle \mathsf{C} \rangle = \langle \mathsf{C}_j \rangle_{j=1}^n$, $\langle P \rangle = \langle P_j \rangle_{j=1}^n$, and $n \geq 0$. Define

$$\mathsf{PFib}^{\mathcal{D}}\big(\langle P \rangle\, ;\, P_0\big) = \mathsf{PFib}^{\mathcal{D}}\big(P_1, \ldots, P_n\, ;\, P_0\big)$$

as the subcategory of $\mathsf{PermCat}^{\mathsf{sg}}\big(\langle \mathsf{C} \rangle\, ;\, \mathsf{C}_0\big)$ with

- opcartesian *strong* n-linear functors (Definition 7.11) as objects and
- opcartesian n-linear transformations (Definition 7.27) as morphisms.

This finishes the definition of $\mathsf{PFib}^{\mathcal{D}}\big(\langle P \rangle\, ;\, P_0\big)$.

Moreover, if the projection functors $\langle P_k \rangle_{k=0}^n$ are unambiguously determined by the context, then we also denote $\mathsf{PFib}^{\mathcal{D}}\big(\langle P \rangle\, ;\, P_0\big)$ by $\mathsf{PFib}^{\mathcal{D}}\big(\langle \mathsf{C} \rangle\, ;\, \mathsf{C}_0\big)$. ◇

First we check that the category $\mathsf{PFib}^{\mathcal{D}}\big(\langle P \rangle\, ;\, P_0\big)$ is well defined.

Lemma 7.35. *In the context of Definition 7.34,*

$$\mathsf{PFib}^{\mathcal{D}}\big(\langle P \rangle\, ;\, P_0\big)$$

is a well-defined subcategory of $\mathsf{PermCat}^{\mathsf{sg}}\big(\langle \mathsf{C} \rangle\, ;\, \mathsf{C}_0\big)$.

Proof. In the arity 0 case, the assertion is that $\mathsf{PFib}^{\mathcal{D}}(\langle\rangle\,;P_0)$ is a subcategory of C_0. This is true because $\mathsf{PFib}^{\mathcal{D}}(\langle\rangle\,;P_0)$ is the P_0-preimage of $1 \in \mathcal{D}$ by

- Explanation 7.18 (2) for objects and
- Explanation 7.30 (2) for morphisms.

Suppose $n > 0$. The identity natural transformation 1_F of each opcartesian n-linear functor F is an opcartesian n-linear transformation. Indeed, in this case the transformation projection axiom (7.29) reduces to the projection axiom (7.13) for F.

It remains to check that opcartesian n-linear transformations are closed under vertical composition. Suppose

$$\big(F\,,\langle F_j^2\rangle_{j=1}^n\big) \xrightarrow{\;\theta\;} \big(G\,,\langle G_j^2\rangle_{j=1}^n\big) \xrightarrow{\;\psi\;} \big(H\,,\langle H_j^2\rangle_{j=1}^n\big)$$

are opcartesian n-linear transformations in $\mathsf{PFib}^{\mathcal{D}}\big(\langle P\rangle\,;P_0\big)$. The vertical composite n-linear natural transformation

$$\big(F\,,\langle F_j^2\rangle_{j=1}^n\big) \xrightarrow{\;\psi\theta\;} \big(H\,,\langle H_j^2\rangle_{j=1}^n\big)$$

satisfies the transformation projection axiom (7.33) by the following computation for each n-tuple of objects $\langle x_j\rangle \in \prod_{j=1}^n \mathsf{C}_j$.

$$\begin{aligned}
P_0(\psi\theta)_{\langle x_j\rangle} &= P_0\big(\psi_{\langle x_j\rangle}\theta_{\langle x_j\rangle}\big) & \\
&= \big(P_0\psi_{\langle x_j\rangle}\big)\big(P_0\theta_{\langle x_j\rangle}\big) && \text{by functoriality of } P_0\\
&= \big(1_{\otimes_{j=1}^n P_j x_j}\big)\big(1_{\otimes_{j=1}^n P_j x_j}\big) && \text{by (7.33) for } \theta \text{ and } \psi\\
&= 1_{\otimes_{j=1}^n P_j x_j}
\end{aligned}$$

This finishes the proof. □

7.3 The Multicategory of Permutative Opfibrations

As in (7.10) we continue to assume that $(\mathcal{D},\oplus,\otimes)$ is a small tight bipermutative category. In Definition 7.34 we constructed the multimorphism categories of $\mathsf{PFib}^{\mathcal{D}}$. Recall from Definition 3.2 that a non-symmetric V-multicategory is a V-multicategory without the symmetric group action and the corresponding axioms. Beyond the objects and the multimorphism categories, a non-symmetric V-multicategory is determined by its colored units and composition, subject to

- the associativity axiom (3.7) and
- the unity axioms (3.8) and (3.9).

In this section we construct the rest of the non-symmetric Cat-multicategory structure on $\mathsf{PFib}^{\mathcal{D}}$. In the rare situation that the multiplicative braiding in $\mathcal{D}$ is the identity, $\mathsf{PFib}^{\mathcal{D}}$ is a Cat-multicategory. This (non-symmetric) Cat-multicategory comes with a (non-symmetric) Cat-multifunctor

$$\mathsf{PFib}^{\mathcal{D}} \xrightarrow{\mathsf{U}} \mathsf{PermCat}^{\mathsf{sg}}$$

that sends each small permutative opfibration over $\mathcal{D}$ to its total category. Here is an outline of this section.

- Definition 7.36 specifies the colored units, composition, and symmetric group action in $\mathsf{PFib}^{\mathcal{D}}$ by restricting those in the Cat-multicategory $\mathsf{PermCat}^{\mathsf{sg}}$ (Theorem 6.31).
- Lemmas 7.37 and 7.42 show that the composition in $\mathsf{PermCat}^{\mathsf{sg}}$ restricts to $\mathsf{PFib}^{\mathcal{D}}$.
- Lemmas 7.44 and 7.48 show that the symmetric group action in $\mathsf{PermCat}^{\mathsf{sg}}$ restricts to $\mathsf{PFib}^{\mathcal{D}}$ *if* the multiplicative braiding $\beta^{\otimes}$ in $\mathcal{D}$ is the identity. The last assumption is used in the diagrams (7.47) and (7.50). A weaker converse holds: if the symmetric group action in $\mathsf{PermCat}^{\mathsf{sg}}$ restricts to $\mathsf{PFib}^{\mathcal{D}}$, then $\otimes = \otimes^{\mathsf{op}}$ in $\mathcal{D}$.
- The main Theorem 7.51 proves the existence of the non-symmetric Cat-multicategory $\mathsf{PFib}^{\mathcal{D}}$, which is, furthermore, a Cat-multicategory if $\beta^{\otimes} = 1$ in $\mathcal{D}$.
- Corollary 7.52 states that there exists a non-symmetric Cat-multifunctor U from $\mathsf{PFib}^{\mathcal{D}}$ to $\mathsf{PermCat}^{\mathsf{sg}}$ that sends each small permutative opfibration over $\mathcal{D}$ to the total category. Moreover, U is a Cat-multifunctor if $\beta^{\otimes} = 1$ in $\mathcal{D}$.

Definition 7.36. For a small tight bipermutative category $\mathcal{D}$ as in (7.10), define the data of a Cat-multicategory $\mathsf{PFib}^{\mathcal{D}}$ as follows.

Objects An object in $\mathsf{PFib}^{\mathcal{D}}$ is a small permutative opfibration over $\mathcal{D}$ (Definition 7.6).

Multimorphisms The multimorphism categories in $\mathsf{PFib}^{\mathcal{D}}$ are the ones in Definition 7.34:

- The objects are opcartesian strong multilinear functors (Definition 7.11).
- The morphisms are opcartesian multilinear transformations (Definition 7.27).
- Identity morphisms are identity natural transformations.
- Composition is vertical composition of natural transformations.

Units For each small permutative opfibration $P : \mathsf{C} \longrightarrow \mathcal{D}$, the P-colored unit is the identity opcartesian 1-linear functor $1_{\mathsf{C}} : \mathsf{C} \longrightarrow \mathsf{C}$ (Example 7.23).

Composition It is the restriction of the composition in $\mathsf{PermCat}^{\mathsf{sg}}$ (6.28) via the subcategory inclusions in Lemma 7.35.

Symmetric Group Action It is the restriction of the symmetric group action in $\mathsf{PermCat}^{\mathsf{sg}}$ (6.24) via the subcategory inclusions in Lemma 7.35.

This finishes the definition of $\mathsf{PFib}^{\mathcal{D}}$. ◇

Composition

To show that the composition in $\mathsf{PFib}^{\mathcal{D}}$ is well defined, we begin with n-ary 1-cells.

Lemma 7.37. *In the context of Definition 7.36, the composition in* $\mathsf{PFib}^{\mathcal{D}}$ *is well defined on opcartesian strong multilinear functors.*

Proof. Suppose given

- small permutative opfibrations over $\mathcal{D}$ (Definition 7.6)
  - $P_k : \mathsf{C}_k \longrightarrow \mathcal{D}$ for $k \in \{0, 1, \ldots, n\}$ and
  - $Q_{ji} : \mathsf{B}_{ji} \longrightarrow \mathcal{D}$ for $j \in \{1, \ldots, n\}$ and $i \in \{1, \ldots, \ell_j\}$,
- an opcartesian n-linear functor (Definition 7.11)

$$\langle P \rangle \xrightarrow{\left(F, \langle F_j^2 \rangle_{j=1}^n\right)} P_0 \tag{7.38}$$

  with $\langle P \rangle = \langle P_j \rangle_{j=1}^n$, and

- for each $j \in \{1, \ldots, n\}$ an opcartesian ℓ_j-linear functor

$$\langle Q_j \rangle \xrightarrow{\left(G_j, \left\langle (G_j)_i^2 \right\rangle_{i=1}^{\ell_j}\right)} P_j \tag{7.39}$$

  with $\langle Q_j \rangle = \langle Q_{ji} \rangle_{i=1}^{\ell_j}$.

With $\ell = \ell_1 + \cdots + \ell_n$, the composite ℓ-linear functor

$$\prod_{j=1}^n \prod_{i=1}^{\ell_j} \mathsf{B}_{ji} \xrightarrow{\prod_j G_j} \prod_{j=1}^n \mathsf{C}_j \xrightarrow{F} \mathsf{C}_0 \tag{7.40}$$

has linearity constraints in (6.30). It is strong if F and each G_j are strong. We must show that the composite ℓ-linear functor in (7.40) satisfies axioms 1 through 4 in Definition 7.11.

If some $\ell_j = 0$, then the argument below needs a slight adjustment using

- Explanation 7.18 (2), which identifies opcartesian 0-linear functors with objects in the preimage of $1 \in \mathcal{D}$;
- the fact that 1 is the strict multiplicative unit in $\mathcal{D}$; and
- the unitarity condition of a split opfibration (Definition 7.3 (8)).

With this in mind, in the rest of this proof we assume that each $\ell_j > 0$.

Axiom 1. This holds because F and each G_j preserve opcartesian morphisms.

Axiom 2. The diagram (7.13) for the composite $F \circ (\prod_j G_j)$ in (7.40) is the boundary of the following diagram.

$$\begin{array}{ccccc}
\prod_{j=1}^n \prod_{i=1}^{\ell_j} \mathsf{B}_{ji} & \xrightarrow{\prod_j G_j} & \prod_{j=1}^n \mathsf{C}_j & \xrightarrow{F} & \mathsf{C}_0 \\
{\scriptstyle \prod_{j,i} Q_{ji}} \downarrow & & {\scriptstyle \prod_j P_j} \downarrow & & \downarrow {\scriptstyle P_0} \\
\mathcal{D}^{\ell_1 + \cdots + \ell_n} & \xrightarrow{\prod_j \otimes} & \mathcal{D}^n & \xrightarrow{\otimes} & \mathcal{D}
\end{array}$$

The left and right squares commute by the projection axiom (7.13) for, respectively, $\langle G_j \rangle_{j=1}^n$ and F.

Axiom 3. This holds because F and each G_j preserve chosen opcartesian lifts.

Axiom 4. By definition (6.30) for each $q \in \{1, \ldots, n\}$ and $p \in \{1, \ldots, \ell_q\}$ with

$$r = \ell_1 + \cdots + \ell_{q-1} + p,$$

the r-th linearity constraint of $(F \circ \prod_j G_j)$ in (7.40) is the composite

$$(F \circ \textstyle\prod_j G_j)_r^2 = F\big(1 \circ_q (G_q)_p^2\big) \circ F_q^2. \tag{7.41}$$

Here

- F_q^2 is the q-th linearity constraint of F, and
- $(G_q)_p^2$ is the p-th linearity constraint of G_q.

Axiom 4 requires each component of $(F \circ \prod_j G_j)_r^2$ to be the chosen opcartesian lift of L_r^{-1}, where L_r is the unique Laplaza coherence isomorphism in $\mathcal{D}$ that distributes over the sum in the r-th tensor factor in its domain. This axiom holds for the following reasons.

1. By axiom 4 for F, each component of F_q^2 is the chosen opcartesian lift of L_q^{-1}, where L_q is the unique Laplaza coherence isomorphism in $\mathcal{D}$ that distributes over the sum in the q-th tensor factor in its domain.
2. By axiom 4 for G_q, each component of $(G_q)_p^2$ is the chosen opcartesian lift of L_p^{-1}, where L_p is the unique Laplaza coherence isomorphism in $\mathcal{D}$ that distributes over the sum in the p-th tensor factor in its domain.

3. By the unitarity axiom of a split opfibration (Definition 7.3 (8)), each component of each entry in the n-tuple

$$\langle 1 \circ_q (G_q)^2_p \rangle = \langle 1, \ldots, (G_q)^2_p, \ldots, 1 \rangle$$

is a chosen opcartesian lift. By

- (2) above,
- the functoriality of each P_j, and
- axiom 3 for F,

each component of

$$F\langle 1 \circ_q (G_q)^2_p \rangle$$

is the chosen opcartesian lift of

$$1 \otimes \cdots \otimes \mathsf{L}_p^{-1} \otimes \cdots \otimes 1.$$

4. By the uniqueness in Theorem 2.90, there is a morphism equality in $\mathcal{D}$

$$\mathsf{L}_q \circ \big(1 \otimes \cdots \otimes \mathsf{L}_p \otimes \cdots \otimes 1\big) = \mathsf{L}_r.$$

5. By the multiplicativity axiom of a split opfibration (Definition 7.3 (8)) and (1), (3), and (4) above, each component of the right-hand side of (7.41) is the chosen opcartesian lift of

$$\big(1 \otimes \cdots \otimes \mathsf{L}_p^{-1} \otimes \cdots \otimes 1\big) \circ \mathsf{L}_q^{-1} = \mathsf{L}_r^{-1}.$$

This finishes the proof. □

Next we consider composition of n-ary 2-cells.

Lemma 7.42. *In the context of Definition 7.36, the composition in* $\mathsf{PFib}^{\mathcal{D}}$ *is well defined on opcartesian multilinear transformations.*

Proof. Reusing the notation in (7.38) through (7.40), suppose given

- an opcartesian n-linear transformation

$$F \xrightarrow{\theta} F'$$

in $\mathsf{PFib}^{\mathcal{D}}\big(\langle P \rangle ; P_0\big)$ and

- for each $j \in \{1, \ldots, n\}$ an opcartesian ℓ_j-linear transformation

$$G_j \xrightarrow{\theta_j} G'_j$$

in $\mathsf{PFib}^{\mathcal{D}}\big(\langle Q_j \rangle ; P_j\big)$.

By the definition (6.29) of the composition, the transformation projection axiom (7.29) for the composite ℓ-linear natural transformation $\gamma\big(\theta\,;\langle\theta_j\rangle\big)$ states that the two boundary whiskered natural transformations in the following diagram are equal.

$$\begin{array}{ccccc}
\prod_{j=1}^{n}\prod_{i=1}^{\ell_j}\mathsf{B}_{ji} & \overset{\prod_j G_j}{\underset{\prod_j G'_j}{\overset{\Downarrow \prod_j\theta_j}{\rightrightarrows}}} & \prod_{j=1}^{n}\mathsf{C}_j & \overset{F}{\underset{F'}{\overset{\Downarrow\theta}{\rightrightarrows}}} & \mathsf{C}_0 \\
{\scriptstyle \prod_{j,i} Q_{ji}}\downarrow & & {\scriptstyle \prod_j P_j}\downarrow & & \downarrow{\scriptstyle P_0} \\
\mathcal{D}^{\ell_1+\cdots+\ell_n} & \xrightarrow{\prod_j \otimes} & \mathcal{D}^n & \xrightarrow{\otimes} & \mathcal{D}
\end{array}$$

The two boundary whiskered natural transformations above are equal by the transformation projection axiom (7.29) for $\langle\theta_j\rangle$ and θ. $\square$

Symmetry

Next we show that the symmetric group action in $\mathsf{PFib}^{\mathcal{D}}$, which is a restriction of the one in $\mathsf{PermCat}^{\mathsf{sg}}$, is well defined when the multiplicative braiding in $\mathcal{D}$ is the identity, together with a weaker converse statement. For a monoidal category $(\mathsf{C},\otimes)$, denote by $\otimes^{\mathsf{op}}$, which is called the *opposite monoidal product*, the composite functor

$$\mathsf{C}^2 \xrightarrow{\xi} \mathsf{C}^2 \xrightarrow{\otimes} \mathsf{C}, \qquad \text{composite } \otimes^{\mathsf{op}} \tag{7.43}$$

where ξ swaps the two arguments.

Lemma 7.44. *In the context of Definition 7.36, the following statements hold.*

1. *If the multiplicative braiding $\beta^{\otimes}$ in $\mathcal{D}$ is the identity natural transformation, then the symmetric group action in $\mathsf{PFib}^{\mathcal{D}}$ is well defined on opcartesian strong multilinear functors.*

2. *If the symmetric group action in $\mathsf{PFib}^{\mathcal{D}}$ is well defined on opcartesian strong multilinear functors, then $\otimes = \otimes^{\mathsf{op}}$ in $\mathcal{D}$.*

Proof. First suppose $\beta^{\otimes} = 1$ in $\mathcal{D}$. We want to show that the symmetric group action in $\mathsf{PermCat}^{\mathsf{sg}}$ restricts to $\mathsf{PFib}^{\mathcal{D}}$. Suppose given small permutative opfibrations over $\mathcal{D}$ (Definition 7.6)

$$\mathsf{C}_k \xrightarrow{P_k} \mathcal{D} \quad \text{for} \quad k \in \{0,1,\ldots,n\}$$

with $\langle \mathsf{C}\rangle = \langle \mathsf{C}_j\rangle_{j=1}^{n}$ and $\langle P\rangle = \langle P_j\rangle_{j=1}^{n}$. For a permutation $\sigma \in \Sigma_n$, to show that the symmetric group action

$$\mathsf{PFib}^{\mathcal{D}}\big(\langle P\rangle\,; P_0\big) \xrightarrow{\sigma} \mathsf{PFib}^{\mathcal{D}}\big(\langle P\rangle\sigma\,; P_0\big) \tag{7.45}$$

is well defined on objects, suppose given an opcartesian n-linear functor (Definition 7.11)

$$\langle P\rangle \xrightarrow{\left(F\,,\,\langle F_j^2\rangle_{j=1}^n\right)} P_0.$$

We must show that the n-linear functor

$$\prod_{j=1}^n \mathsf{C}_{\sigma(j)} \xrightarrow{\;\sigma\;} \prod_{j=1}^n \mathsf{C}_j \xrightarrow{\;F\;} \mathsf{C}_0 \tag{7.46}$$

with the linearity constraints in (6.26) satisfies axioms 1 through 4 in Definition 7.11.

Axiom 1. This axiom holds for $F^\sigma = F \circ \sigma$ because

- the permutation σ sends an n-tuple of opcartesian morphisms to an n-tuple of opcartesian morphisms and
- F preserves opcartesian morphisms by axiom 1.

Note that axiom 1 is valid even without the assumption $\beta^\otimes = 1$.

Axiom 2. The diagram (7.13) for F^σ in (7.46) is the boundary of the following diagram.

$$\begin{array}{ccccc}
\prod_{j=1}^n \mathsf{C}_{\sigma(j)} & \xrightarrow{\;\sigma\;} & \prod_{j=1}^n \mathsf{C}_j & \xrightarrow{\;F\;} & \mathsf{C}_0 \\
{\scriptstyle \prod_j P_{\sigma(j)}}\big\downarrow & & {\scriptstyle \prod_j P_j}\big\downarrow & & \big\downarrow{\scriptstyle P_0} \\
\mathcal{D}^n & \xrightarrow{\;\sigma\;} & \mathcal{D}^n & \xrightarrow{\;\otimes\;} & \mathcal{D} \\
& & \beta^\otimes \Uparrow \cong & & \\
& & \underbrace{\hspace{6em}}_{\otimes} & &
\end{array} \tag{7.47}$$

The detail of the diagram (7.47) is as follows.

- The left square commutes by the naturality of the braiding in $(\mathsf{Cat}, \times)$.
- The right square commutes by axiom 2 for F.

In the bottom region, the natural isomorphism

$$\otimes \xrightarrow[\cong]{\;\beta^\otimes\;} \otimes \circ \sigma$$

is an iterate of the multiplicative braiding $\beta^\otimes$ in $\mathcal{D}$. Under the current assumption that $\beta^\otimes = 1$, the bottom region in (7.47) strictly commutes, proving axiom 2 for F^σ.

Axiom 3. This axiom for F^σ follows from

- axiom 2 for F^σ and
- axiom 3 for F.

Axiom 4. This axiom for F^σ follows from

- the definition (6.26) of the i-th linearity constraint $(F^\sigma)^2_i = F^2_{\sigma(i)}$,
- the assumption that $\beta^\otimes = 1$ in $\mathcal{D}$,
- axiom 2 for F^σ, and
- axiom 4 for F.

This finishes the proof of assertion (1).

For assertion (2), suppose that the symmetric group action in $\mathsf{PFib}^{\mathcal{D}}$ is well defined on opcartesian strong multilinear functors. For the opcartesian strong 2-linear functor (7.26)

$$\langle 1_{\mathcal{D}^\oplus} \rangle_{j=1}^2 \xrightarrow{\left(\otimes\, ,\, \langle \mathsf{L}_j^{-1} \rangle_{j=1}^2\right)} 1_{\mathcal{D}^\oplus}$$

and $\sigma = (1,2) \in \Sigma_2$, the boundary of the commutative diagram (7.47) yields the equality

$$\otimes = \otimes \circ \xi = \otimes^{\mathsf{op}}.$$

This proves assertion (2). $\square$

Next we consider opcartesian multilinear transformations.

Lemma 7.48. *In the context of Definition 7.36, the following statements hold.*

1. *If the multiplicative braiding $\beta^\otimes$ in $\mathcal{D}$ is the identity natural transformation, then the symmetric group action in $\mathsf{PFib}^{\mathcal{D}}$ is well defined on opcartesian multilinear transformations.*
2. *If the symmetric group action in $\mathsf{PFib}^{\mathcal{D}}$ is well defined on opcartesian multilinear transformations, then $\otimes = \otimes^{\mathsf{op}}$ in $\mathcal{D}$.*

Proof. First suppose $\beta^\otimes = 1$ in $\mathcal{D}$. Reusing the notation in the proof of Lemma 7.44, suppose given an opcartesian n-linear transformation θ in $\mathsf{PFib}^{\mathcal{D}}\big(\langle P \rangle\, ;\, P_0\big)$ as follows.

$$\prod_{j=1}^n \mathsf{C}_j \overset{(F, \langle F_j^2 \rangle)}{\underset{(G, \langle G_j^2 \rangle)}{\Downarrow \theta}} \mathsf{C}_0$$

By definition (6.25), the axiom (7.29) for $\theta^\sigma = \theta * 1_\sigma$ is the equality

$$1_{P_0} * \theta * 1_\sigma = 1_{\otimes \circ (\prod_j P_{\sigma(j)})} \tag{7.49}$$

as in the boundary of the following diagram.

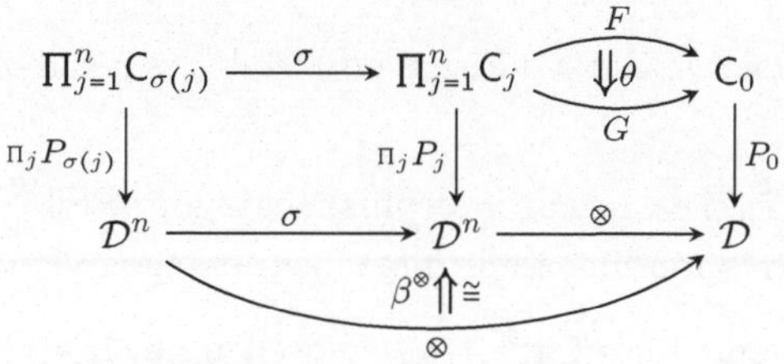

(7.50)

The detail of the diagram (7.50) is as follows.

- The left square commutes by the naturality of the braiding in $(\mathsf{Cat}, \times)$.
- To its right, the equality

$$1_{P_0} * \theta = 1_{\otimes \circ (\prod_j P_j)}$$

  holds by the axiom (7.29) for θ.
- Under the current assumption that $\beta^{\otimes} = 1$ in $\mathcal{D}$, the bottom region strictly commutes.

This proves the equality (7.49).

For assertion (2), suppose that the symmetric group action in $\mathsf{PFib}^{\mathcal{D}}$ is well defined on opcartesian multilinear transformations. Similar to the proof of Lemma 7.44 (2), the desired equality, $\otimes = \otimes^{\mathsf{op}}$ in $\mathcal{D}$, follows from considering the diagram (7.50) in the case $n = 2$ with

- $\sigma \in \Sigma_2$ the nonidentity permutation,
- $F = G$ the opcartesian strong 2-linear functor (7.26)

$$\langle 1_{\mathcal{D}^{\oplus}} \rangle_{j=1}^2 \xrightarrow{\left(\otimes, \langle \mathsf{L}_j^{-1} \rangle_{j=1}^2\right)} 1_{\mathcal{D}^{\oplus}},$$

  and
- $\theta = 1_F$ the identity opcartesian 2-linear transformation of F.

This finishes the proof. □

Multicategory Structure

Next is the main result of this section. Recall the Cat-multicategory $\mathsf{PermCat}^{\mathsf{sg}}$ (Theorem 6.31).

Theorem 7.51. *Suppose $\mathcal{D}$ is a small tight bipermutative category.*

1. $\mathsf{PFib}^{\mathcal{D}}$ *in Definition 7.36 is a non-symmetric* Cat*-multicategory.*

2. *If the multiplicative braiding* $\beta^{\otimes}$ *in* $\mathcal{D}$ *is the identity natural transformation, then* $\mathsf{PFib}^{\mathcal{D}}$ *is a* Cat*-multicategory.*

3. *If the non-symmetric* Cat*-multicategory* $\mathsf{PFib}^{\mathcal{D}}$ *is a* Cat*-multicategory, then* $\otimes = \otimes^{\mathsf{op}}$ *in* $\mathcal{D}$.

Proof. By Lemma 7.35 each multimorphism category $\mathsf{PFib}^{\mathcal{D}}\big(\langle P\rangle\,;\,P_0\big)$ is a subcategory of the corresponding multimorphism category $\mathsf{PermCat}^{\mathsf{sg}}\big(\langle \mathsf{C}\rangle\,;\,\mathsf{C}_0\big)$.

- In both $\mathsf{PFib}^{\mathcal{D}}$ and $\mathsf{PermCat}^{\mathsf{sg}}$, the colored units are given by identity 1-linear functors.
- The composition in $\mathsf{PFib}^{\mathcal{D}}$, which is a restriction of the composition in $\mathsf{PermCat}^{\mathsf{sg}}$, is well defined by Lemmas 7.37 and 7.42.

Therefore, the associativity and unity axioms, (3.7) through (3.9), hold in $\mathsf{PFib}^{\mathcal{D}}$ as they do in the Cat-multicategory $\mathsf{PermCat}^{\mathsf{sg}}$. This proves assertion 1.

For assertion 2, under the assumption that $\beta^{\otimes} = 1$ in $\mathcal{D}$, the symmetric group action in $\mathsf{PFib}^{\mathcal{D}}$ is well defined by Lemma 7.44 (1) and Lemma 7.48 (1). The symmetry group action axiom, (3.6), and the equivariance axioms, (3.10) and (3.12), hold in $\mathsf{PFib}^{\mathcal{D}}$ as they do in the Cat-multicategory $\mathsf{PermCat}^{\mathsf{sg}}$.

Assertion 3 follows from Lemma 7.44 (2) and Lemma 7.48 (2). □

Recall from Definition 3.19 that a non-symmetric V-multifunctor is a V-multifunctor without the symmetric group action axiom (3.20). In other words, a non-symmetric V-multifunctor is only required to preserve the colored units (3.21) and the composition (3.22). The following observation follows from Theorem 7.51.

Corollary 7.52. *Suppose* $\mathcal{D}$ *is a small tight bipermutative category.*

1. *There exists a non-symmetric* Cat*-multifunctor*

$$\mathsf{PFib}^{\mathcal{D}} \xrightarrow{\;\mathsf{U}\;} \mathsf{PermCat}^{\mathsf{sg}} \tag{7.53}$$

   defined by

   - *the object assignment*

   $$\mathsf{U}(P : \mathsf{C} \longrightarrow \mathcal{D}) = \mathsf{C}$$

   for each small permutative opfibration P *over* $\mathcal{D}$ *and*

   - *the subcategory inclusions in Lemma 7.35 for multimorphism categories.*

2. *If the multiplicative braiding* $\beta^{\otimes}$ *in* $\mathcal{D}$ *is the identity natural transformation, then* U *is a* Cat*-multifunctor.*

Explanation 7.54. Although U in (7.53) consists of subcategory inclusions between multimorphism categories, $\mathsf{PFib}^{\mathcal{D}}$ is generally *not* a sub-Cat-multicategory of $\mathsf{PermCat}^{\mathsf{sg}}$. The reason is that U is generally not injective on objects because a small permutative category C may admit several different permutative opfibrations $\mathsf{C} \longrightarrow \mathcal{D}$. ◇

Example 7.55. Consider the small tight bipermutative categories

$$\mathsf{Fin}^{\mathsf{sk}}, \quad \Sigma, \quad \mathcal{F}, \quad \text{and} \quad \mathcal{A}$$

in, respectively, Examples 2.48, 2.57, 2.58, and 2.63. With $\mathcal{D}$ denoting any one of them,

- $\mathsf{PFib}^{\mathcal{D}}$ is a non-symmetric Cat-multicategory by Theorem 7.51 (1), and
- there is a non-symmetric Cat-multifunctor U as in (7.53).

Since $\otimes \neq \otimes^{\mathsf{op}}$ in $\mathcal{D}$, $\mathsf{PFib}^{\mathcal{D}}$ does *not* have a well-defined symmetric group action by Lemma 7.44 (2). Therefore, $\mathsf{PFib}^{\mathcal{D}}$ is *not* a Cat-multicategory. ◇

7.4 Permutative Opfibrations via the Grothendieck Construction

As in (7.10) we continue to assume that $(\mathcal{D}, \oplus, \otimes)$ is a small tight bipermutative category. For an additive symmetric monoidal functor (Definition 5.31)

$$(X, X^2, X^0) : \mathcal{D}^{\oplus} = (\mathcal{D}, \oplus, 0, \beta^{\oplus}) \longrightarrow \mathsf{Cat},$$

its Grothendieck construction $\int_{\mathcal{D}} X$ (Definition 6.33) is a small permutative category (Definition 6.37 and Lemma 6.42)

$$\left(\textstyle\int_{\mathcal{D}} X, \square, (0, X^0*), \beta^{\square}\right).$$

By Theorem 6.110 $X \longmapsto \int_{\mathcal{D}} X$ is the object assignment of a pseudo symmetric Cat-multifunctor (Definition 4.1)

$$\mathcal{D}\mathsf{Cat} \xrightarrow{\;\int_{\mathcal{D}}\;} \mathsf{PermCat}^{\mathsf{sg}}.$$

It is, furthermore, a Cat-multifunctor if the multiplicative braiding $\beta^{\otimes}$ in $\mathcal{D}$ is the identity natural transformation.

In this section we show that $\int_{\mathcal{D}}$ lifts to a non-symmetric Cat-multifunctor against the non-symmetric Cat-multifunctor U in (7.53) as follows.

$$\begin{array}{ccc} & \overset{\int_{\mathcal{D}}}{\dashrightarrow} & \mathsf{PFib}^{\mathcal{D}} \\ \mathcal{D}\mathsf{Cat} \nearrow & & \big\downarrow \mathsf{U} \\ \mathcal{D}\mathsf{Cat} & \xrightarrow{\;\int_{\mathcal{D}}\;} & \mathsf{PermCat}^{\mathsf{sg}} \end{array}$$

Moreover, if $\beta^{\otimes} = 1$ in $\mathcal{D}$, then this lift is a Cat-multifunctor. In other words, it strictly preserves the symmetric group action.

Extension of Object Assignment

First we extend the Grothendieck construction $\int_{\mathcal{D}} X$ to a permutative opfibration U_X over $\mathcal{D}$ (Definition 7.6).

Lemma 7.56. *For each additive symmetric monoidal functor*

$$(X, X^2, X^0) : \mathcal{D}^{\oplus} \longrightarrow \mathsf{Cat},$$

the first-factor projection

$$\int_{\mathcal{D}} X \xrightarrow{U_X} \mathcal{D} \tag{7.57}$$

defined by

- $U_X(a, x) = a$ *for each object* (a, x) *in* $\int_{\mathcal{D}} X$ *and*
- $U_X(f, p) = f$ *for each morphism* (f, p) *in* $\int_{\mathcal{D}} X$

is a small permutative opfibration over $\mathcal{D}$.

Proof. Given

- an object $(a, x) \in \int_{\mathcal{D}} X$ and
- a morphism $f : a \longrightarrow b \in \mathcal{D}$,

the chosen opcartesian lift (Definition 7.3 (7)) of the fore-lift $\langle (a, x) ; f \rangle$ is defined as the morphism in $\int_{\mathcal{D}} X$

$$(a, x) \xrightarrow{(f, 1_{f_* x})} (b, f_* x) \tag{7.58}$$

with

$$f_* = Xf : Xa \longrightarrow Xb.$$

This definition makes U_X into a split opfibration over $\mathcal{D}$ by [JY21, 10.1.11]. Axioms 1 through 3 in Definition 7.6 follow from

- the definition of the permutative structure $\big(\square, (0, X^0 *), \beta^{\square}\big)$ in (6.38) through (6.41), and
- the functoriality of each component functor of the monoidal constraint X^2 (5.34).

This finishes the proof. □

The assignment $X \longmapsto U_X$, with $U_X : \int_{\mathcal{D}} X \longrightarrow \mathcal{D}$ in (7.57), is the object assignment of the lifted Grothendieck construction.

Lifting Multimorphism Functors

Next we observe that each multimorphism functor of $\int_{\mathcal{D}}$ lands in $\mathsf{PFib}^{\mathcal{D}}$. Recall that the multimorphism functors of $\int_{\mathcal{D}}$ are given in

- Definition 6.43 for arity 0 inputs and
- Definitions 6.45 and 6.61 and Lemma 6.69 for positive arity inputs.

For the next several results, suppose given additive symmetric monoidal functors

$$(Z, Z^2, Z^0) \quad \text{and} \quad \big\langle (X_j, X_j^2, X_j^0) \big\rangle_{j=1}^{n} : \mathcal{D}^{\oplus} \longrightarrow \mathsf{Cat} \tag{7.59}$$

with $n \geq 0$. We consider the solid-arrow functors

$$\begin{array}{ccc} & & \mathsf{PFib}^{\mathcal{D}}\big(\langle U_X \rangle ; U_Z\big) \\ & \nearrow & \big\downarrow \mathsf{U} \\ \mathcal{D}\mathsf{Cat}\big(\langle X \rangle ; Z\big) & \xrightarrow{\int_{\mathcal{D}}} & \mathsf{PermCat}^{\mathsf{sg}}\big(\langle \int_{\mathcal{D}} X \rangle ; \int_{\mathcal{D}} Z\big) \end{array} \tag{7.60}$$

with

- $\langle X \rangle = \langle X_j \rangle_{j=1}^{n}$, $\langle \int_{\mathcal{D}} X \rangle = \langle \int_{\mathcal{D}} X_j \rangle_{j=1}^{n}$, and $\langle U_X \rangle = \langle U_{X_j} \rangle_{j=1}^{n}$,
- each

  $$\int_{\mathcal{D}} X_j \xrightarrow{U_{X_j}} \mathcal{D} \quad \text{and} \quad \int_{\mathcal{D}} Z \xrightarrow{U_Z} \mathcal{D}$$

  as in (7.57),
- $\int_{\mathcal{D}}$ the functor in (6.44) if $n = 0$ and (6.70) if $n > 0$, and
- U the subcategory inclusion in Lemma 7.35.

Lemmas 7.61 and 7.62 show that $\int_{\mathcal{D}}$ lifts against U as the dotted arrow. First we consider the arity 0 case.

Lemma 7.61. *In the context of* (7.60) *with* $n = 0$, *the functor*

$$\mathcal{D}\mathsf{Cat}\big(\langle\rangle ; Z\big) \xrightarrow{\int_{\mathcal{D}}} \mathsf{PermCat}^{\mathsf{sg}}\big(\langle\rangle ; \int_{\mathcal{D}} Z\big)$$

factors through the subcategory $\mathsf{PFib}^{\mathcal{D}}\big(\langle\rangle ; U_Z\big)$ *and yields an isomorphism*

$$\mathcal{D}\mathsf{Cat}\big(\langle\rangle ; Z\big) \xrightarrow[\cong]{\int_{\mathcal{D}}} \mathsf{PFib}^{\mathcal{D}}\big(\langle\rangle ; U_Z\big).$$

Proof. The subcategory

$$\mathsf{PFib}^{\mathcal{D}}\big(\langle\rangle ; U_Z\big) \subset \int_{\mathcal{D}} Z$$

is the U_Z-preimage of $1 \in \mathcal{D}$ by

- Explanation 7.18 (2) for objects and
- Explanation 7.30 (2) for morphisms.

In other words, in $\mathsf{PFib}^{\mathcal{D}}(\langle\rangle\,; U_Z)$

- each object has the form $(1, x)$ for some object $x \in Z1$ and
- each morphism has the form

$$(1, x) \xrightarrow{(1_1, p)} (1, y)$$

for some morphism $p : x \longrightarrow y$ in $Z1$.

Therefore, the multimorphism functor in (6.44) factors through the subcategory

$$\mathsf{PFib}^{\mathcal{D}}(\langle\rangle\,; U_Z) \subset \textstyle\int_{\mathcal{D}} Z = \mathsf{PermCat}^{\mathsf{sg}}(\langle\rangle\,; \int_{\mathcal{D}} Z)$$

and yields an isomorphism

$$Z1 = \mathcal{D}\mathsf{Cat}(\langle\rangle\,; Z) \xrightarrow[\cong]{\int_{\mathcal{D}}} \mathsf{PFib}^{\mathcal{D}}(\langle\rangle\,; U_Z)$$

of categories. □

Next we consider the positive arity case. In some diagrams below, the symbols $\exists$ and ! mean, respectively, *there exists* and *unique*.

Lemma 7.62. *In the context of* (7.60) *with* $n > 0$*, the image of the functor*

$$\mathcal{D}\mathsf{Cat}(\langle X\rangle\,; Z) \xrightarrow{\int_{\mathcal{D}}} \mathsf{PermCat}^{\mathsf{sg}}(\langle \textstyle\int_{\mathcal{D}} X\rangle\,; \int_{\mathcal{D}} Z)$$

lies in the subcategory $\mathsf{PFib}^{\mathcal{D}}(\langle U_X\rangle\,; U_Z)$ *of the codomain.*

Proof. First we consider objects, which are additive natural transformations, followed by morphisms, which are additive modifications.

Objects. Given an additive natural transformation (5.40)

$$\big\langle (X_j, X_j^2, X_j^0)\big\rangle_{j=1}^{n} \xrightarrow{\ \phi\ } (Z, Z^2, Z^0) \quad \text{in} \quad \mathcal{D}\mathsf{Cat}(\langle X\rangle\,; Z),$$

we must show that the strong n-linear functor (Lemma 6.57)

$$\textstyle\prod_{j=1}^{n} \big(\int_{\mathcal{D}} X_j\big) \xrightarrow{(\int_{\mathcal{D}}\phi,\, \langle (\int_{\mathcal{D}}\phi)_i^2\rangle_{i=1}^{n})} \int_{\mathcal{D}} Z \tag{7.63}$$

is opcartesian. Using Lemma 7.21 we need to verify axioms 2 through 4 in Definition 7.11. We use the notation in Definition 6.45.

Axiom 2. The diagram (7.13) for $\int_{\mathcal{D}}\phi$ is as follows.

$$\begin{array}{ccc} \prod_{j=1}^{n}(\int_{\mathcal{D}} X_j) & \xrightarrow{\int_{\mathcal{D}}\phi} & \int_{\mathcal{D}} Z \\ {\scriptstyle \prod_j U_{X_j}} \downarrow & & \downarrow {\scriptstyle U_Z} \\ \mathcal{D}^n & \xrightarrow{\otimes} & \mathcal{D} \end{array}$$

This diagram is commutative by

- the definition (7.57) of $U_?$ as the first-factor projection and
- the definitions (6.49) and (6.52) of $\int_{\mathcal{D}}\phi$ on, respectively, objects and morphisms.

This proves axiom 2.

Axiom 3. For each $j \in \{1,\ldots,n\}$, suppose given a fore-lift $\langle (a_j, x_j); f_j\rangle$ with respect to $U_{X_j} : \int_{\mathcal{D}} X_j \longrightarrow \mathcal{D}$. Its chosen opcartesian lift is the morphism (7.58)

$$(a_j, x_j) \xrightarrow{\left(f_j\,,\,1_{(f_j)_* x_j}\right)} \left(b_j\,,\,(f_j)_* x_j\right) \quad \text{in} \quad \textstyle\int_{\mathcal{D}} X_j.$$

By

- the definition (6.52) of $\int_{\mathcal{D}}\phi$ on morphisms and
- the functoriality of $\phi_{\langle b\rangle}$,

there are morphism equalities in $\int_{\mathcal{D}} Z$ as follows.

$$\begin{aligned} &(\textstyle\int_{\mathcal{D}}\phi)\big\langle (f_j\,,\,1_{(f_j)_* x_j})\big\rangle_{j=1}^{n} : (\textstyle\int_{\mathcal{D}}\phi)\langle (a_j, x_j)\rangle \longrightarrow (\textstyle\int_{\mathcal{D}}\phi)\big\langle b_j, (f_j)_* x_j\big\rangle \\ &= \Big(f\,,\,\phi_{\langle b\rangle}\big\langle 1_{(f_j)_* x_j}\big\rangle_{j=1}^{n}\Big) : \big(a\,,\,\phi_{\langle a\rangle}\langle x\rangle\big) \longrightarrow \big(b\,,\,\phi_{\langle b\rangle}\langle (f_j)_* x_j\rangle\big) \\ &= \big(f\,,\,1_{f_*\phi_{\langle a\rangle}\langle x\rangle}\big) : \big(a\,,\,\phi_{\langle a\rangle}\langle x\rangle\big) \longrightarrow \big(b\,,\,f_*\phi_{\langle a\rangle}\langle x\rangle\big) \end{aligned}$$

By definition (7.58) the last morphism is the chosen opcartesian lift of the fore-lift

$$\big\langle (\textstyle\int_{\mathcal{D}}\phi)\langle (a_j, x_j)\rangle\,;\,\otimes_{j=1}^{n} f_j\big\rangle = \big\langle \big(a\,,\,\phi_{\langle a\rangle}\langle x\rangle\big)\,;\,f\big\rangle$$

with respect to $U_Z : \int_{\mathcal{D}} Z \longrightarrow \mathcal{D}$. Thus $\int_{\mathcal{D}}\phi$ preserves chosen opcartesian lifts, proving axiom 3.

Axiom 4. This axiom follows from

- the definition (6.56) of each linearity constraint $(\int_{\mathcal{D}}\phi)_i^2$ and
- the definition (7.58) of the chosen opcartesian lift of each fore-lift with respect to $U_Z : \int_{\mathcal{D}} Z \longrightarrow \mathcal{D}$.

We have shown that $\int_{\mathcal{D}}\phi$ in (7.63) is opcartesian, so it is an object in the subcategory $\mathsf{PFib}^{\mathcal{D}}\big(\langle U_X\rangle\,;\,U_Z\big)$.

Morphisms. Given an additive modification (5.53)

$$\big\langle (X_j, X_j^2, X_j^0)\big\rangle_{j=1}^n \underset{\psi}{\overset{\phi}{\Downarrow\Phi}} (Z, Z^2, Z^0) \quad \text{in} \quad \mathcal{D}\mathsf{Cat}\big(\langle X\rangle\,;\,Z\big),$$

we must prove the axiom (7.29) for $\int_{\mathcal{D}}\Phi$ (Lemma 6.67), which asserts that the following two whiskered natural transformations are equal.

$$\begin{array}{ccc}
\prod_{j=1}^n\big(\int_{\mathcal{D}}X_j\big) & \underset{\int_{\mathcal{D}}\psi}{\overset{\int_{\mathcal{D}}\phi}{\Downarrow \int_{\mathcal{D}}\Phi}} & \int_{\mathcal{D}}Z \\
{\scriptstyle \prod_j U_{X_j}}\big\downarrow & & \big\downarrow{\scriptstyle U_Z} \\
\mathcal{D}^n & \xrightarrow{\quad\otimes\quad} & \mathcal{D}
\end{array}$$

These two natural transformations are equal by

- the definition (6.63) of each component

$$\big(\textstyle\int_{\mathcal{D}}\Phi\big)_{\langle (a_j,x_j)\rangle} = \big(1_a\,,\,(\Phi_{\langle a\rangle})_{\langle x\rangle}\big)$$

  and

- the definition (7.57) of $U_?$ as the first-factor projection.

This shows that $\int_{\mathcal{D}}\Phi$ is a morphism in the subcategory $\mathsf{PFib}^{\mathcal{D}}\big(\langle U_X\rangle\,;\,U_Z\big)$. □

We are now ready for the main result of this chapter.

Theorem 7.64. *Suppose $\mathcal{D}$ is a small tight bipermutative category.*

1. *There is a non-symmetric* Cat*-multifunctor*

$$\mathcal{D}\mathsf{Cat} \xrightarrow{\ \int_{\mathcal{D}}\ } \mathsf{PFib}^{\mathcal{D}} \tag{7.65}$$

   defined by

   - *the object assignment*

$$X \longmapsto \big(U_X : \textstyle\int_{\mathcal{D}}X \longrightarrow \mathcal{D}\big)$$

     in (7.57) and

   - *the multimorphism functors from Lemmas 7.61 and 7.62.*

2. *The diagram of non-symmetric* Cat*-multifunctors*

$$\begin{array}{ccc} & \overset{\int_{\mathcal{D}}}{\nearrow} & \mathsf{PFib}^{\mathcal{D}} \\ & & \big\downarrow \mathsf{U} \\ \mathcal{D}\mathsf{Cat} & \xrightarrow{\int_{\mathcal{D}}} & \mathsf{PermCat}^{\mathsf{sg}} \end{array}$$

   is commutative, with the bottom $\int_{\mathcal{D}}$ *and* U *from, respectively,* (6.111) *and* (7.53).

3. *If the multiplicative braiding* $\beta^{\otimes}$ *in* $\mathcal{D}$ *is the identity natural transformation, then* $\int_{\mathcal{D}}$ *in* (7.65) *is a* Cat*-multifunctor.*

Proof. By Theorem 7.51 $\mathsf{PFib}^{\mathcal{D}}$ is a non-symmetric Cat-multicategory whose colored units and composition (Definition 7.36) are both inherited from $\mathsf{PermCat}^{\mathsf{sg}}$ (Theorem 6.31) via the subcategory inclusions in Lemma 7.35. The latter are the multimorphism functors of U in (7.53). If $\beta^{\otimes} = 1$ in $\mathcal{D}$, then $\mathsf{PFib}^{\mathcal{D}}$ is a Cat-multicategory whose symmetric group action is inherited from $\mathsf{PermCat}^{\mathsf{sg}}$.

Assertions 1 and 2 follow from Theorem 6.110 (1), Theorem 7.51 (1), and Lemmas 7.56, 7.61, and 7.62.

Assertion 3 follows from Theorem 6.110 (2) and Theorem 7.51 (2). □

Example 7.66. Suppose $\mathcal{D}$ is any one of the small tight bipermutative categories

$$\mathsf{Fin}^{\mathsf{sk}}, \quad \Sigma, \quad \mathcal{F}, \quad \text{and} \quad \mathcal{A}$$

in, respectively, Examples 2.48, 2.57, 2.58, and 2.63. Then the following statements hold.

- $\mathsf{PFib}^{\mathcal{D}}$ is a non-symmetric Cat-multicategory and *not* a Cat-multicategory by Example 7.55.
- Theorem 7.64 (1) and (2) apply to $\mathcal{D}$.
- $\int_{\mathcal{D}}$ in (7.65) is *not* a Cat-multifunctor because $\mathsf{PFib}^{\mathcal{D}}$ is not a Cat-multicategory. ◇

8

The Grothendieck Construction is a Cat-*Multiequivalence*

Context

Throughout this chapter, we assume that

$$\big(\mathcal{D}, (\oplus, 0, \beta^{\oplus}), (\otimes, 1, \beta^{\otimes}), (\partial^l, \partial^r)\big)$$

is a small tight bipermutative category (Definition 2.41) with

- additive structure $\mathcal{D}^{\oplus} = (\mathcal{D}, \oplus, 0, \beta^{\oplus})$ and
- multiplicative structure $\mathcal{D}^{\otimes} = (\mathcal{D}, \otimes, 1, \beta^{\otimes})$.

In Theorem 7.64 (2) we observe that the pseudo symmetric Cat-multifunctor

$$\mathcal{D}\mathsf{Cat} \xrightarrow{\int_{\mathcal{D}}} \mathsf{PermCat}^{\mathsf{sg}}$$

in Theorem 6.110 factors through the non-symmetric Cat-multicategory $\mathsf{PFib}^{\mathcal{D}}$ (Theorem 7.51) into two non-symmetric Cat-multifunctors, $\int_{\mathcal{D}}$ (7.65) and U (7.53).

$$\begin{array}{ccc} \mathcal{D}\mathsf{Cat} & \xrightarrow{\int_{\mathcal{D}}} & \mathsf{PermCat}^{\mathsf{sg}} \\ & {\scriptstyle\int_{\mathcal{D}}}\searrow \quad \nearrow{\scriptstyle \mathsf{U}} & \\ & \mathsf{PFib}^{\mathcal{D}} & \end{array}$$

Purpose

Recall from Definition 3.42 that a Cat-*multiequivalence* is an equivalence in the 2-category Cat-Multicat of small Cat-multicategories. The non-symmetric variant is defined similarly. By Theorem 3.69 a Cat-multifunctor is a Cat-multiequivalence if and only if it is essentially surjective on objects and an isomorphism on each multimorphism category.

The main purpose of this chapter is to strengthen the factorization above by showing that the non-symmetric Cat-multifunctor

$$\mathcal{D}\mathsf{Cat} \xrightarrow{\int_{\mathcal{D}}} \mathsf{PFib}^{\mathcal{D}} \tag{8.1}$$

DOI: 10.1201/9781003451235-8

is a non-symmetric Cat-*multiequivalence.* Moreover, if the multiplicative braiding in $\mathcal{D}$ is the identity natural transformation, then this $\int_{\mathcal{D}}$ is a Cat-multiequivalence. See Theorem 8.76. These assertions are proved using Theorem 3.69. Below is a summary of the main properties of the two $\int_{\mathcal{D}}$ and U.

- $\int_{\mathcal{D}} : \mathcal{D}\mathsf{Cat} \longrightarrow \mathsf{PermCat}^{\mathsf{sg}}$ is a pseudo symmetric Cat-multifunctor (Theorem 6.110).
- $\mathsf{U} : \mathsf{PFib}^{\mathcal{D}} \longrightarrow \mathsf{PermCat}^{\mathsf{sg}}$ is a non-symmetric Cat-multifunctor (Corollary 7.52).
- $\int_{\mathcal{D}} : \mathcal{D}\mathsf{Cat} \xrightarrow{\sim} \mathsf{PFib}^{\mathcal{D}}$ is a non-symmetric Cat-multiequivalence (Theorem 8.76).

Implication for Inverse K-Theory

The inverse K-theory functor (Chapters 9 and 10) is the composite $\int_{\mathcal{A}} \circ A$ of two functors, with $\mathcal{A}$ the small tight bipermutative category in Example 2.63. Applied to $\mathcal{A}$, Theorem 8.76 implies that the pseudo symmetric Cat-multifunctor $\int_{\mathcal{A}}$ in inverse K-theory factors through $\mathsf{PFib}^{\mathcal{A}}$ as displayed below.

$$\begin{array}{ccc} \mathcal{A}\mathsf{Cat} & \xrightarrow{\int_{\mathcal{A}}} & \mathsf{PermCat}^{\mathsf{sg}} \\ & {\scriptstyle \int_{\mathcal{A}}} \searrow^{\sim} \quad \nearrow {\scriptstyle \mathsf{U}} & \\ & \mathsf{PFib}^{\mathcal{A}} & \end{array}$$

Its first factor is the non-symmetric Cat-multiequivalence

$$\mathcal{A}\mathsf{Cat} \xrightarrow[\sim]{\int_{\mathcal{A}}} \mathsf{PFib}^{\mathcal{A}}$$

given by the Grothendieck construction.

Organization

As we mentioned above, we prove that $\int_{\mathcal{D}}$ in (8.1) is a non-symmetric Cat-multiequivalence using Theorem 3.69. In other words, we must show that $\int_{\mathcal{D}}$ is essentially surjective on objects and an isomorphism on each multimorphism category. In Section 8.1 we first show that $\int_{\mathcal{D}}$ is essentially surjective on objects. This means that each small permutative opfibration over $\mathcal{D}$ is, up to an opcartesian 1-linear isomorphism, the Grothendieck construction of some additive symmetric monoidal functor.

In Section 8.2 we show that each multimorphism functor of $\int_{\mathcal{D}}$ is injective on both objects and morphism sets. Injectivity on objects means that each additive natural transformation is uniquely determined by its Grothendieck construction as a strong multilinear functor (Lemma 8.20). Injectivity on morphism sets means that each additive modification is uniquely determined by its Grothendieck construction as a multilinear natural transformation (Lemma 8.28).

In Section 8.3 we show that each multimorphism functor of $\int_{\mathcal{D}}$ is surjective on objects. This means that each opcartesian strong n-linear functor in $\mathsf{PFib}^{\mathcal{D}}$ is equal to the Grothendieck construction of some additive natural transformation in $\mathcal{D}\mathsf{Cat}$.

In Section 8.4 we show that each multimorphism functor of $\int_{\mathcal{D}}$ is surjective on morphism sets, finishing the proof of Theorem 8.76. This means that each opcartesian n-linear transformation is equal to the Grothendieck construction of some additive modification. The proof that $\int_{\mathcal{D}}$ is, furthermore, a Cat-multiequivalence if $\beta^{\otimes} = 1$ in $\mathcal{D}$ requires no extra effort because the lemmas in this chapter do not depend on any special property of $\beta^{\otimes}$. Thus the symmetric case of Theorem 3.69 applies when $\beta^{\otimes} = 1$ in $\mathcal{D}$.

We remind the reader of our left normalized bracketing Convention 1.24 for iterated monoidal product.

8.1 Essential Surjectivity on Objects

This section contains the first main step toward showing that the Grothendieck construction $\int_{\mathcal{D}}$ in Theorem 7.64 (1) is a non-symmetric Cat-multiequivalence. We show that $\int_{\mathcal{D}}$ is essentially surjective on objects between the underlying 1-ary categories. To clarify the presentation, we separate the proof of essential surjectivity into the following two parts:

1. Lemma 8.2 shows that any split opfibration $P : \mathsf{E} \longrightarrow \mathsf{B}$ can be realized by a Grothendieck construction as $U_X : \int_{\mathsf{B}} X \longrightarrow \mathsf{B}$, up to an isomorphism $\varphi : \int_{\mathsf{B}} X \longrightarrow \mathsf{E}$ that preserves both opcartesian morphisms and chosen opcartesian lifts. This step is only about split opfibrations and does not involve any permutative structure.

2. The essential surjectivity of $\int_{\mathcal{D}}$ is proved in Lemma 8.9. For a small permutative opfibration $P : \mathsf{E} \longrightarrow \mathcal{D}$, we first equip $X : \mathcal{D} \longrightarrow \mathsf{Cat}$ in Lemma 8.2 with the structure of an additive symmetric monoidal functor $\mathcal{D}^{\oplus} \longrightarrow \mathsf{Cat}$. Then we show that φ is an opcartesian 1-linear isomorphism in $\mathsf{PFib}^{\mathcal{D}}$.

As stated in the first paragraph of Section 7.1, results in [JY21, Ch. 9 and 10] about fibrations and the Grothendieck construction are applicable in our current context of opfibrations by considering the opposite functor between the opposite categories.

Lemma 8.2. *Suppose $P : \mathsf{E} \longrightarrow \mathsf{B}$ is a split opfibration. Then there exist*

- *a functor $X : \mathsf{B} \longrightarrow \mathsf{Cat}$ and*
- *an isomorphism of categories*

$$\int_{\mathsf{B}} X \xrightarrow[\cong]{\varphi} \mathsf{E} \tag{8.3}$$

that

(i) makes the diagram

$$\begin{array}{ccc} \int_{\mathsf{B}} X & \xrightarrow[\cong]{\varphi} & \mathsf{E} \\ & {\scriptstyle U_X} \searrow \quad \swarrow {\scriptstyle P} & \\ & \mathsf{B} & \end{array}$$

commutative and

(ii) preserves opcartesian morphisms and chosen opcartesian lifts.

Proof. First we construct the functor $X : \mathsf{B} \longrightarrow \mathsf{Cat}$.

Objects. Given an object $a \in \mathsf{B}$, define the subcategory

$$Xa = P^{-1}(a) \subset \mathsf{E}. \tag{8.4}$$

In other words,

- an object $x \in Xa$ is an object in E such that $Px = a$, and
- a morphism $p : x \longrightarrow y$ in Xa is a morphism in E such that $Pp = 1_a$ in B.

Morphisms. Given a morphism $f : a \longrightarrow b$ in B, the functor

$$P^{-1}(a) = Xa \xrightarrow{Xf} Xb = P^{-1}(b)$$

is defined as follows.

- Each object $y \in P^{-1}(a)$ is sent to the codomain

$$(Xf)y = y_f \in P^{-1}(b) \tag{8.5}$$

of the *chosen* opcartesian lift $\overline{f} : y \longrightarrow y_f$ of the fore-lift $\langle y\,; f\rangle$ with respect to P (Definition 7.3 (7)).

- Each morphism $p : y \longrightarrow y'$ in $P^{-1}(a)$ is sent to the *unique* raise

$$(Xf)p = p_f : y_f \longrightarrow y'_f \quad \text{in} \quad P^{-1}(b) \tag{8.6}$$

of the fore-raise $\langle \overline{f}, \overline{f}'p; 1_b\rangle$ (Definition 7.3 (4)), as depicted below.

$$\begin{array}{ccc} y'_f & \xleftarrow{\overline{f}'} & y' \\ {\scriptstyle p_f} \uparrow & & \uparrow {\scriptstyle p} \\ y_f & \xleftarrow{\overline{f}} & y \end{array} \qquad \stackrel{P}{\longmapsto} \qquad \begin{array}{ccc} b & \xleftarrow{f} & a \\ {\scriptstyle 1_b} \uparrow & & \uparrow {\scriptstyle 1_a} \\ b & \xleftarrow{f} & a \end{array}$$

Here $\overline{f}$ and $\overline{f}'$ are the chosen opcartesian lifts of, respectively, $\langle y\,;f\rangle$ and $\langle y'\,;f\rangle$. The unique existence of p_f follows from the fact that $\overline{f}$ is an opcartesian morphism.

Functoriality. The functoriality of $X : \mathsf{B} \longrightarrow \mathsf{Cat}$ follows from

- [JY21, 10.4.7], which shows that X is a strictly unitary pseudofunctor, and
- the assumption that P is a *split* opfibration, which forces the lax functoriality constraint of X to be the identity.

Moreover, by [JY21, 10.1.11] the first-factor projection functor (7.57)

$$\int_{\mathsf{B}} X \xrightarrow{U_X} \mathsf{B}$$

is a split opfibration.

The Functor $\varphi : \int_{\mathsf{B}} X \longrightarrow \mathsf{E}$. For each object $(a,x) \in \int_{\mathsf{B}} X$ with $a \in \mathsf{B}$ and $x \in P^{-1}(a)$, define the object

$$\varphi(a,x) = x \in \mathsf{E}. \tag{8.7}$$

Suppose given a morphism

$$(a,x) \xrightarrow{(f,p)} (b,y) \quad \text{in} \quad \int_{\mathsf{B}} X$$

with

- $f : a \longrightarrow b$ in B,
- $p : (Xf)x = x_f \longrightarrow y$ in $P^{-1}(b)$, and
- $\overline{f} : x \longrightarrow x_f$ the chosen opcartesian lift of $\langle x\,;f\rangle$ with respect to P.

Define $\varphi(f,p)$ as the following composite in E.

$$\begin{array}{ccccc} \varphi(a,x) & & \xrightarrow{\quad\varphi(f,p)\quad} & & \varphi(b,y) \\ \| & & & & \| \\ x & \xrightarrow{\overline{f}} & x_f & \xrightarrow{p} & y \end{array} \tag{8.8}$$

Properties of φ. The assignments (8.7) and (8.8) have the following properties.

- By [JY21, 10.4.12] $\varphi : \int_{\mathsf{B}} X \longrightarrow \mathsf{E}$ is a functor such that
$$U_X = P\varphi : \int_{\mathsf{B}} X \longrightarrow \mathsf{B}.$$
- By [JY21, 10.4.23] φ is an isomorphism and preserves opcartesian morphisms.

To see that φ preserves chosen opcartesian lifts, consider a fore-lift $\langle (a,x)\,;f\rangle$ with respect to U_X. Its chosen opcartesian lift is (7.58)

$$(a,x) \xrightarrow{(f,1_{x_f})} (b, f_*x = x_f) \quad \text{in} \quad \textstyle\int_{\mathsf{B}} X.$$

By (8.8) its image under φ is

$$\varphi(f, 1_{x_f}) = 1_{x_f} \circ \overline{f} = \overline{f} \quad \text{in} \quad \mathsf{E}.$$

This is the chosen opcartesian lift of $\langle \varphi(a,x) = x\,; f\rangle$ with respect to P. □

Restricting to 1-ary multimorphism categories, each non-symmetric Cat-multicategory has an underlying 2-category (Example 3.18 and Proposition 1.35). By the same restriction, a non-symmetric Cat-multifunctor has an underlying 2-functor. Next we observe that the underlying 2-functor of the Grothendieck construction $\int_{\mathcal{D}}$ is essentially surjective on objects.

Lemma 8.9. *Suppose $\mathcal{D}$ is a small tight bipermutative category. Then the non-symmetric* Cat*-multifunctor in* (7.65)

$$\mathcal{D}\mathsf{Cat} \xrightarrow{\int_{\mathcal{D}}} \mathsf{PFib}^{\mathcal{D}}$$

is essentially surjective on objects.

Proof. Suppose given an object in $\mathsf{PFib}^{\mathcal{D}}$, that is, a small permutative opfibration (Definition 7.6)

$$(\mathsf{E}, \oplus, e, \beta) \xrightarrow{P} \mathcal{D}^{\oplus} = (\mathcal{D}, \oplus, 0, \beta^{\oplus}).$$

We must show that there exist

- an object in $\mathcal{D}\mathsf{Cat}$, that is, an additive symmetric monoidal functor (Definition 5.31)

  $$(X, X^2, X^0) : \mathcal{D}^{\oplus} \longrightarrow \mathsf{Cat}$$

  and

- an invertible opcartesian 1-linear functor in $\mathsf{PFib}^{\mathcal{D}}$ (Definition 7.11)

  $$\Big(\textstyle\int_{\mathcal{D}} X \xrightarrow{U_X} \mathcal{D}\Big) \xrightarrow[\cong]{\varphi} \Big(\mathsf{E} \xrightarrow{P} \mathcal{D}\Big).$$

By Example 7.24 φ is precisely a strict symmetric monoidal isomorphism (7.8)

$$\textstyle\int_{\mathcal{D}} X \xrightarrow[\cong]{\varphi} \mathsf{E}$$

that

- satisfies $U_X = P\varphi$ and
- preserves chosen opcartesian lifts.

We use Lemma 7.56 and the functors in Lemma 8.2 with $\mathsf{B} = \mathcal{D}$, namely

$$\mathcal{D} \xrightarrow{X} \mathsf{Cat} \quad \text{and} \quad \int_{\mathcal{D}} X \xrightarrow[\cong]{\varphi} \mathsf{E}.$$

It remains to

- equip the functor X, defined in (8.4) through (8.6), with the structure (X^2, X^0) of a symmetric monoidal functor and
- show that the functor φ, defined in (8.7) and (8.8), is strict symmetric monoidal.

To clarify the presentation, we divide the rest of this proof into several steps, starting with the constructions of X^0 and X^2.

Step 1: The Unit Constraint of X

Define the unit constraint

$$\mathbf{1} \xrightarrow{X^0} X\mathsf{0} = P^{-1}(\mathsf{0}) \subset \mathsf{E}$$

by the unit object (5.33)

$$X^0 * = e \in \mathsf{E}. \tag{8.10}$$

This is well defined because P is strict symmetry monoidal (7.8), which includes the property

$$Pe = \mathsf{0} \quad \text{in} \quad \mathcal{D}.$$

Step 2: The Monoidal Constraint of X

For objects $a, b \in \mathcal{D}$ the (a,b)-component functor of the monoidal constraint X^2 (5.34),

$$P^{-1}(a) \times P^{-1}(b) = Xa \times Xb \xrightarrow{X^2_{a,b}} X(a \oplus b) = P^{-1}(a \oplus b), \tag{8.11}$$

is defined by

$$X^2_{a,b}(x,y) = x \oplus y \in \mathsf{E} \tag{8.12}$$

for both objects and morphisms $(x,y) \in P^{-1}(a) \times P^{-1}(b)$.

- The assignment $X^2_{a,b}$ is well defined because P strictly preserves the monoidal product (7.8).
- The functoriality of $X^2_{a,b}$ follows from the functoriality of $\oplus$ in E.

Step 3: The Naturality of the Monoidal Constraint

For morphisms

$$a \xrightarrow{f} b \quad \text{and} \quad a' \xrightarrow{f'} b' \quad \text{in } \mathcal{D},$$

the naturality diagram for X^2 (8.11) is as follows, with $(-)_* = X(-)$.

$$\begin{array}{ccc} P^{-1}(a)\times P^{-1}(a') & \xrightarrow{\oplus} & P^{-1}(a\oplus a') \\ {\scriptstyle f_*\times f'_*}\big\downarrow & & \big\downarrow{\scriptstyle (f\oplus f')_*} \\ P^{-1}(b)\times P^{-1}(b') & \xrightarrow{\oplus} & P^{-1}(b\oplus b') \end{array} \tag{8.13}$$

Objects. The following equalities in E show that the diagram (8.13) commutes for any pair of objects $(x, x') \in P^{-1}(a) \times P^{-1}(a')$.

$$\begin{aligned} (f\oplus f')_*(x\oplus x') &= (x\oplus x')_{f\oplus f'} && \text{by (8.5)} \\ &= x_f \oplus x'_{f'} && \text{by Definition 7.6 (2)} \\ &= f_*x \oplus f'_*x' && \text{by (8.5)} \end{aligned}$$

In the computation above, we remind the reader of our notation:

- $\overline{f} : x \longrightarrow x_f$ is the chosen opcartesian lift of $\langle x\,; f\rangle$ with respect to P.
- $\overline{f}' : x' \longrightarrow x'_{f'}$ is the chosen opcartesian lift of $\langle x'\,; f'\rangle$.
- $\overline{f\oplus f'} = \overline{f} \oplus \overline{f}' : x\oplus x' \longrightarrow (x\oplus x')_{f\oplus f'}$ is the chosen opcartesian lift of $\langle x\oplus x'\,; f\oplus f'\rangle$, in which the equality follows from axiom 2 in Definition 7.6.

Morphisms. To check the commutativity of (8.13) on morphisms, suppose given a pair of morphisms

$$x \xrightarrow{p} y \in P^{-1}(a) \quad \text{and} \quad x' \xrightarrow{p'} y' \in P^{-1}(a'). \tag{8.14}$$

By Definition 7.6 (2) and (8.6), the morphism in $P^{-1}(b\oplus b')$

$$(f\oplus f')_*(p\oplus p') = (p\oplus p')_{f\oplus f'} : (x\oplus x')_{f\oplus f'} \longrightarrow (y\oplus y')_{f\oplus f'}$$

is the *unique* raise of the fore-raise

$$\langle \overline{f}\oplus\overline{f}', (\overline{f}\oplus\overline{f}')(p\oplus p'); 1_{b\oplus b'}\rangle \tag{8.15}$$

as depicted below.

$$\begin{array}{ccc} (y\oplus y')_{f\oplus f'} = y_f\oplus y'_{f'} & \xleftarrow{\overline{f}\oplus\overline{f}'} & y\oplus y' \\ {\scriptstyle (p\oplus p')_{f\oplus f'}}\,\vdots\!\uparrow & & \big\uparrow{\scriptstyle p\oplus p'} \\ (x\oplus x')_{f\oplus f'} = x_f\oplus x'_{f'} & \xleftarrow{\overline{f}\oplus\overline{f}'} & x\oplus x' \end{array} \quad \overset{P}{\longmapsto} \quad \begin{array}{ccc} b\oplus b' & \xleftarrow{f\oplus f'} & a\oplus a' \\ {\scriptstyle 1_{b\oplus b'}}\big\uparrow & & \big\uparrow{\scriptstyle 1_{a\oplus a'}} \\ b\oplus b' & \xleftarrow{f\oplus f'} & a\oplus a' \end{array}$$

By (8.6) for p_f and $p'_{f'}$ the sum

$$p_f \oplus p'_{f'} = f_* p \oplus f'_* p'$$

is also a raise of the fore-raise (8.15). So uniqueness implies that

$$f_* p \oplus f'_* p' = (f \oplus f')_* (p \oplus p').$$

This proves that the diagram (8.13) is commutative.

Step 4: Symmetric Monoidal Functor Axioms

Next we check that (X, X^2, X^0) is a symmetric monoidal functor.

Unity. Since $\mathcal{D}^{\oplus}$ is a permutative category, by definitions (8.10) and (8.12), the unity diagrams (1.18) are as follows.

$$\begin{array}{ccc} \mathbf{1} \times P^{-1}(b) & \xrightarrow{\cong} & P^{-1}(b) \\ {\scriptstyle e \times 1}\downarrow & & \| \\ P^{-1}(0) \times P^{-1}(b) & \xrightarrow{\oplus} & P^{-1}(0 \oplus b) \end{array} \qquad \begin{array}{ccc} P^{-1}(a) \times \mathbf{1} & \xrightarrow{\cong} & P^{-1}(a) \\ {\scriptstyle 1 \times e}\downarrow & & \| \\ P^{-1}(a) \times P^{-1}(0) & \xrightarrow{\oplus} & P^{-1}(a \oplus 0) \end{array}$$

These two diagrams commute because e is the monoidal unit in the permutative category E.

Associativity. The associativity diagram (1.19) of (X, X^2) is as follows for objects $a, b, c \in \mathcal{D}$.

$$\begin{array}{ccc} (P^{-1}(a) \times P^{-1}(b)) \times P^{-1}(c) & \xrightarrow{\cong} & P^{-1}(a) \times (P^{-1}(b) \times P^{-1}(c)) \\ {\scriptstyle \oplus \times 1}\downarrow & & \downarrow{\scriptstyle 1 \times \oplus} \\ P^{-1}(a \oplus b) \times P^{-1}(c) & & P^{-1}(a) \oplus P^{-1}(b \oplus c) \\ {\scriptstyle \oplus}\downarrow & & \downarrow{\scriptstyle \oplus} \\ P^{-1}(a \oplus b \oplus c) & = & P^{-1}(a \oplus b \oplus c) \end{array}$$

This diagram commutes because $\oplus$ in E is strictly associative.

Braiding. Denoting by

- $\beta^{\oplus} : a \oplus a' \longrightarrow a' \oplus a$ the braiding in $\mathcal{D}$,
- $\beta^{\oplus}_* = X\beta^{\oplus} : P^{-1}(a \oplus a') \longrightarrow P^{-1}(a' \oplus a)$ as in (8.4) through (8.6), and
- ξ the braiding in $(\mathsf{Cat}, \times)$,

the compatibility diagram (1.20) of (X, X^2) with the braidings is as follows.

$$\begin{array}{ccc} P^{-1}(a) \times P^{-1}(a') & \xrightarrow{\oplus} & P^{-1}(a \oplus a') \\ {\scriptstyle \xi}\downarrow & & \downarrow{\scriptstyle \beta^{\oplus}_*} \\ P^{-1}(a') \times P^{-1}(a) & \xrightarrow{\oplus} & P^{-1}(a' \oplus a) \end{array} \tag{8.16}$$

The following equalities in E show that the diagram (8.16) commutes for any pair of objects $(x, x') \in P^{-1}(a) \times P^{-1}(a')$.

$$\begin{aligned} \beta_*^{\oplus}(x \oplus x') &= (x \oplus x')_{\beta^{\oplus}} && \text{by (8.5)} \\ &= x' \oplus x && \text{by Definition 7.6 (3)} \\ &= (\oplus \circ \xi)(x, x') \end{aligned}$$

To check the commutativity of (8.16) on morphisms, suppose given a pair of morphisms $(p, p') \in P^{-1}(a) \times P^{-1}(a')$ as in (8.14). By Definition 7.6 (3) and (8.6), the morphism in $P^{-1}(a' \oplus a)$

$$\beta_*^{\oplus}(p \oplus p') = (p \oplus p')_{\beta^{\oplus}} : (x \oplus x')_{\beta^{\oplus}} \longrightarrow (y \oplus y')_{\beta^{\oplus}}$$

is the *unique* raise of the fore-raise

$$\left\langle \beta_{x,x'}, \beta_{y,y'}(p \oplus p'); 1_{a' \oplus a} \right\rangle \tag{8.17}$$

as depicted below.

$$\begin{array}{ccc} (y \oplus y')_{\beta^{\oplus}} = y' \oplus y & \xleftarrow{\beta_{y,y'}} & y \oplus y' \\ {\scriptstyle (p \oplus p')_{\beta^{\oplus}}} \uparrow & & \uparrow {\scriptstyle p \oplus p'} \\ (x \oplus x')_{\beta^{\oplus}} = x' \oplus x & \xleftarrow{\beta_{x,x'}} & x \oplus x' \end{array} \quad \overset{P}{\longmapsto} \quad \begin{array}{ccc} a' \oplus a & \xleftarrow{\beta^{\oplus}_{a,a'}} & a \oplus a' \\ {\scriptstyle 1_{a' \oplus a}} \uparrow & & \uparrow {\scriptstyle 1_{a \oplus a'}} \\ a' \oplus a & \xleftarrow{\beta^{\oplus}_{a,a'}} & a \oplus a' \end{array}$$

By the naturality of β, the sum $p' \oplus p$ is also a raise of the fore-raise (8.17). So uniqueness implies that

$$\beta_*^{\oplus}(p \oplus p') = (p \oplus p')_{\beta^{\oplus}} = p' \oplus p.$$

This proves that the diagram (8.16) is commutative. We have shown that (X, X^2, X^0) is a symmetric monoidal functor.

Step 5: Strict Symmetric Monoidality of φ

Next we check that the categorical isomorphism

$$\left(\textstyle\int_{\mathcal{D}} X, \square, (0, e), \beta^{\square} \right) \xrightarrow[\cong]{\varphi} (\mathsf{E}, \oplus, e, \beta),$$

defined in (8.7) and (8.8), is a strict symmetric monoidal functor as in (7.8).

Monoidal Unit. The functor φ preserves the monoidal unit because, by definition (8.7),

$$\varphi(0, e) = e \in \mathsf{E}.$$

Objects. The functor φ preserves the monoidal product on objects by the following equalities in E for objects (a, x) and $(b, y) \in \int_{\mathcal{D}} X$.

$$\begin{aligned} \varphi\big((a, x) \square (b, y)\big) &= \varphi(a \oplus b, x \oplus y) && \text{by (6.39) and (8.12)} \\ &= x \oplus y && \text{by (8.7)} \\ &= \varphi(a, x) \oplus \varphi(b, y) && \text{by (8.7)} \end{aligned}$$

Morphisms. To show that φ preserves the monoidal product on morphisms, suppose given two morphisms in $\int_{\mathcal{D}} X$,

$$(a,x) \xrightarrow{(f,p)} (a',x') \quad \text{and} \quad (b,y) \xrightarrow{(g,q)} (b',y'),$$

with

- $f : a \longrightarrow a'$ and $g : b \longrightarrow b'$ in $\mathcal{D}$,
- $p : f_* x = x_f \longrightarrow x'$ in $Xa' = P^{-1}(a')$ by (8.4) and (8.5), and
- $q : g_* y = y_g \longrightarrow y'$ in $Xb' = P^{-1}(b')$.

Here

$$x \xrightarrow{\overline{f}} x_f \quad \text{and} \quad y \xrightarrow{\overline{g}} y_g \quad \text{in} \quad \mathsf{E}$$

are the chosen opcartesian lifts of, respectively, the fore-lifts $\langle x\,; f\rangle$ and $\langle y\,; g\rangle$ with respect to P. The functor φ preserves the monoidal product on morphisms by the following morphism equalities in E.

$$\begin{aligned}
\varphi\big((f,p) \,\square\, (g,q)\big) &= \varphi(f \oplus g, p \oplus q) && \text{by (6.40) and (8.12)}\\
&= (p \oplus q)\overline{(f \oplus g)} && \text{by (8.8)}\\
&= (p \oplus q)(\overline{f} \oplus \overline{g}) && \text{by Definition 7.6 (2)}\\
&= p\overline{f} \oplus q\overline{g} && \text{by functoriality of } \oplus\\
&= \varphi(f,p) \oplus \varphi(g,q) && \text{by (8.8)}
\end{aligned}$$

Braiding. The functor φ preserves the braiding by the following morphism equalities in E.

$$\begin{aligned}
\varphi\beta^{\square}_{(a,x),(b,y)} &= \varphi(\beta^{\oplus}_{a,b}, 1_{y\oplus x}) && \text{by (6.41) and (8.12)}\\
&= 1_{y\oplus x}\overline{\beta^{\oplus}_{a,b}} && \text{by (8.8)}\\
&= \beta_{x,y} && \text{by Definition 7.6 (3)}\\
&= \beta_{\varphi(a,x),\varphi(b,y)} && \text{by (8.7)}
\end{aligned}$$

Therefore, we may define the unit constraint φ^0 and the monoidal constraint φ^2 of φ to be the identities, and $(\varphi, \varphi^2, \varphi^0)$ is a strict symmetric monoidal functor. □

8.2 Faithfulness on Multimorphism Categories

For the rest of this chapter, suppose

$$(Z, Z^2, Z^0) \quad \text{and} \quad \big\langle (X_j, X_j^2, X_j^0)\big\rangle_{j=1}^{n} : \mathcal{D}^{\oplus} \longrightarrow \mathsf{Cat} \tag{8.18}$$

are additive symmetric monoidal functors (Definition 5.31) with $n \geq 0$. We use the notation in (7.60), so

$$\langle X \rangle = \langle X_j \rangle_{j=1}^n, \quad \langle \textstyle\int_{\mathcal{D}} X \rangle = \langle \int_{\mathcal{D}} X_j \rangle_{j=1}^n, \quad \text{and} \quad \langle U_X \rangle = \langle U_{X_j} \rangle_{j=1}^n \tag{8.19}$$

with each $U_{X_j} : \int_{\mathcal{D}} X_j \longrightarrow \mathcal{D}$ the small permutative opfibration in Lemma 7.56.

By Lemma 7.61 the arity 0 multimorphism functor is an isomorphism

$$Z1 = \mathcal{D}\mathsf{Cat}(\langle\rangle\,;\,Z) \xrightarrow[\cong]{\int_{\mathcal{D}}} \mathsf{PFib}^{\mathcal{D}}(\langle\rangle\,;\,U_Z) = U_Z^{-1}(1)$$

with $U_Z : \int_{\mathcal{D}} Z \longrightarrow \mathcal{D}$ (Lemma 7.56). Our next objective is to show that the multimorphism functors of $\int_{\mathcal{D}}$ are also isomorphisms for positive arity inputs. In this section, we show that they are injective on objects and morphisms; see Lemmas 8.20 and 8.28. By Definition 5.58 the objects and morphisms in $\mathcal{D}\mathsf{Cat}(\langle X \rangle\,;\,Z)$ are, respectively, additive natural transformations (5.40) and additive modifications (5.53).

Lemma 8.20. *For $n > 0$ the functor from Lemma 7.62*

$$\mathcal{D}\mathsf{Cat}(\langle X \rangle\,;\,Z) \xrightarrow{\int_{\mathcal{D}}} \mathsf{PFib}^{\mathcal{D}}(\langle U_X \rangle\,;\,U_Z)$$

is injective on objects.

Proof. Suppose given additive natural transformations (5.40)

$$\big\langle (X_j, X_j^2, X_j^0) \big\rangle_{j=1}^n \overset{\phi}{\underset{\psi}{\rightrightarrows}} (Z, Z^2, Z^0) \tag{8.21}$$

such that the following two strong n-linear functors (Lemma 6.57) are equal.

$$\textstyle\prod_{j=1}^n (\int_{\mathcal{D}} X_j) \overset{(\int_{\mathcal{D}}\phi\,,\,\langle(\int_{\mathcal{D}}\phi)_i^2\rangle_{i=1}^n)}{\underset{(\int_{\mathcal{D}}\psi\,,\,\langle(\int_{\mathcal{D}}\psi)_i^2\rangle_{i=1}^n)}{\rightrightarrows}} \int_{\mathcal{D}} Z \tag{8.22}$$

We must show that $\phi = \psi$. These additive natural transformations consist of component functors (5.7)

$$\textstyle\prod_{j=1}^n X_j a_j \overset{\phi_{\langle a \rangle}}{\underset{\psi_{\langle a \rangle}}{\rightrightarrows}} Za \tag{8.23}$$

for each n-tuple of objects

$$\langle a \rangle = \langle a_j \rangle_{j=1}^n \in \mathcal{D}^n \quad \text{with} \quad a = \otimes_{j=1}^n a_j \in \mathcal{D}. \tag{8.24}$$

We must show that $\phi_{\langle a \rangle} = \psi_{\langle a \rangle}$ as functors.

Objects. For each n-tuple of objects

$$\langle x\rangle = \langle x_j\rangle_{j=1}^n \in \textstyle\prod_{j=1}^n X_j a_j \tag{8.25}$$

the desired equality,

$$\phi_{\langle a\rangle}\langle x\rangle = \psi_{\langle a\rangle}\langle x\rangle \quad\text{in}\quad Za,$$

follows from the following equalities in $\int_{\mathcal{D}} Z$.

$$\begin{aligned}\big(a\,,\,\phi_{\langle a\rangle}\langle x\rangle\big) &= (\textstyle\int_{\mathcal{D}}\phi)\langle(a_j,x_j)\rangle && \text{by (6.49)}\\ &= (\textstyle\int_{\mathcal{D}}\psi)\langle(a_j,x_j)\rangle && \text{by (8.22)}\\ &= \big(a\,,\,\psi_{\langle a\rangle}\langle x\rangle\big) && \text{by (6.49)}\end{aligned}$$

Morphisms. For each n-tuple of morphisms

$$\langle p\rangle = \big\langle p_j : x_j \longrightarrow y_j\big\rangle_{j=1}^n \in \textstyle\prod_{j=1}^n X_j a_j \tag{8.26}$$

consider the n-tuple of morphisms

$$\big\langle (1_{a_j}, p_j) : (a_j, x_j) \longrightarrow (a_j, y_j)\big\rangle_{j=1}^n \in \textstyle\prod_{j=1}^n (\int_{\mathcal{D}} X_j).$$

The desired equality,

$$\phi_{\langle a\rangle}\langle p\rangle = \psi_{\langle a\rangle}\langle p\rangle \quad\text{in}\quad Za,$$

follows from the following morphism equalities in $\int_{\mathcal{D}} Z$.

$$\begin{aligned}\big(1_a\,,\,\phi_{\langle a\rangle}\langle p\rangle\big) &= (\textstyle\int_{\mathcal{D}}\phi)\langle(1_{a_j},p_j)\rangle && \text{by (6.52)}\\ &= (\textstyle\int_{\mathcal{D}}\psi)\langle(1_{a_j},p_j)\rangle && \text{by (8.22)}\\ &= \big(1_a\,,\,\psi_{\langle a\rangle}\langle p\rangle\big) && \text{by (6.52)}\end{aligned}$$

This proves that $\phi_{\langle a\rangle} = \psi_{\langle a\rangle}$ in (8.23). □

Remark 8.27. The proof of Lemma 8.20 uses only the assignments of $\int_{\mathcal{D}}\phi$ and $\int_{\mathcal{D}}\psi$ on objects and morphisms of the form $\langle(1_{a_j}, p_j)\rangle$. Their linearity constraints, $\langle(\int_{\mathcal{D}}\phi)_i^2\rangle_{i=1}^n$ and $\langle(\int_{\mathcal{D}}\psi)_i^2\rangle_{i=1}^n$, are not used in the proof. ◇

Lemma 8.28. *For $n > 0$ the functor from Lemma 7.62*

$$\mathcal{D}\mathsf{Cat}\big(\langle X\rangle; Z\big) \xrightarrow{\;\int_{\mathcal{D}}\;} \mathsf{PFib}^{\mathcal{D}}\big(\langle U_X\rangle; U_Z\big)$$

is injective on morphism sets.

Proof. Suppose given additive natural transformations ϕ and ψ as in (8.21) and additive modifications (5.53)

$$\big\langle (X_j, X_j^2, X_j^0)\big\rangle_{j=1}^n \underset{\psi}{\overset{\phi}{\rightrightarrows}}\ \Phi\Downarrow\ \Downarrow\Theta\ \ (Z, Z^2, Z^0) \tag{8.29}$$

such that the following two n-linear natural transformations (6.62) are equal.

$$\prod_{j=1}^{n}(\textstyle\int_{\mathcal{D}} X_j) \underset{\int_{\mathcal{D}}\psi}{\overset{\int_{\mathcal{D}}\phi}{\rightrightarrows}} \textstyle\int_{\mathcal{D}} Z \qquad \int_{\mathcal{D}}\Phi \Downarrow \ \Downarrow \int_{\mathcal{D}}\Theta \tag{8.30}$$

We must show that $\Phi = \Theta$.

With the notation in (8.23) and (8.24), Φ and Θ consist of the following component natural transformations (5.11).

$$\prod_{j=1}^{n} X_j a_j \underset{\psi_{\langle a\rangle}}{\overset{\phi_{\langle a\rangle}}{\rightrightarrows}} Za \qquad \Phi_{\langle a\rangle} \Downarrow \ \Downarrow \Theta_{\langle a\rangle} \tag{8.31}$$

For each n-tuple of objects $\langle x\rangle \in \prod_{j=1}^{n} X_j a_j$ as in (8.25), the desired equality,

$$(\Phi_{\langle a\rangle})_{\langle x\rangle} = (\Theta_{\langle a\rangle})_{\langle x\rangle} : \phi_{\langle a\rangle}\langle x\rangle \longrightarrow \psi_{\langle a\rangle}\langle x\rangle \quad \text{in} \quad Za,$$

follows from the following morphism equalities in $\int_{\mathcal{D}} Z$.

$$\begin{aligned} \big(1_a\,,\,(\Phi_{\langle a\rangle})_{\langle x\rangle}\big) &= \big(\textstyle\int_{\mathcal{D}}\Phi\big)_{\langle (a_j, x_j)\rangle} && \text{by (6.63)} \\ &= \big(\textstyle\int_{\mathcal{D}}\Theta\big)_{\langle (a_j, x_j)\rangle} && \text{by (8.30)} \\ &= \big(1_a\,,\,(\Theta_{\langle a\rangle})_{\langle x\rangle}\big) && \text{by (6.63)} \end{aligned}$$

This proves that $\Phi_{\langle a\rangle} = \Theta_{\langle a\rangle}$ in (8.31). □

8.3 Object Fullness on Multimorphism Categories

We continue to use the notation in (8.18), (8.19), and (8.24) through (8.26) with $n > 0$. In Lemmas 8.20 and 8.28 we observe that the component functors (Lemma 7.62)

$$\mathcal{D}\mathsf{Cat}(\langle X\rangle; Z) \xrightarrow{\int_{\mathcal{D}}} \mathsf{PFib}^{\mathcal{D}}(\langle U_X\rangle; U_Z)$$

are injective on objects and morphism sets. In this section we show that these component functors are surjective on objects. Recall from Definition 7.34 that

$$\mathsf{PFib}^{\mathcal{D}}(\langle U_X\rangle; U_Z) \subset \mathsf{PermCat}^{\mathsf{sg}}(\langle \textstyle\int_{\mathcal{D}} X\rangle; \int_{\mathcal{D}} Z)$$

is the subcategory with

- opcartesian strong n-linear functors (Definition 7.11) as objects and
- opcartesian n-linear transformations (Definition 7.27) as morphisms.

Here is an outline of this section.

- In Definition 8.32 we associate to each opcartesian strong n-linear functor F in $\mathsf{PFib}^{\mathcal{D}}\big(\langle U_X\rangle\,;\,U_Z\big)$ the data of an additive natural transformation ϕ in $\mathcal{D}\mathsf{Cat}\big(\langle X\rangle\,;\,Z\big)$. The projection axiom (7.13) plays a crucial role in this definition, as we discuss in Explanation 8.41.
- Lemmas 8.43 and 8.44 show that ϕ is a natural transformation (5.5).
- Lemmas 8.51 and 8.53 show that ϕ is an *additive* natural transformation (Definition 5.39).
- Lemma 8.57 shows that $\int_{\mathcal{D}}\phi = F$ as strong n-linear functors. This is sufficient to show the object surjectivity of the component functors of $\int_{\mathcal{D}}$ because an opcartesian n-linear functor is an n-linear functor with extra properties but no extra structure (Explanation 7.18 (1)).

Definition 8.32. Suppose given an opcartesian strong n-linear functor

$$\langle U_X\rangle \xrightarrow{(F\,,\,\langle F_j^2\rangle_{j=1}^n)} U_Z. \tag{8.33}$$

This means a strong n-linear functor (Definition 6.4)

$$\textstyle\prod_{j=1}^n\big(\int_{\mathcal{D}}X_j\big) \xrightarrow{\left(F\,,\,\langle F_j^2\rangle_{j=1}^n\right)} \int_{\mathcal{D}}Z \tag{8.34}$$

that satisfies axioms 1 through 4 in Definition 7.11. Define the data of an additive natural transformation (5.40)

$$\big\langle (X_j, X_j^2, X_j^0)\big\rangle_{j=1}^n \xrightarrow{\phi} (Z, Z^2, Z^0) \tag{8.35}$$

with component functors

$$\textstyle\prod_{j=1}^n X_j a_j \xrightarrow{\phi_{\langle a\rangle}} Za \tag{8.36}$$

as follows, where we use the notation in (8.24) through (8.26).

Objects For each n-tuple of objects $\langle x\rangle \in \prod_{j=1}^n X_j a_j$, define the object

$$\phi_{\langle a\rangle}\langle x\rangle = F_2\langle x\rangle \quad \text{in} \quad Za \tag{8.37}$$

where $F_2\langle x\rangle$ is the second entry in the object

$$F\langle (a_j, x_j)\rangle = \big(a, F_2\langle x\rangle\big) \quad \text{in} \quad \textstyle\int_{\mathcal{D}}Z. \tag{8.38}$$

Morphisms For each n-tuple of morphisms $\langle p\rangle \in \prod_{j=1}^{n} X_j a_j$, define the morphism

$$\phi_{\langle a\rangle}\langle p\rangle = F_2\langle p\rangle : F_2\langle x\rangle \longrightarrow F_2\langle y\rangle \quad \text{in} \quad Za \tag{8.39}$$

where $F_2\langle p\rangle$ is the second entry in the morphism

$$F\langle(1_{a_j}, p_j)\rangle = \big(1_a, F_2\langle p\rangle\big) \quad \text{in} \quad \textstyle\int_{\mathcal{D}} Z. \tag{8.40}$$

This finishes the definition of ϕ in (8.35). ◇

Explanation 8.41. The projection axiom (7.13) for F in (8.34) is the following commutative diagram of functors.

$$\begin{array}{ccc} \prod_{j=1}^{n}(\int_{\mathcal{D}} X_j) & \xrightarrow{\;F\;} & \int_{\mathcal{D}} Z \\ {\scriptstyle \prod_j U_{X_j}} \big\downarrow & & \big\downarrow {\scriptstyle U_Z} \\ \mathcal{D}^n & \xrightarrow{\;\otimes\;} & \mathcal{D} \end{array} \tag{8.42}$$

Since each $U_?$ (7.57) is the first-factor projection, evaluating the commutative diagram (8.42) at the object

$$\langle(a_j, x_j)\rangle \in \textstyle\prod_{j=1}^{n}(\int_{\mathcal{D}} X_j)$$

yields the equalities

$$U_Z F\langle(a_j, x_j)\rangle = \textstyle\bigotimes_{j=1}^{n} U_{X_j}(a_j, x_j) = \bigotimes_{j=1}^{n} a_j = a.$$

So the first entry of the object

$$F\langle(a_j, x_j)\rangle \in \textstyle\int_{\mathcal{D}} Z$$

is $a \in \mathcal{D}$. This is the meaning of (8.38).

Likewise, evaluating the commutative diagram (8.42) at the morphism

$$\big\langle(1_{a_j}, p_j) : (a_j, x_j) \longrightarrow (a_j, y_j)\big\rangle \in \textstyle\prod_{j=1}^{n}(\int_{\mathcal{D}} X_j)$$

yields the morphism equalities

$$U_Z F\langle(1_{a_j}, p_j)\rangle = \textstyle\bigotimes_{j=1}^{n} U_{X_j}(1_{a_j}, p_j) = \bigotimes_{j=1}^{n} 1_{a_j} = 1_a.$$

So the first entry of the morphism

$$F\langle(1_{a_j}, p_j)\rangle \in \textstyle\int_{\mathcal{D}} Z$$

is 1_a. This is the meaning of (8.40). ◇

In the rest of this section we show that ϕ in (8.35) is an additive natural transformation (Definition 5.39) such that

$$\big(\textstyle\int_{\mathcal{D}}\phi\,,\,\langle(\int_{\mathcal{D}}\phi)^2_j\rangle_{j=1}^{n}\big) = \big(F\,,\,\langle F^2_j\rangle_{j=1}^{n}\big)$$

as strong n-linear functors, with the left-hand side from Lemma 6.57.

Naturality

We begin by showing that each component of ϕ is well defined.

Lemma 8.43. *Each component* $\phi_{\langle a\rangle}$ *in* (8.36) *is a functor.*

Proof. There are morphism equalities in $\int_{\mathcal{D}} Z$ as follows.

$$\begin{aligned}
\big(1_a, F_2\langle 1_{x_j}\rangle\big) &= F\langle (1_{a_j}, 1_{x_j})\rangle && \text{by (8.40)}\\
&= F\langle 1_{(a_j,x_j)}\rangle && \text{by (6.34)}\\
&= 1_{F\langle (a_j,x_j)\rangle} && \text{by functoriality of } F\\
&= 1_{(a,F_2\langle x\rangle)} && \text{by (8.38)}\\
&= \big(1_a, 1_{F_2\langle x\rangle}\big) && \text{by (6.34)}
\end{aligned}$$

So $\phi_{\langle a\rangle}$ preserves identity morphisms by definitions (8.37) and (8.39).

For morphisms

$$x_j \xrightarrow{p_j} y_j \xrightarrow{q_j} z_j \quad \text{in} \quad X_j a_j$$

there are morphism equalities in $\int_{\mathcal{D}} Z$ as follows.

$$\begin{aligned}
\big(1_a, F_2\langle q_j p_j\rangle\big) &= F\langle (1_{a_j}, q_j p_j)\rangle && \text{by (8.40)}\\
&= F\big\langle (1_{a_j}, q_j)\circ(1_{a_j}, p_j)\big\rangle && \text{by (6.35)}\\
&= F\langle (1_{a_j}, q_j)\rangle \circ F\langle (1_{a_j}, p_j)\rangle && \text{by functoriality of } F\\
&= \big(1_a, F_2\langle q\rangle\big)\circ\big(1_a, F_2\langle p\rangle\big) && \text{by (8.40)}\\
&= \big(1_a, F_2\langle q\rangle \circ F_2\langle p\rangle\big) && \text{by (6.35)}
\end{aligned}$$

So $\phi_{\langle a\rangle}$ preserves composition of morphisms by definition (8.39). □

Next we show that

$$\textstyle\bigotimes_{j=1}^n X_j \xrightarrow{\phi} Z$$

is a natural transformation (5.5). Below we write Zf as f_* and $X_j f_j$ as $(f_j)_*$.

Lemma 8.44. *With the notation in* (5.8) *and* (5.9) *and the component functors* $\phi_{\langle a\rangle}$ *in* (8.36), *the following naturality diagram commutes.*

$$\begin{array}{ccc}
\prod_{j=1}^n X_j a_j & \xrightarrow{\phi_{\langle a\rangle}} & Za\\
{\scriptstyle \prod_{j=1}^n (f_j)_*}\big\downarrow & & \big\downarrow{\scriptstyle f_*}\\
\prod_{j=1}^n X_j a'_j & \xrightarrow{\phi_{\langle a'\rangle}} & Za'
\end{array} \tag{8.45}$$

Proof. We check the commutativity of (8.45) first on objects and then on morphisms.

Objects. Suppose given an n-tuple of objects $\langle x\rangle \in \prod_{j=1}^{n} X_j a_j$. Consider the fore-lift with respect to $U_Z : \int_{\mathcal{D}} Z \longrightarrow \mathcal{D}$

$$\big\langle F\langle (a_j, x_j)\rangle \,;\, \otimes_{j=1}^{n} f_j\big\rangle = \big\langle (a, \phi_{\langle a\rangle}\langle x\rangle) \,;\, f\big\rangle \tag{8.46}$$

with the equality from (8.37) and (8.38). There are two ways to obtain its chosen opcartesian lift as follows.

- By definition (7.58) the chosen opcartesian lift of (8.46) is the morphism

$$\big(a\,,\,\phi_{\langle a\rangle}\langle x\rangle\big) \xrightarrow{\big(f\,,\,1_{f_*\phi_{\langle a\rangle}\langle x\rangle}\big)} \big(a'\,,\,f_*\phi_{\langle a\rangle}\langle x\rangle\big) \quad \text{in} \quad \int_{\mathcal{D}} Z. \tag{8.47}$$

- On the other hand, for each $j \in \{1,\ldots,n\}$, the chosen opcartesian lift of the fore-lift $\langle (a_j, x_j)\,;\, f_j\rangle$ with respect to $U_{X_j} : \int_{\mathcal{D}} X_j \longrightarrow \mathcal{D}$ is, by definition (7.58), the morphism

$$(a_j, x_j) \xrightarrow{\big(f_j\,,\,1_{(f_j)_* x_j}\big)} \big(a'_j\,,\,(f_j)_* x_j\big) \quad \text{in} \quad \int_{\mathcal{D}} X_j. \tag{8.48}$$

  Since F preserves chosen opcartesian lifts (Definition 7.11 (3)), the chosen opcartesian lift of (8.46) is also given by

$$\begin{array}{ccc} F\langle (a_j, x_j)\rangle & \xrightarrow{F\langle (f_j, 1)\rangle} & F\big\langle \big(a'_j\,,\,(f_j)_* x_j\big)\big\rangle \\ \| & & \| \\ \big(a\,,\,\phi_{\langle a\rangle}\langle x\rangle\big) & & \big(a'\,,\,\phi_{\langle a'\rangle}\langle (f_j)_* x_j\rangle\big) \end{array} \tag{8.49}$$

  with each equality from (8.37) and (8.38).

The codomains of (8.47) and (8.49) yield the object equality

$$f_*\phi_{\langle a\rangle}\langle x\rangle = \phi_{\langle a'\rangle}\langle (f_j)_* x_j\rangle \quad \text{in} \quad Za'.$$

This proves the commutativity of (8.45) on objects.

Morphisms. Suppose given an n-tuple of morphisms $\langle p\rangle \in \prod_{j=1}^{n} X_j a_j$ as in (8.26). For each $j \in \{1,\ldots,n\}$, the following diagram in $\int_{\mathcal{D}} X_j$ is commutative by (6.35).

$$\begin{array}{ccc} (a_j, x_j) & \xrightarrow{\big(f_j\,,\,1_{(f_j)_* x_j}\big)} & \big(a'_j\,,\,(f_j)_* x_j\big) \\ {\scriptstyle (1_{a_j}, p_j)}\big\downarrow & & \big\downarrow{\scriptstyle (1_{a'_j}\,,\,(f_j)_* p_j)} \\ (a_j, y_j) & \xrightarrow{\big(f_j\,,\,1_{(f_j)_* y_j}\big)} & \big(a'_j\,,\,(f_j)_* y_j\big) \end{array} \tag{8.50}$$

Indeed, in each of the two composites in (8.50),

- the first component is $f_j : a_j \longrightarrow a'_j$ in $\mathcal{D}$, and
- the second component is

$$(f_j)_* x_j \xrightarrow{(f_j)_* p_j} (f_j)_* y_j \quad \text{in} \quad X_j a'_j.$$

In the commutative diagram (8.50), as in (8.48),

- the top $\left(f_j\,,\,1_{(f_j)_* x_j}\right)$ is the chosen opcartesian lift of $\langle (a_j, x_j)\,;f_j\rangle$ and
- the bottom $\left(f_j\,,\,1_{(f_j)_* y_j}\right)$ is the chosen opcartesian lift of $\langle (a_j, y_j)\,;f_j\rangle$

with respect to $U_{X_j} : \int_{\mathcal{D}} X_j \longrightarrow \mathcal{D}$.

Applying F to the commutative diagrams (8.50) for $j \in \{1,\ldots,n\}$ and using (8.39), (8.40), (8.47), and (8.49) yield the following commutative diagram in $\int_{\mathcal{D}} Z$.

$$\begin{array}{ccc} (a\,,\,\phi_{\langle a\rangle}\langle x\rangle) & \xrightarrow{(f\,,\,1_{f_*\phi_{\langle a\rangle}\langle x\rangle})} & (a'\,,\,\phi_{\langle a'\rangle}\langle (f_j)_* x_j\rangle) \\ {\scriptstyle (1_a\,,\,\phi_{\langle a\rangle}\langle p\rangle)}\big\downarrow & & \big\downarrow{\scriptstyle (1_{a'}\,,\,\phi_{\langle a'\rangle}\langle (f_j)_* p_j\rangle)} \\ (a\,,\,\phi_{\langle a\rangle}\langle y\rangle) & \xrightarrow[]{(f\,,\,1_{f_*\phi_{\langle a\rangle}\langle y\rangle})} & (a'\,,\,\phi_{\langle a'\rangle}\langle (f_j)_* y_j\rangle) \end{array}$$

The second components of these two equal composites yield the morphism equality

$$f_*\phi_{\langle a\rangle}\langle p\rangle = \phi_{\langle a'\rangle}\langle (f_j)_* p_j\rangle \quad \text{in} \quad Za'.$$

This proves the commutativity of (8.45) on morphisms. □

Unity and Additivity

Lemmas 8.43 and 8.44 show that

$$\textstyle\bigotimes_{j=1}^n X_j \xrightarrow{\ \phi\ } Z,$$

with component functors $\phi_{\langle a\rangle}$ in (8.36), is a natural transformation as in (5.5). To show that ϕ is an additive natural transformation, next we check the two axioms in Definition 5.39, starting with the unity axiom.

Lemma 8.51. *The natural transformation* (8.36)

$$\textstyle\bigotimes_{j=1}^n X_j \xrightarrow{\ \phi\ } Z$$

satisfies the unity axiom (5.43).

Proof. With $\circ_i$ as in (6.2), the unity axiom asserts the equalities in $Z0$

$$\begin{cases} \phi_{\langle a\rangle \circ_i 0}\big(\langle x\rangle \circ_i (X_i^0 *)\big) = Z^0 * \\ \phi_{\langle a\rangle \circ_i 0}\big(\langle p\rangle \circ_i 1_{X_i^0 *}\big) = 1_{Z^0 *} \end{cases}$$

for objects $\langle x\rangle$ and morphisms $\langle p\rangle$ in $\prod_{j=1}^n X_j a_j$. The unity axiom for objects is proved by the following equalities in $\int_{\mathcal{D}} Z$.

$$\begin{cases} (0, Z^0 *) & \\ = F\big\langle (a_j, x_j) \circ_i (0, X_i^0 *)\big\rangle & \text{by (6.7) and (6.38)} \\ = \big(a_1 \otimes \cdots \otimes 0 \otimes \cdots \otimes a_n\,,\,\phi_{\langle a\rangle \circ_i 0}(\langle x\rangle \circ_i X_i^0 *)\big) & \text{by (8.37) and (8.38)} \\ = \big(0\,,\,\phi_{\langle a\rangle \circ_i 0}(\langle x\rangle \circ_i X_i^0 *)\big) & \text{by (2.31)} \end{cases} \tag{8.52}$$

The unity axiom for morphisms is proved by the computation (8.52) by

- replacing $(0, Z^0*)$, $\langle(a_j, x_j)\rangle$, and $(0, X_i^0*)$ with, respectively,

$$\big(1_0\,,\,1_{Z^0*}\big), \quad \langle(1_{a_j}, p_j)\rangle, \quad \text{and} \quad \big(1_0\,,\,1_{X_i^0*}\big)$$

  and

- using (6.34)

$$\big(1_0\,,\,1_{Z^0*}\big) = 1_{(0,Z^0*)} \quad \text{in} \quad \textstyle\int_{\mathcal{D}} Z$$

  and the fact that F preserves identity morphisms.

This finishes the proof of the unity axiom for ϕ. □

Lemma 8.53. *The natural transformation* (8.36)

$$\textstyle\bigotimes_{j=1}^n X_j \xrightarrow{\ \phi\ } Z$$

satisfies the additivity axiom (5.51).

Proof. We first prove the additivity axiom on objects and then on morphisms.

Objects. Using the notation in (5.44), (5.46), (5.49), and (5.50) for objects, we compute the (co)domain objects in $\int_{\mathcal{D}} Z$ of the i-th linearity constraint, F_i^2, as follows.

$$\begin{cases} F\langle(a_j, x_j)\rangle \,\square\, F\big\langle(a_j, x_j) \circ_i (a_i', x_i')\big\rangle & \\ \quad = \big(a\,,\,\phi_{\langle a\rangle}\langle x\rangle\big) \,\square\, \big(a'\,,\,\phi_{\langle a'\rangle}\langle x'\rangle\big) & \text{by (8.37) and (8.38)} \\ \quad = \big(a \oplus a'\,,\,Z^2_{a,a'}\big(\phi_{\langle a\rangle}\langle x\rangle\,,\,\phi_{\langle a'\rangle}\langle x'\rangle\big)\big) & \text{by (6.39)} \\ F\big\langle(a_j, x_j) \circ_i \big((a_i, x_i) \,\square\, (a_i', x_i')\big)\big\rangle & \\ \quad = F\big\langle(a_j, x_j) \circ_i (a_i''\,,\,x_i'')\big\rangle & \text{by (6.39)} \\ \quad = \big(a''\,,\,\phi_{\langle a''\rangle}\langle x''\rangle\big) & \text{by (8.37) and (8.38)} \end{cases} \tag{8.54}$$

By the constraint lift axiom 4 in Definition 7.11, this component of F_i^2 is the chosen opcartesian lift of the fore-lift

$$\big\langle F\langle(a_j, x_j)\rangle \,\square\, F\big\langle(a_j, x_j) \circ_i (a_i', x_i')\big\rangle\,;\, \mathsf{L}^{-1}\big\rangle$$

with respect to $U_Z : \int_{\mathcal{D}} Z \longrightarrow \mathcal{D}$. By (7.58) and (8.54) this chosen opcartesian lift, which is F_i^2, is also given by the following morphism in $\int_{\mathcal{D}} Z$.

$$\big(a \oplus a'\,,\,Z^2_{a,a'}\big(\phi_{\langle a\rangle}\langle x\rangle\,,\,\phi_{\langle a'\rangle}\langle x'\rangle\big)\big) \xrightarrow{\ (\mathsf{L}^{-1}, 1)\ } \big(a''\,,\,\phi_{\langle a''\rangle}\langle x''\rangle\big) \tag{8.55}$$

Its second entry is the identity morphism in Za''

$$\mathsf{L}_*^{-1} Z^2_{a,a'}\big(\phi_{\langle a\rangle}\langle x\rangle\,,\,\phi_{\langle a'\rangle}\langle x'\rangle\big) \xrightarrow{\ 1\ } \phi_{\langle a''\rangle}\langle x''\rangle.$$

Applying L_* yields the equality in $Z(a \oplus a')$

$$\mathsf{L}_* \phi_{\langle a'' \rangle} \langle x'' \rangle = Z^2_{a,a'} \big(\phi_{\langle a \rangle} \langle x \rangle \, , \, \phi_{\langle a' \rangle} \langle x' \rangle \big). \tag{8.56}$$

This proves the additivity axiom (5.51) for objects.

Morphisms. For the additivity axiom (5.51) for morphisms, suppose given morphisms

$$\begin{aligned} \langle p \rangle = \big\langle p_j : x_j \longrightarrow y_j \big\rangle_{j=1}^n \in \textstyle\prod_{j=1}^n X_j a_j \quad \text{and} \\ p'_i : x'_i \longrightarrow y'_i \in X_i a'_i. \end{aligned}$$

We extend the notation in (5.50) and (5.51) to morphisms by replacing each occurrence of x with p. By (8.54) and (8.55), the naturality diagram of F_i^2 for these morphisms is the following commutative diagram in $\int_{\mathcal{D}} Z$.

$$\begin{array}{ccc} \big(a \oplus a' \, , \, Z^2_{a,a'}(\phi_{\langle a \rangle}\langle x \rangle \, , \, \phi_{\langle a' \rangle}\langle x' \rangle)\big) & \xrightarrow{(\mathsf{L}^{-1},1)} & \big(a'' \, , \, \phi_{\langle a'' \rangle}\langle x'' \rangle\big) \\ {\scriptstyle (1 \, , \, Z^2_{a,a'}(\phi_{\langle a \rangle}\langle p \rangle \, , \, \phi_{\langle a' \rangle}\langle p' \rangle))} \big\downarrow & & \big\downarrow {\scriptstyle (1 \, , \, \phi_{\langle a'' \rangle}\langle p'' \rangle)} \\ \big(a \oplus a' \, , \, Z^2_{a,a'}(\phi_{\langle a \rangle}\langle y \rangle \, , \, \phi_{\langle a' \rangle}\langle y' \rangle)\big) & \xrightarrow{(\mathsf{L}^{-1},1)} & \big(a'' \, , \, \phi_{\langle a'' \rangle}\langle y'' \rangle\big) \end{array}$$

The second component of this commutative diagram is the morphism equality in Za''

$$\mathsf{L}_*^{-1} Z^2_{a,a'} \big(\phi_{\langle a \rangle} \langle p \rangle \, , \, \phi_{\langle a' \rangle} \langle p' \rangle \big) = \phi_{\langle a'' \rangle} \langle p'' \rangle.$$

Applying L_* yields the morphism equality in $Z(a \oplus a')$

$$\mathsf{L}_* \phi_{\langle a'' \rangle} \langle p'' \rangle = Z^2_{a,a'} \big(\phi_{\langle a \rangle} \langle p \rangle \, , \, \phi_{\langle a' \rangle} \langle p' \rangle \big).$$

This proves the additivity axiom (5.51) for morphisms. □

Preimage of F

By Lemmas 8.43, 8.44, 8.51, and 8.53,

$$\big\langle (X_j, X_j^2, X_j^0) \big\rangle_{j=1}^n \xrightarrow{\ \phi\ } (Z, Z^2, Z^0)$$

in (8.35) is an additive natural transformation (Definition 5.39). Next we show that its Grothendieck construction, in the sense of Lemma 6.57, is equal to F.

Lemma 8.57. *In the context of Definition 8.32, there is an equality*

$$\big(\textstyle\int_{\mathcal{D}} \phi \, , \, \langle (\int_{\mathcal{D}} \phi)_j^2 \rangle_{j=1}^n \big) = \big(F \, , \, \langle F_j^2 \rangle_{j=1}^n \big)$$

of strong n-linear functors.

Proof. We must prove the following two statements.

1. There is an equality of functors

$$\textstyle\int_{\mathcal{D}}\phi = F : \prod_{j=1}^{n}(\int_{\mathcal{D}} X_j) \longrightarrow \int_{\mathcal{D}} Z.$$

2. In the context of (8.54) through (8.56), the following two morphisms in $\int_{\mathcal{D}} Z$ are equal.

$$\big(a \oplus a'\,,\, \mathsf{L}_*\phi_{\langle a''\rangle}\langle x''\rangle\big) \underset{(\int_{\mathcal{D}}\phi)_i^2}{\overset{F_i^2 = (\mathsf{L}^{-1},1)}{\Longrightarrow}} \big(a''\,,\, \phi_{\langle a''\rangle}\langle x''\rangle\big)$$

Once (1) is proved, (2) follows from the definition (6.56) of $(\int_{\mathcal{D}}\phi)_i^2$. We prove (1) first on objects and then on morphisms.

Objects. For each n-tuple of objects

$$\langle (a_j, x_j)\rangle \in \textstyle\prod_{j=1}^{n}(\int_{\mathcal{D}} X_j),$$

the following equalities in $\int_{\mathcal{D}} Z$ prove (1) on objects.

$$\begin{aligned} (\textstyle\int_{\mathcal{D}}\phi)\langle (a_j, x_j)\rangle &= \big(a\,,\, \phi_{\langle a\rangle}\langle x\rangle\big) && \text{by (6.49)}\\ &= F\langle (a_j, x_j)\rangle && \text{by (8.37) and (8.38)} \end{aligned}$$

Morphisms. Suppose given an n-tuple of morphisms

$$\big\langle (f_j, p_j) : (a_j, x_j) \longrightarrow (b_j, y_j)\big\rangle \in \textstyle\prod_{j=1}^{n}(\int_{\mathcal{D}} X_j).$$

We must show that its images under $\int_{\mathcal{D}}\phi$ and F are equal. By definition (6.35) each morphism (f_j, p_j) in $\int_{\mathcal{D}} X_j$ factors as follows.

$$\begin{array}{ccc} (a_j, x_j) & \xrightarrow{\quad (f_j, p_j) \quad} & (b_j, y_j) \\ {\scriptstyle (f_j\,,\,1_{(f_j)_* x_j})} \searrow & & \nearrow {\scriptstyle (1_{b_j}\,,\,p_j)} \\ & (b_j\,,\,(f_j)_* x_j) & \end{array} \tag{8.58}$$

Now we observe the following for the first morphism $\big(f_j\,,\,1_{(f_j)_* x_j}\big)$ in (8.58).

- By definition (7.58) $(f_j, 1)$ is the chosen opcartesian lift of the fore-lift $\langle (a_j, x_j)\,;\, f_j\rangle$ with respect to $U_{X_j} : \int_{\mathcal{D}} X_j \longrightarrow \mathcal{D}$.
- Since F preserves chosen opcartesian lifts (Definition 7.11 (3)), the image

  $$F\langle (a_j, x_j)\rangle \xrightarrow{F\langle (f_j,1)\rangle} F\big\langle (b_j\,,\,(f_j)_* x_j)\big\rangle \quad \text{in} \quad \textstyle\int_{\mathcal{D}} Z$$

  is the chosen opcartesian lift of the fore-lift

  $$\big\langle F\langle (a_j, x_j)\rangle\,;\, \textstyle\bigotimes_{j=1}^{n} f_j\big\rangle = \big\langle \big(a\,,\, \phi_{\langle a\rangle}\langle x\rangle\big)\,;\, f\big\rangle$$

  with respect to $U_Z : \int_{\mathcal{D}} Z \longrightarrow \mathcal{D}$. By definition (7.58) this chosen opcartesian lift, which is $F\langle (f_j, 1)\rangle$, is also given by

  $$\big(a\,,\, \phi_{\langle a\rangle}\langle x\rangle\big) \xrightarrow{\big(f\,,\, 1_{f_*\phi_{\langle a\rangle}\langle x\rangle}\big)} \big(b\,,\, f_*\phi_{\langle a\rangle}\langle x\rangle\big). \tag{8.59}$$

The following morphism equalities in $\int_{\mathcal{D}} Z$ prove (1) on morphisms.

$$\begin{aligned}
&(\textstyle\int_{\mathcal{D}}\phi)\langle(f_j, p_j)\rangle \\
&= \big(f\,,\,\phi_{\langle b\rangle}\langle p\rangle\big) && \text{by (6.52)}\\
&= \big(1_b\,,\,\phi_{\langle b\rangle}\langle p\rangle\big)\circ\big(f\,,\,1_{f_*\phi_{\langle a\rangle}\langle x\rangle}\big) && \text{by (6.35)}\\
&= F\langle(1_{b_j}, p_j)\rangle \circ F\big\langle\big(f_j\,,\,1_{(f_j)_* x_j}\big)\big\rangle && \text{by (8.39), (8.40), and (8.59)}\\
&= F\langle(f_j, p_j)\rangle && \text{by (8.58)}
\end{aligned}$$

This finishes the proof of (1). □

8.4 Morphism Fullness and the Main Theorem

We continue to use the notation in (8.18), (8.19), and (8.24) through (8.26) with $n > 0$. In Lemmas 8.20, 8.28, and 8.57 we observe that the component functors (Lemma 7.62)

$$\mathcal{D}\mathsf{Cat}(\langle X\rangle\,;\,Z) \xrightarrow{\int_{\mathcal{D}}} \mathsf{PFib}^{\mathcal{D}}(\langle U_X\rangle\,;\,U_Z)$$

are (i) injective on objects and morphism sets and (ii) surjective on objects. In this section we show that these component functors are also surjective on morphism sets.

Recall the following.

- $\mathcal{D}\mathsf{Cat}(\langle X\rangle\,;\,Z)$ has additive natural transformations as objects and additive modifications as morphisms (Definitions 5.39 and 5.52).
- $\mathsf{PFib}^{\mathcal{D}}(\langle U_X\rangle\,;\,U_Z)$ has opcartesian strong n-linear functors as objects and opcartesian n-linear transformations as morphisms (Definitions 7.11 and 7.27).

Here is an outline of this section.

- In Definition 8.60 we associate to each opcartesian n-linear transformation $\omega : \int_{\mathcal{D}}\phi \longrightarrow \int_{\mathcal{D}}\psi$ between the Grothendieck constructions of two additive natural transformations the data of an additive modification $\Phi : \phi \longrightarrow \psi$.
- Lemmas 8.68 and 8.70 show that Φ is a modification.
- Lemmas 8.73 and 8.74 show that Φ is an additive modification.
- Lemma 8.75 shows that $\int_{\mathcal{D}}\Phi$ and ω are equal as opcartesian n-linear transformations. This proves that the multimorphism functors of $\int_{\mathcal{D}}$ are surjective on morphism sets.

- Using the Multiequivalence Theorem 3.69 and the lemmas in this chapter, Theorem 8.76 proves that

$$\mathcal{D}\mathsf{Cat} \xrightarrow{\int_{\mathcal{D}}} \mathsf{PFib}^{\mathcal{D}}$$

is a non-symmetric Cat-multiequivalence. If, moreover, $\beta^{\otimes} = 1$ in $\mathcal{D}$, then $\int_{\mathcal{D}}$ is a Cat-multiequivalence.

Definition 8.60. Suppose given additive natural transformations

$$\langle (X_j, X_j^2, X_j^0) \rangle_{j=1}^n \overset{\phi}{\underset{\psi}{\rightrightarrows}} (Z, Z^2, Z^0)$$

and an opcartesian n-linear transformation

$$\left(\textstyle\int_{\mathcal{D}} \phi \,,\, \langle (\int_{\mathcal{D}} \phi)_j^2 \rangle_{j=1}^n \right) \xrightarrow{\ \omega\ } \left(\textstyle\int_{\mathcal{D}} \psi \,,\, \langle (\int_{\mathcal{D}} \psi)_j^2 \rangle_{j=1}^n \right) \tag{8.61}$$

with the (co)domain given by Lemma 6.57. The structure ω is given by an n-linear natural transformation (Definition 6.16)

$$\prod_{j=1}^n (\textstyle\int_{\mathcal{D}} X_j) \underset{(\int_{\mathcal{D}} \psi \,,\, \langle (\int_{\mathcal{D}} \psi)_j^2 \rangle_{j=1}^n)}{\overset{(\int_{\mathcal{D}} \phi \,,\, \langle (\int_{\mathcal{D}} \phi)_j^2 \rangle_{j=1}^n)}{\overset{\Downarrow \omega}{\rightrightarrows}}} \textstyle\int_{\mathcal{D}} Z \tag{8.62}$$

that satisfies the transformation projection axiom (7.29). Define the data of an additive modification

$$\langle (X_j, X_j^2, X_j^0) \rangle_{j=1}^n \underset{\psi}{\overset{\phi}{\overset{\Downarrow \Phi}{\rightrightarrows}}} (Z, Z^2, Z^0) \tag{8.63}$$

with $\langle a \rangle$-component natural transformations (5.11)

$$\prod_{j=1}^n X_j a_j \underset{\psi_{\langle a \rangle}}{\overset{\phi_{\langle a \rangle}}{\overset{\Downarrow \Phi_{\langle a \rangle}}{\rightrightarrows}}} Za \tag{8.64}$$

as follows. For each n-tuple of objects $\langle x \rangle \in \prod_{j=1}^n X_j a_j$, define the morphism

$$(\Phi_{\langle a \rangle})_{\langle x \rangle} = \omega_{\langle x \rangle} : \phi_{\langle a \rangle} \langle x \rangle \longrightarrow \psi_{\langle a \rangle} \langle x \rangle \quad \text{in} \quad Za \tag{8.65}$$

where $\omega_{\langle x \rangle}$ is the second entry in the $\langle (a_j, x_j) \rangle$-component morphism

$$\omega_{\langle (a_j, x_j) \rangle} = \left(1_a \,,\, \omega_{\langle x \rangle} \right) : (\textstyle\int_{\mathcal{D}} \phi) \langle (a_j, x_j) \rangle \longrightarrow (\textstyle\int_{\mathcal{D}} \psi) \langle (a_j, x_j) \rangle \tag{8.66}$$

in $\int_{\mathcal{D}} Z$. This finishes the definition of Φ in (8.63). ◇

Explanation 8.67. In (8.66) the (co)domain objects are

$$\big(\textstyle\int_{\mathcal{D}}\phi\big)\langle(a_j,x_j)\rangle = \big(a\,,\,\phi_{\langle a\rangle}\langle x\rangle\big) \quad \text{and}$$
$$\big(\textstyle\int_{\mathcal{D}}\psi\big)\langle(a_j,x_j)\rangle = \big(a\,,\,\psi_{\langle a\rangle}\langle x\rangle\big)$$

in $\int_{\mathcal{D}} Z$ by definition (6.49). The transformation projection axiom (7.29) for ω in (8.61) means that

$$U_Z\omega_{\langle(a_j,x_j)\rangle} = 1_{\otimes_{j=1}^n U_{X_j}(a_j,x_j)} = 1_a \quad \text{in} \quad \mathcal{D}.$$

So the first entry in the morphism

$$\big(a\,,\,\phi_{\langle a\rangle}\langle x\rangle\big) \xrightarrow{\omega_{\langle(a_j,x_j)\rangle}} \big(a\,,\,\psi_{\langle a\rangle}\langle x\rangle\big) \quad \text{in} \quad \textstyle\int_{\mathcal{D}} Z$$

is the identity morphism $1_a \in \mathcal{D}$. This is the meaning of (8.66). ◇

In the rest of this section we show that $\Phi : \phi \longrightarrow \psi$ in (8.63) is an additive modification such that

$$\textstyle\int_{\mathcal{D}}\Phi = \omega$$

as n-linear natural transformations, with $\int_{\mathcal{D}}\Phi$ given by Lemma 6.67.

Naturality and the Modification Axiom

To show that Φ is an additive modification, first we check that its components are natural transformations.

Lemma 8.68. *In* (8.64) *each component*

$$\phi_{\langle a\rangle} \xrightarrow{\Phi_{\langle a\rangle}} \psi_{\langle a\rangle}$$

is a natural transformation between functors $\prod_{j=1}^n X_j a_j \longrightarrow Za$.

Proof. For each n-tuple of morphisms

$$\langle p\rangle = \big\langle p_j : x_j \longrightarrow y_j\big\rangle_{j=1}^n \in \textstyle\prod_{j=1}^n X_j a_j$$

the desired naturality diagram for $\Phi_{\langle a\rangle}$ is the following diagram in Za.

$$\begin{array}{ccc} \phi_{\langle a\rangle}\langle x\rangle & \xrightarrow{(\Phi_{\langle a\rangle})_{\langle x\rangle}} & \psi_{\langle a\rangle}\langle x\rangle \\ {\scriptstyle \phi_{\langle a\rangle}\langle p\rangle}\downarrow & & \downarrow{\scriptstyle \psi_{\langle a\rangle}\langle p\rangle} \\ \phi_{\langle a\rangle}\langle y\rangle & \xrightarrow{(\Phi_{\langle a\rangle})_{\langle y\rangle}} & \psi_{\langle a\rangle}\langle y\rangle \end{array} \tag{8.69}$$

To show that (8.69) is commutative, consider the n-tuple of morphisms

$$\big\langle (1_{a_j}, p_j) : (a_j, x_j) \longrightarrow (a_j, y_j)\big\rangle_{j=1}^n \in \textstyle\prod_{j=1}^n\big(\int_{\mathcal{D}} X_j\big).$$

The naturality of $\omega : \int_{\mathcal{D}}\phi \longrightarrow \int_{\mathcal{D}}\psi$ in (8.62) yields the following commutative diagram in $\int_{\mathcal{D}} Z$.

$$\begin{array}{ccc}
(\int_{\mathcal{D}}\phi)\langle(a_j,x_j)\rangle & \xrightarrow{\ \omega_{\langle(a_j,x_j)\rangle}\ } & (\int_{\mathcal{D}}\psi)\langle(a_j,x_j)\rangle \\
{\scriptstyle (\int_{\mathcal{D}}\phi)\langle(1_{a_j},p_j)\rangle}\Big\downarrow & & \Big\downarrow{\scriptstyle (\int_{\mathcal{D}}\psi)\langle(1_{a_j},p_j)\rangle} \\
(\int_{\mathcal{D}}\phi)\langle(a_j,y_j)\rangle & \xrightarrow{\ \omega_{\langle(a_j,y_j)\rangle}\ } & (\int_{\mathcal{D}}\psi)\langle(a_j,y_j)\rangle
\end{array}$$

By the definitions

- (6.49) and (6.52) of $\int_{\mathcal{D}}\phi$ on, respectively, objects and morphisms and
- (8.65) and (8.66) of $\Phi_{\langle a\rangle}$,

the previous commutative diagram is equal to the following diagram.

$$\begin{array}{ccc}
\big(a\,,\,\phi_{\langle a\rangle}\langle x\rangle\big) & \xrightarrow{\ (1_a\,,\,(\Phi_{\langle a\rangle})_{\langle x\rangle})\ } & \big(a\,,\,\psi_{\langle a\rangle}\langle x\rangle\big) \\
{\scriptstyle (1_a\,,\,\phi_{\langle a\rangle}\langle p\rangle)}\Big\downarrow & & \Big\downarrow{\scriptstyle (1_a\,,\,\psi_{\langle a\rangle}\langle p\rangle)} \\
\big(a\,,\,\phi_{\langle a\rangle}\langle y\rangle\big) & \xrightarrow{\ (1_a\,,\,(\Phi_{\langle a\rangle})_{\langle y\rangle})\ } & \big(a\,,\,\psi_{\langle a\rangle}\langle y\rangle\big)
\end{array}$$

The second component of this commutative diagram yields the desired diagram (8.69). □

Next we show that Φ is a modification.

Lemma 8.70. *In* (8.63) $\Phi : \phi \longrightarrow \psi$ *is a modification.*

Proof. The desired modification axiom (5.12) states that, for each n-tuple of objects $\langle x\rangle \in \prod_{j=1}^{n} X_j a_j$, the following morphism equality holds in Za'.

$$f_*(\Phi_{\langle a\rangle})_{\langle x\rangle} = (\Phi_{\langle a'\rangle})_{\langle (f_j)_* x_j\rangle} : f_*\phi_{\langle a\rangle}\langle x\rangle \longrightarrow \psi_{\langle a'\rangle}\langle (f_j)_* x_j\rangle \tag{8.71}$$

To prove the equality in (8.71), consider the n-tuple of morphisms

$$\Big\langle \big(f_j\,,\,1_{(f_j)_* x_j}\big) : (a_j,x_j) \longrightarrow \big(a'_j\,,\,(f_j)_* x_j\big)\Big\rangle \in \textstyle\prod_{j=1}^{n}(\int_{\mathcal{D}} X_j).$$

The naturality of $\omega : \int_{\mathcal{D}}\phi \longrightarrow \int_{\mathcal{D}}\psi$ in (8.62) yields the following commutative diagram in $\int_{\mathcal{D}} Z$.

$$\begin{array}{ccc}
(\int_{\mathcal{D}}\phi)\langle(a_j,x_j)\rangle & \xrightarrow{\ (\int_{\mathcal{D}}\phi)\langle(f_j,1)\rangle\ } & (\int_{\mathcal{D}}\phi)\big\langle\big(a'_j\,,\,(f_j)_* x_j\big)\big\rangle \\
{\scriptstyle \omega_{\langle(a_j,x_j)\rangle}}\Big\downarrow & & \Big\downarrow{\scriptstyle \omega_{\langle(a'_j\,,\,(f_j)_* x_j)\rangle}} \\
(\int_{\mathcal{D}}\psi)\langle(a_j,x_j)\rangle & \xrightarrow{\ (\int_{\mathcal{D}}\psi)\langle(f_j,1)\rangle\ } & (\int_{\mathcal{D}}\psi)\big\langle\big(a'_j\,,\,(f_j)_* x_j\big)\big\rangle
\end{array}$$

By (6.49), (6.52), (8.65), and (8.66) the previous commutative diagram is equal to the following diagram.

$$\begin{array}{ccc} \big(a\,,\,\phi_{\langle a\rangle}\langle x\rangle\big) & \xrightarrow{(f\,,\,\phi_{\langle a'\rangle}\langle 1\rangle)} & \big(a'\,,\,\phi_{\langle a'\rangle}\langle (f_j)_* x_j\rangle\big) \\ {\scriptstyle (1_a\,,\,(\Phi_{\langle a\rangle})_{\langle x\rangle})}\downarrow & & \downarrow{\scriptstyle (1_{a'}\,,\,(\Phi_{\langle a'\rangle})_{\langle (f_j)_* x_j\rangle})} \\ \big(a\,,\,\psi_{\langle a\rangle}\langle x\rangle\big) & \xrightarrow{(f\,,\,\psi_{\langle a'\rangle}\langle 1\rangle)} & \big(a'\,,\,\psi_{\langle a'\rangle}\langle (f_j)_* x_j\rangle\big) \end{array} \tag{8.72}$$

By the functoriality of $\phi_{\langle a'\rangle}$ and $\psi_{\langle a'\rangle}$, there are identity morphisms

$$\phi_{\langle a'\rangle}\langle 1\rangle = 1_{\phi_{\langle a'\rangle}\langle (f_j)_* x_j\rangle} \quad \text{and}$$
$$\psi_{\langle a'\rangle}\langle 1\rangle = 1_{\psi_{\langle a'\rangle}\langle (f_j)_* x_j\rangle}$$

in Za'. Thus the second component of the commutative diagram (8.72) yields the desired equality (8.71). □

Unity and Additivity

Next we show that Φ is an *additive* modification (Definition 5.52).

Lemma 8.73. $\Phi : \phi \longrightarrow \psi$ *in* (8.63) *satisfies the unity axiom* (5.54).

Proof. With $\circ_i$ as in (6.2), there are morphism equalities in $\int_{\mathcal{D}} Z$ as follows.

$$\begin{aligned} &\big(1_0\,,\,(\Phi_{\langle a\rangle \circ_i 0})_{\langle x\rangle \circ_i (X_i^0 *)}\big) \\ &= \omega_{\langle (a_j, x_j)\rangle \circ_i (0, X_i^0 *)} && \text{by (2.31), (8.65), and (8.66)} \\ &= 1_{(0, Z^0 *)} && \text{by (6.18) and (6.38)} \\ &= \big(1_0\,,\,1_{Z^0 *}\big) && \text{by (6.34)} \end{aligned}$$

The morphism equality

$$(\Phi_{\langle a\rangle \circ_i 0})_{\langle x\rangle \circ_i (X_i^0 *)} = 1_{Z^0 *} \quad \text{in} \quad Z0$$

is the unity axiom for Φ. □

Lemma 8.74. $\Phi : \phi \longrightarrow \psi$ *in* (8.63) *satisfies the additivity axiom* (5.57).

Proof. In the context of (6.53) through (6.56), (8.65), and (8.66), the constraint compatibility axiom (6.19) for $\omega : \int_{\mathcal{D}} \phi \longrightarrow \int_{\mathcal{D}} \psi$ in (8.62) is the following commutative diagram in $\int_{\mathcal{D}} Z$.

$$\begin{array}{ccc} \big(a \oplus a'\,,\,Z^2_{a,a'}(\phi_{\langle a\rangle}\langle x\rangle\,,\,\phi_{\langle a'\rangle}\langle x'\rangle)\big) & \xrightarrow{(\mathsf{L}^{-1},1)} & \big(a''\,,\,\phi_{\langle a''\rangle}\langle x''\rangle\big) \\ {\scriptstyle (1_{a\oplus a'}\,,\,Z^2_{a,a'}((\Phi_{\langle a\rangle})_{\langle x\rangle}\,,\,(\Phi_{\langle a'\rangle})_{\langle x'\rangle}))}\downarrow & & \downarrow{\scriptstyle (1_{a''}\,,\,(\Phi_{\langle a''\rangle})_{\langle x''\rangle})} \\ \big(a \oplus a'\,,\,Z^2_{a,a'}(\psi_{\langle a\rangle}\langle x\rangle\,,\,\psi_{\langle a'\rangle}\langle x'\rangle)\big) & \xrightarrow{(\mathsf{L}^{-1},1)} & \big(a''\,,\,\psi_{\langle a''\rangle}\langle x''\rangle\big) \end{array}$$

The second entry of this commutative diagram yields the morphism equality

$$(\Phi_{\langle a'' \rangle})_{\langle x'' \rangle} = \mathsf{L}_*^{-1} Z^2_{a,a'}\big((\Phi_{\langle a \rangle})_{\langle x \rangle}\,,\, (\Phi_{\langle a' \rangle})_{\langle x' \rangle}\big) \quad \text{in} \quad Za''.$$

Applying L_* yields the equality

$$\mathsf{L}_*(\Phi_{\langle a'' \rangle})_{\langle x'' \rangle} = Z^2_{a,a'}\big((\Phi_{\langle a \rangle})_{\langle x \rangle}\,,\, (\Phi_{\langle a' \rangle})_{\langle x' \rangle}\big) \quad \text{in} \quad Z(a \oplus a').$$

This proves the additivity axiom for Φ. □

Preimage of ω

Lemmas 8.70, 8.73, and 8.74 show that $\Phi : \phi \longrightarrow \psi$ in (8.63) is an additive modification (Definition 5.52). By Lemmas 6.67 and 7.62 its Grothendieck construction $\int_{\mathcal{D}} \Phi$ is an opcartesian n-linear transformation.

Lemma 8.75. *In the context of Definition 8.60, there is an equality*

$$\textstyle\int_{\mathcal{D}} \Phi = \omega : \int_{\mathcal{D}} \phi \longrightarrow \int_{\mathcal{D}} \psi$$

of opcartesian n-linear transformations.

Proof. Opcartesian n-linear transformations are n-linear natural transformations with the extra property (7.29) but no extra structure (Explanation 7.30 (1)). Thus it suffices to show that $\int_{\mathcal{D}} \Phi$ and ω are equal as natural transformations. For each n-tuple of objects

$$\langle (a_j, x_j) \rangle \in \textstyle\prod_{j=1}^n (\int_{\mathcal{D}} X_j)$$

there are equalities in $\int_{\mathcal{D}} Z$ as follows.

$$\begin{aligned} (\textstyle\int_{\mathcal{D}} \Phi)_{\langle (a_j, x_j) \rangle} &= \big(1_a\,,\, (\Phi_{\langle a \rangle})_{\langle x \rangle}\big) && \text{by (6.63)} \\ &= \omega_{\langle (a_j, x_j) \rangle} && \text{by (8.65) and (8.66)} \end{aligned}$$

This proves $\int_{\mathcal{D}} \Phi = \omega$ as natural transformations. □

Multiequivalence

We are now ready for the main result of this chapter.

Theorem 8.76. *Suppose $\mathcal{D}$ is a small tight bipermutative category.*

1. *The non-symmetric* Cat*-multifunctor in* (7.65)

$$\mathcal{D}\mathsf{Cat} \xrightarrow{\ \int_{\mathcal{D}}\ } \mathsf{PFib}^{\mathcal{D}}$$

is a non-symmetric Cat*-multiequivalence.*

2. *If the multiplicative braiding $\beta^{\otimes}$ in $\mathcal{D}$ is the identity natural transformation, then $\int_{\mathcal{D}}$ is a* Cat-*multiequivalence.*

Proof. We have proved the following regarding $\int_{\mathcal{D}}$.

- It is essentially surjective on objects (Lemma 8.9).
- It is an isomorphism on arity 0 multimorphism categories (Lemma 7.61).
- Between positive arity multimorphism categories, $\int_{\mathcal{D}}$ is
  - injective on objects (Lemma 8.20),
  - injective on morphism sets (Lemma 8.28),
  - surjective on objects (Lemma 8.57), and
  - surjective on morphism sets (Lemma 8.75).

Assertion (1) now follows from the Multiequivalence Theorem 3.69.

Under the assumption that $\beta^{\otimes} = 1$, $\int_{\mathcal{D}}$ is a Cat-multifunctor between Cat-multicategories by Theorem 7.64 (3). Thus assertion (2) follows from Theorem 3.69 and the lemmas stated in the previous paragraph. $\square$

Example 8.77. Suppose $\mathcal{D}$ is any one of the small tight bipermutative categories

$$\mathsf{Fin}^{\mathsf{sk}}, \quad \Sigma, \quad \mathcal{F}, \quad \text{and} \quad \mathcal{A}$$

in, respectively, Examples 2.48, 2.57, 2.58, and 2.63. Then $\mathsf{PFib}^{\mathcal{D}}$ is a non-symmetric Cat-multicategory and *not* a Cat-multicategory by Example 7.55. The commutative diagram of non-symmetric Cat-multifunctors

$$\begin{array}{ccc} & & \mathsf{PFib}^{\mathcal{D}} \\ & \overset{\int_{\mathcal{D}}}{\underset{\sim}{\nearrow}} & \Big\downarrow \mathsf{U} \\ \mathcal{D}\mathsf{Cat} & \xrightarrow{\quad\int_{\mathcal{D}}\quad} & \mathsf{PermCat}^{\mathsf{sg}} \end{array}$$

in Theorem 7.64 (2) shows that the pseudo symmetric Cat-multifunctor

$$\mathcal{D}\mathsf{Cat} \xrightarrow{\quad\int_{\mathcal{D}}\quad} \mathsf{PermCat}^{\mathsf{sg}}$$

in (6.111), regarded as a non-symmetric Cat-multifunctor, factors as

- the non-symmetric Cat-multiequivalence $\int_{\mathcal{D}}$ in Theorem 8.76
- followed by the non-symmetric Cat-multifunctor U in (7.53).

The case $\mathcal{D} = \mathcal{A}$ is used in Theorem 10.12 (3) in the context of inverse K-theory. $\diamond$

Part III

Pseudo Symmetric Enriched Multifunctorial Inverse K-Theory

9

The Cat-*Multifunctor A*

The main topic of Part III is the pseudo symmetric Cat-multifunctorial inverse K-theory

$$\Gamma\text{-}\mathsf{Cat} \xrightarrow{A} \mathcal{A}\mathsf{Cat} \xrightarrow{\int_{\mathcal{A}}} \mathsf{PermCat}^{\mathsf{sg}}. \qquad (\mathcal{P} = \textstyle\int_{\mathcal{A}} \circ A)$$

The Grothendieck construction

$$\mathcal{A}\mathsf{Cat} \xrightarrow{\int_{\mathcal{A}}} \mathsf{PermCat}^{\mathsf{sg}}$$

is the pseudo symmetric Cat-multifunctor in (6.111) with the following data:

- $\mathcal{A}$ is the small tight bipermutative category in Example 2.63.
- $\mathcal{A}\mathsf{Cat}$ is the Cat-multicategory in Theorem 5.71.
- $\mathsf{PermCat}^{\mathsf{sg}}$ is the Cat-multicategory in Theorem 6.31.

In this chapter we construct the Cat-multifunctor (Theorem 9.120)

$$\Gamma\text{-}\mathsf{Cat} \xrightarrow{A} \mathcal{A}\mathsf{Cat}.$$

We emphasize that—unlike the pseudo symmetric Cat-multifunctor $\int_{\mathcal{A}}$ that is *not* a Cat-multifunctor—A is a Cat-multifunctor in the symmetric sense (Definition 3.19). In other words, A strictly preserves the symmetric group action, colored units, and composition in Γ-Cat and $\mathcal{A}$Cat.

Summary

Below is a summary of the Cat-multicategories Γ-Cat and $\mathcal{A}$Cat and the Cat-multifunctor A.

- Γ-Cat (Definition 9.11).
  - Its objects are Γ-categories (Definition 9.11).
  - Its n-ary 1-cells are pointed natural transformations (Explanation 9.28 and (9.33) and (9.45)).
  - Its n-ary 2-cells are pointed modifications (Explanation 9.28 and (9.33) and (9.49)).

DOI: 10.1201/9781003451235-9

  - Its units are identity natural transformations (9.51).
  - Its symmetry and composition are given in (9.52) and (9.59).

- $\mathcal{A}$Cat (Theorem 5.71).

  - Its objects are additive symmetric monoidal functors (Definition 5.31).
  - Its n-ary 1-cells are additive natural transformations (Definition 5.39).
  - Its n-ary 2-cells are additive modifications (Definition 5.52).
  - Its units are identity natural transformations (5.13).
  - Its symmetry is given in (5.14).
  - Its composition is given in (5.19) and Explanation 5.21.
  - It is *not* induced by a symmetric monoidal structure (Explanation 5.64).

- A (Theorem 9.120).

  - Its object assignment is given in Definition 9.60 and Lemma 9.66.
  - Its n-ary 1-cell assignment is given in (9.83) and (9.94).
  - Its n-ary 2-cell assignment is given in (9.84) and (9.104).
  - It is a Cat-multifunctor by Lemmas 9.112 and 9.116 and Theorem 9.120.

Organization

To prepare for the discussion of A and inverse K-theory, in Section 9.1 we review the symmetric monoidal closed structure on the category of Γ-categories. While Γ-categories are relatively simple to define, the main subtlety of the symmetric monoidal closed structure on Γ-Cat is that it uses the *pointed* Day convolution (9.18), along with the pointed version of the monoidal unit (9.19) and pointed hom diagram (9.20). This symmetric monoidal closed structure induces a Cat-multicategory structure on Γ-Cat, which is explained in detail in Section 9.2.

To make the discussion of the Cat-multifunctoriality of A easier to understand, in Section 9.3 we first construct the underlying functor of A. In particular, this section contains the object assignment of A, which sends a Γ-category to an object in $\mathcal{A}$Cat. The Cat-multifunctor A is constructed in two stages.

- In Section 9.4 we construct the multimorphism functors of A and observe the following.

  - A is an isomorphism of categories in arity 0 (Lemma 9.85).

   – In positive arity A is injective on both objects and morphism sets (Lemma 9.110).

- In Section 9.5 we prove that the object assignment in Section 9.3 and the multimorphism functors in Section 9.4 define a Cat-multifunctor A from Γ-Cat to $\mathcal{A}\mathsf{Cat}$.

We remind the reader of our left normalized bracketing Convention 1.24 for iterated monoidal product.

9.1 The Symmetric Monoidal Category of Γ-Categories

The purpose of this section is to discuss the symmetric monoidal closed category of Γ-categories. In Section 9.2 we discuss its induced Cat-multicategory structure. Here is an outline of this section.

- We begin with a discussion of pointed categories and the symmetric monoidal closed structure on the category Cat_* of small pointed categories (Theorem 9.7).
- Then we define Γ-categories (Definition 9.11) and discuss its symmetric monoidal closed structure (Theorem 9.23).
- We describe explicitly the monoidal unit, the pointed Day convolution, and the pointed mapping object for Γ-categories in, respectively, Explanations 9.24, 9.26, and 9.28.

We emphasize that the symmetric monoidal structure on Γ-Cat uses the *pointed* Day convolution in (9.18), which is different from the unpointed version in (1.51). As a result, the associated monoidal unit (9.19) and internal hom (9.20) are also different from the unpointed versions in (1.54) and (1.55). Much more detailed and general discussion of pointed categories and pointed Day convolution can be found in [JY∞, Ch. 4].

Pointed Categories

Definition 9.1 (Pointed Diagrams). A *pointed category* is a pair $(\mathsf{C}, *)$ consisting of a category C and a distinguished object $* \in \mathsf{C}$, which is called the *basepoint*. The basepoint is also denoted by the functor $\iota^{\mathsf{C}} : \mathbf{1} \longrightarrow \mathsf{C}$ with $\iota^{\mathsf{C}}(*) = *$.

- For pointed categories $(\mathsf{C}, *)$ and $(\mathsf{D}, *)$, a *pointed functor*

$$(\mathsf{C}, *) \xrightarrow{\;F\;} (\mathsf{D}, *)$$

  is a functor $F : \mathsf{C} \longrightarrow \mathsf{D}$ such that $F(*) = *$.

- For pointed functors $F, G : (\mathsf{C}, *) \longrightarrow (\mathsf{D}, *)$, a *pointed natural transformation*

$$F \xrightarrow{\theta} G$$

is a natural transformation $\theta : F \longrightarrow G$ such that

$$\theta_* = 1_*.$$

Denote by Cat_* the category with

- small pointed categories as objects and
- pointed functors as morphisms.

This finishes the definition. ◇

For an elementary discussion of (co)limits in Cat, the reader is referred to [Yau20, Section 1.4]. For categories C and D with C small, recall that the *diagram category* $[\mathsf{C}, \mathsf{D}]$ has functors $\mathsf{C} \longrightarrow \mathsf{D}$ as objects and natural transformations as morphisms (Definition 1.15).

Definition 9.2. Suppose $(\mathsf{C}, \iota^{\mathsf{C}})$ and $(\mathsf{D}, \iota^{\mathsf{D}})$ are small pointed categories. Define the following small pointed categories.

- The *smash unit* is defined as the coproduct in Cat

$$E = \mathbf{1} \amalg \mathbf{1} \tag{9.3}$$

with the basepoint given by the first copy of $\mathbf{1}$.

- The *wedge product* $\mathsf{C} \vee \mathsf{D}$ is defined as the pushout in Cat of the span

$$\mathsf{C} \xleftarrow{\iota^{\mathsf{C}}} \mathbf{1} \xrightarrow{\iota^{\mathsf{D}}} \mathsf{D} \tag{9.4}$$

with the basepoint given by the composite

$$\mathbf{1} \longrightarrow \mathsf{C} \longrightarrow \mathsf{C} \vee \mathsf{D}.$$

- The *smash product* $\mathsf{C} \wedge \mathsf{D}$ is defined as the following pushout in Cat.

$$\begin{array}{ccc} (\mathsf{C} \times \mathbf{1}) \amalg (\mathbf{1} \times \mathsf{D}) & \xrightarrow{(1_{\mathsf{C}} \times \iota^{\mathsf{D}}) \amalg (\iota^{\mathsf{C}} \times 1_{\mathsf{D}})} & \mathsf{C} \times \mathsf{D} \\ \downarrow & & \downarrow \mathsf{pr} \\ \mathbf{1} & \xrightarrow{\iota^{\mathsf{C} \wedge \mathsf{D}}} & \mathsf{C} \wedge \mathsf{D} \end{array} \tag{9.5}$$

The functor $\iota^{\mathsf{C} \wedge \mathsf{D}}$ makes $\mathsf{C} \wedge \mathsf{D}$ into a pointed category.

- The *pointed diagram category* $[\mathsf{C},\mathsf{D}]_*$ is defined as the following pullback in Cat.

$$\begin{array}{ccc} [\mathsf{C},\mathsf{D}]_* & \longrightarrow & \mathbf{1} \\ \downarrow & & \downarrow \iota^{\mathsf{D}} \\ [\mathsf{C},\mathsf{D}] & \xrightarrow{[\iota^{\mathsf{C}},\mathsf{D}]} & [\mathbf{1},\mathsf{D}] \end{array} \tag{9.6}$$

  The composite functor

$$\mathbf{1} \xrightarrow{\cong} [\mathsf{C},\mathbf{1}] \xrightarrow{[\mathsf{C},\iota^{\mathsf{D}}]} [\mathsf{C},\mathsf{D}]$$

  induces a canonical functor

$$\mathbf{1} \longrightarrow [\mathsf{C},\mathsf{D}]_*$$

  making $[\mathsf{C},\mathsf{D}]_*$ into a pointed category.

This finishes the definition. ◇

The following result is a special case of [JY∞, 4.2.3].

Theorem 9.7. *In the context of Definition 9.2,*

$$\big(\mathsf{Cat}_*, \wedge, E, [,]_*\big)$$

is a complete and cocomplete symmetric monoidal closed category.

Explanation 9.8 (Pointed Diagram Category)**.** Unraveling the pullback (9.6), the pointed diagram category $[\mathsf{C},\mathsf{D}]_*$ is the category with

- pointed functors $(\mathsf{C},*) \longrightarrow (\mathsf{D},*)$ as objects,
- pointed natural transformations as morphisms,
- identity natural transformations as identity morphisms, and
- vertical composition of natural transformations as composition.

The basepoint in $[\mathsf{C},\mathsf{D}]_*$ is given by the constant functor $\mathsf{C} \longrightarrow \mathsf{D}$ at the basepoint in D. Using this explicit description, the pointed diagram category $[\mathsf{C},\mathsf{D}]_*$ is well defined as long as C is small. ◇

Γ-Categories

To define Γ-categories, first recall from Example 2.58 the small tight bipermutative category

$$\big(\mathcal{F}, (\vee, \underline{0}, \xi^{\vee}), (\wedge, \underline{1}, \xi^{\wedge}), \partial^{l} = (\delta^{r})^{-1}, \partial^{r} = 1\big).$$

It has

- pointed finite sets $\underline{n} = \{0, \ldots, n\}$ for $n \geq 0$ as objects and
- pointed functions as morphisms.

The basepoint of $\underline{n}$ is 0.

- The additive structure $\vee$ is given by the wedge of pointed finite sets.
- The multiplicative structure $\wedge$ is given by the smash product of pointed finite sets.

The following two statements hold by Theorem 2.45.

- The right factorization $\partial^r = (\delta^l)^{-1}$ is the identity, since $\delta^l = 1$ by (2.61).
- The left factorization $\partial^l = (\delta^r)^{-1}$, with δ^r in (2.62), is the pointed bijection given by the inverse of the tetris permutation (2.55) for nonzero elements.

Definition 9.9. For objects $\underline{m}, \underline{n} \in \mathcal{F}$, the *zero morphism* $\underline{m} \longrightarrow \underline{n}$ is the composite

$$\underline{m} \longrightarrow \underline{0} \longrightarrow \underline{n}.$$

A *nonzero morphism* in $\mathcal{F}$ is a morphism that does not factor through $\underline{0}$. The subset of nonzero morphisms $\underline{m} \longrightarrow \underline{n}$ in $\mathcal{F}$ is denoted

$$\mathcal{F}^\flat(\underline{m}; \underline{n}). \tag{9.10}$$

Note that $\mathcal{F}^\flat(\underline{m}; \underline{n})$ is empty if either $\underline{m}$ or $\underline{n}$ is $\underline{0}$. ◇

Definition 9.11 (Γ-Categories)**.** For the pointed categories $(\mathsf{Cat}, \mathbf{1})$ and $(\mathcal{F}, \underline{0})$, define the pointed diagram category (Explanation 9.8)

$$\Gamma\text{-}\mathsf{Cat} = [\mathcal{F}, \mathsf{Cat}]_*.$$

- An object in Γ-Cat is a pointed functor

$$(\mathcal{F}, \underline{0}) \longrightarrow (\mathsf{Cat}, \mathbf{1})$$

  and is called a Γ-*category*.

- A morphism in Γ-Cat is a pointed natural transformation and is called a Γ-*category morphism*. ◇

Explanation 9.12 (Canonical Basepoints)**.** Suppose $X : (\mathcal{F}, \underline{0}) \longrightarrow (\mathsf{Cat}, \mathbf{1})$ is a Γ-category.

- For each $n \geq 0$, the unique pointed function $\iota_n : \underline{0} \longrightarrow \underline{n}$ yields a functor

$$\mathbf{1} = X\underline{0} \xrightarrow{X\iota_n} X\underline{n}.$$

  The image of the object $* \in \mathbf{1}$ in $X\underline{n}$ under $X\iota_n$ is also denoted by $*$. Each $X\underline{n} \in \mathsf{Cat}$ is regarded as a pointed category with basepoint $*$.

- For each morphism $f : \underline{m} \longrightarrow \underline{n}$ in $\mathcal{F}$, the functor

$$X\underline{m} \xrightarrow{Xf} X\underline{n}$$

is a pointed functor, since $f\iota_m = \iota_n$ in $\mathcal{F}$.

Therefore, X canonically yields a pointed functor

$$(\mathcal{F}, \underline{0}) \xrightarrow{X} (\mathsf{Cat}_*, \mathbf{1}) \tag{9.13}$$

with Cat_* the category of small pointed categories and pointed functors (Definition 9.1).

Furthermore, suppose $F : X \longrightarrow Y$ is a morphism of Γ-categories, that is, a pointed natural transformation. It consists of a component functor

$$X\underline{n} \xrightarrow{F_{\underline{n}}} Y\underline{n} \quad \text{for each} \quad \underline{n} \in \mathcal{F}$$

that is natural in $\underline{n}$. By naturality each $F_{\underline{n}}$ is a pointed functor. Therefore, $F : X \longrightarrow Y$ is a pointed natural transformation when X and Y are regarded as pointed functors with codomain $(\mathsf{Cat}_*, \mathbf{1})$ as in (9.13). This discussion implies that there is a canonical isomorphism of categories

$$\Gamma\text{-}\mathsf{Cat} \cong [\mathcal{F}, \mathsf{Cat}_*]_*.$$

In other words, in Definition 9.11 we can equivalently use $(\mathsf{Cat}_*, \mathbf{1})$ instead of $(\mathsf{Cat}, \mathbf{1})$ to define the category of Γ-categories. ◇

Symmetric Monoidal Closed Structure on Γ-Cat

We use the notion of a V-category (Definition 1.25) for $(\mathsf{Cat}_*, \wedge, E, [,]_*)$ in Theorem 9.7. In (9.15), (9.18), and (9.19) below, the empty wedge is defined as the chosen terminal category $\mathbf{1}$.

Definition 9.14. Define $\widehat{\mathcal{F}}$ as the Cat_*-category with the same objects as $\mathcal{F}$ (Example 2.58) and hom objects

$$\widehat{\mathcal{F}}(\underline{m}; \underline{n}) = \bigvee_{\mathcal{F}^\flat(\underline{m}; \underline{n})} E \tag{9.15}$$

in Cat_* for $\underline{m}, \underline{n} \in \mathcal{F}$. ◇

Explanation 9.16. The wedge in (9.15) is defined in (9.4), with

- $E = \mathbf{1} \coprod \mathbf{1}$ the smash unit (9.3) and
- $\mathcal{F}^\flat(\underline{m}; \underline{n})$ the set (9.10) of nonzero morphisms $\underline{m} \longrightarrow \underline{n}$.

Since $\mathcal{F}^\flat(\underline{0}; \underline{n})$ and $\mathcal{F}^\flat(\underline{m}; \underline{0})$ are empty sets, there are pointed categories

$$\widehat{\mathcal{F}}(\underline{0}; \underline{n}) = \mathbf{1} = \widehat{\mathcal{F}}(\underline{m}; \underline{0})$$

for $\underline{m}, \underline{n} \in \mathcal{F}$. ◇

Definition 9.17. Suppose X and Y are Γ-categories. Define the following Γ-categories.

- The *pointed Day convolution* of X and Y is the Γ-category given by the Cat-coend

$$X \wedge Y = \int^{(\underline{m},\underline{n}) \in \widehat{\mathcal{F}} \wedge \widehat{\mathcal{F}}} \bigvee_{\mathcal{F}^\flat(\underline{m} \wedge \underline{n}\,;\,-)} (X\underline{m} \wedge Y\underline{n}). \tag{9.18}$$

  - $\widehat{\mathcal{F}} \wedge \widehat{\mathcal{F}}$ is the tensor product (Definition 1.31) of two copies of the Cat_*-category $\widehat{\mathcal{F}}$ in Definition 9.14.
  - The smash product $X\underline{m} \wedge Y\underline{n}$ is defined in (9.5), with each of X and Y regarded as a pointed functor $(\mathcal{F}, \underline{0}) \longrightarrow (\mathsf{Cat}_*, \mathbf{1})$ as in (9.13).
  - $\mathcal{F}^\flat(-\,;\,-)$ is the set of nonzero morphisms (9.10).

- The *monoidal unit diagram* is the Γ-category given by the wedge

$$J = \widehat{\mathcal{F}}(\underline{1}\,;\,-) = \bigvee_{\mathcal{F}^\flat(\underline{1}\,;\,-)} E. \tag{9.19}$$

- The *pointed hom diagram* for X and Y is the Γ-category given by the Cat-end

$$\mathsf{Hom}_*(X,Y) = \int_{\underline{n} \in \widehat{\mathcal{F}}} \big[X\underline{n},\, Y(- \wedge \underline{n})\big]_* \tag{9.20}$$

  with $[,]_*$ the pointed diagram category in (9.6).

Both $\wedge$ and Hom_* are extended to Γ-category morphisms componentwise. Moreover, we define the following.

- The *braiding* for the pointed Day convolution

$$X \wedge Y \xrightarrow{\xi^{\mathsf{Day}}} Y \wedge X \tag{9.21}$$

  is the pointed natural transformation induced by the braiding for the smash product in Cat_*

$$X\underline{m} \wedge Y\underline{n} \xrightarrow{\xi} Y\underline{n} \wedge X\underline{m},$$

  the multiplicative braiding $\xi^\wedge$ in $\mathcal{F}$ (2.60), and the universal properties of coends.

- Evaluation at the object $\underline{1} \in \mathcal{F}$ defines a symmetric monoidal functor $\mathsf{ev}_{\underline{1}}$ that has a strong symmetric monoidal left adjoint $L_{\underline{1}}$:

$$\mathsf{Cat}_* \underset{\mathsf{ev}_{\underline{1}}}{\overset{L_{\underline{1}}}{\rightleftarrows}} \Gamma\text{-}\mathsf{Cat}.$$

- The *pointed mapping object* is the pointed category

$$\begin{aligned}\mathsf{Map}_*(X,Y) &= \big(\mathsf{Hom}_*(X,Y)\big)(\underline{1})\\ &= \int_{\underline{n}\in\widehat{\mathcal{F}}} [X\underline{n}, Y\underline{n}]_*\end{aligned} \tag{9.22}$$

  given by evaluating $\mathsf{Hom}_*(X,Y)$ in (9.20) at $\underline{1}\in\mathcal{F}$.

This finishes the definition. ◇

Recall the notions of *(co)tensored* and a V-*multicategory* in, respectively, Definitions 1.64 and 3.2. The following result is the special case of [JY∞, 4.3.37 and 6.3.6] for Γ-Cat.

Theorem 9.23. *In the context of Definition 9.17, the following statements hold.*

1. $(\Gamma\text{-}\mathsf{Cat}, \wedge, J, \mathsf{Hom}_*)$ *is a complete and cocomplete symmetric monoidal closed category.*
2. *The adjunction* $(L_{\underline{1}}, \mathsf{ev}_{\underline{1}})$ *makes* Γ-Cat *enriched, tensored, and cotensored over* Cat_* *with hom object* Map_*.
3. *The structure in* (1) *and* (2) *induce a* Cat_**-multicategory structure on* Γ-Cat.

We also regard Γ-Cat as enriched over Cat by forgetting the basepoint in each pointed category. This is valid because the basepoint-forgetting functor

$$\mathsf{Cat}_* \longrightarrow \mathsf{Cat}$$

is symmetric monoidal. Section 9.2 contains a detailed description of the Cat-multicategory structure on Γ-Cat. We end this section with an explicit description of each of the monoidal unit diagram J, the pointed Day convolution $\wedge$, and the pointed mapping object Map_*.

Explanation 9.24 (Monoidal Unit Diagram)**.** Nonzero morphisms $\underline{1}\longrightarrow\underline{n}$ bijectively correspond to elements in the unpointed finite set with n elements

$$\{1,\ldots,n\} = \overline{n}.$$

So for the monoidal unit diagram J in (9.19), there is a discrete pointed category

$$J\underline{n} = \widehat{\mathcal{F}}(\underline{1}\,;\,\underline{n}) = \bigvee_{i\in\overline{n}} (\mathbf{1}\amalg\mathbf{1}) \cong \underline{n} \tag{9.25}$$

for each $\underline{n}\in\mathcal{F}$, with $\vee$ the wedge product in (9.4). ◇

Explanation 9.26 (Pointed Day Convolution)**.** The pointed Day convolution (9.18) is characterized by the following universal property. For Γ-categories X, Y, and Z, a Γ-category morphism

$$X\wedge Y \xrightarrow{\ \psi\ } Z$$

consists of precisely a collection of pointed functors

$$X\underline{m} \wedge Y\underline{n} \xrightarrow{\psi_{\underline{m},\underline{n}}} Z\underline{mn} \quad \text{for} \quad \underline{m}, \underline{n} \in \mathcal{F}$$

such that, for each pair of nonzero morphisms

$$\underline{m} \xrightarrow{f} \underline{p} \quad \text{and} \quad \underline{n} \xrightarrow{g} \underline{q}$$

in $\mathcal{F}$, the diagram

$$\begin{array}{ccc} X\underline{m} \wedge Y\underline{n} & \xrightarrow{\psi_{\underline{m},\underline{n}}} & Z\underline{mn} \\ {\scriptstyle Xf \wedge Yg}\downarrow & & \downarrow{\scriptstyle Z(f \wedge g)} \\ X\underline{p} \wedge Y\underline{q} & \xrightarrow{\psi_{\underline{p},\underline{q}}} & Z\underline{pq} \end{array} \tag{9.27}$$

commutes. The smash product $X\underline{m} \wedge Y\underline{n}$ is the pushout in (9.5) using the canonical basepoints in Explanation 9.12. ◇

Explanation 9.28 (Pointed Mapping Object)**.** Unraveling the Cat-end in (9.22) and using Explanation 9.8 to interpret each pointed diagram category $[X\underline{n}, Y\underline{n}]_*$, the pointed category $\mathsf{Map}_*(X,Y)$ can be described explicitly as follows.

An *object* F in $\mathsf{Map}_*(X,Y)$ is a Γ-category morphism $F : X \longrightarrow Y$, that is, a pointed natural transformation. More explicitly, it consists of pointed functors

$$X\underline{n} \xrightarrow{F_{\underline{n}}} Y\underline{n} \quad \text{for} \quad \underline{n} \in \mathcal{F}$$

such that, for each nonzero morphism $f : \underline{m} \longrightarrow \underline{n}$ in $\mathcal{F}$, the following diagram of functors commutes.

$$\begin{array}{ccc} X\underline{m} & \xrightarrow{F_{\underline{m}}} & Y\underline{m} \\ {\scriptstyle Xf}\downarrow & & \downarrow{\scriptstyle Yf} \\ X\underline{n} & \xrightarrow{F_{\underline{n}}} & Y\underline{n} \end{array} \tag{9.29}$$

The basepoint in $\mathsf{Map}_*(X,Y)$ consists of the constant functors $X\underline{n} \longrightarrow Y\underline{n}$ at the basepoint of $Y\underline{n}$ for $\underline{n} \in \mathcal{F}$.

For objects $F, G \in \mathsf{Map}_*(X,Y)$, a *morphism* $\theta : F \longrightarrow G$ is a pointed modification. More explicitly, it consists of pointed natural transformations

$$X\underline{n} \underset{G_{\underline{n}}}{\overset{F_{\underline{n}}}{\Longrightarrow}} Y\underline{n} \;\; (\Downarrow \theta_{\underline{n}}) \quad \text{for} \quad \underline{n} \in \mathcal{F}$$

such that, for each nonzero morphism $f : \underline{m} \longrightarrow \underline{n}$ in $\mathcal{F}$, the following two whiskered natural transformations are equal.

$$\begin{array}{ccc} X\underline{m} & \overset{F_{\underline{m}}}{\underset{G_{\underline{m}}}{\Downarrow \theta_{\underline{m}}}} & Y\underline{m} \\ {\scriptstyle Xf}\downarrow & & \downarrow{\scriptstyle Yf} \\ X\underline{n} & \overset{F_{\underline{n}}}{\underset{G_{\underline{n}}}{\Downarrow \theta_{\underline{n}}}} & Y\underline{n} \end{array} \tag{9.30}$$

This axiom is called the *modification axiom.*

The *identity morphism* 1_F of an object F consists of identity natural transformations

$$(1_F)_{\underline{n}} = 1_{F_{\underline{n}}}.$$

Composition of morphisms is given componentwise by vertical composition of natural transformations.

There is a similar description for $\mathsf{Hom}_*(X,Y)$ in (9.20). ◇

9.2 The Cat-Multicategory of Γ-Categories

By Theorem 9.23 (3) Γ-Cat has the structure of a Cat_*-multicategory, hence also a Cat-multicategory by forgetting the basepoint in each pointed category. In this section we explain in detail the Cat-multicategory structure on Γ-Cat. This discussion is needed for the discussion of inverse K-theory, whose domain is Γ-Cat, in Chapter 10.

The following discussion of Γ-Cat is analogous to Explanations 5.2 and 5.70 for $\mathcal{D}$-Cat. The main difference is that, on the one hand, $\mathcal{D}$-Cat is entirely in the unpointed setting. In particular, the Day convolution (1.51), the hom diagram (1.54), and the unit diagram (1.55) in $\mathcal{D}$-Cat are all unpointed. On the other hand, in the current setting of Γ-Cat (Definition 9.11), we use the *pointed* Day convolution and associated monoidal unit, pointed hom diagram, and pointed mapping object in Definition 9.17.

Objects and Multimorphism Categories

An object in Γ-Cat is a Γ-category (Definition 9.11), that is, a pointed functor

$$(\mathcal{F}, \underline{0}) \xrightarrow{\;Z\;} (\mathsf{Cat}, \mathbf{1}). \tag{9.31}$$

For Γ-categories Z and $\langle X \rangle = \langle X_j \rangle_{j=1}^n$, the n-ary multimorphism category is

$$\Gamma\text{-}\mathsf{Cat}\big(\langle X \rangle\, ;\, Z\big) = \mathsf{Map}_*\big(X_1 \wedge \cdots \wedge X_n\, ,\, Z\big) \tag{9.32}$$

with

- $\wedge$ the pointed Day convolution (9.18) and
- Map_* the pointed mapping object (9.22).

To unravel $\Gamma\text{-}\mathsf{Cat}\big(\langle X\rangle\,;\,Z\big)$ further, we consider the following cases.

Multimorphism Categories in Arity 1

The 1-ary multimorphism category

$$\Gamma\text{-}\mathsf{Cat}\big(X_1\,;\,Z\big) = \mathsf{Map}_*(X_1, Z)$$

was discussed in Explanation 9.28.

Multimorphism Categories in Arity 0

The 0-ary multimorphism category is

$$\Gamma\text{-}\mathsf{Cat}\big(\langle\rangle\,;\,Z\big) = \mathsf{Map}_*(J, Z) \tag{9.33}$$

with J the monoidal unit diagram (9.19). By Explanations 9.24 and 9.28, an *object* F in $\mathsf{Map}_*(J, Z)$ consists of pointed functors

$$J\underline{n} \cong \underline{n} \xrightarrow{F_{\underline{n}}} Z\underline{n} \quad \text{for} \quad \underline{n} \in \mathcal{F}$$

that are natural for nonzero morphisms in $\mathcal{F}$ as in (9.29). Moreover, since $\underline{n} = \{0, \ldots, n\}$ is a *discrete* pointed category, $F_{\underline{n}}$ is determined by the objects

$$F_{\underline{n}}(j) \in Z\underline{n} \qquad \text{for } j \in \underline{n} \text{ such that} \qquad F_{\underline{n}}(0) = *, \tag{9.34}$$

the basepoint in $Z\underline{n}$.

In fact, the objects $F_{\underline{n}}(j)$ in (9.34) are determined by the object $F_{\underline{1}}(1) \in Z\underline{1}$. In more detail, for each $1 \le j \le n$, consider the morphism in $\mathcal{F}$

$$\underline{1} \xrightarrow{\nu_j} \underline{n} \quad \text{such that} \quad \nu_j(1) = j. \tag{9.35}$$

The naturality diagram (9.29) for ν_j is the following commutative diagram.

$$\begin{array}{ccc} J\underline{1} \cong \underline{1} & \xrightarrow{F_{\underline{1}}} & Z\underline{1} \\ {\scriptstyle \nu_j}\big\downarrow & & \big\downarrow{\scriptstyle Z\nu_j} \\ J\underline{n} \cong \underline{n} & \xrightarrow{F_{\underline{n}}} & Z\underline{n} \end{array} \tag{9.36}$$

Evaluating at the object $1 \in \underline{1}$ yields the object equality

$$F_{\underline{n}}(j) = (Z\nu_j)\big(F_{\underline{1}}(1)\big) \quad \text{in} \quad Z\underline{n}. \tag{9.37}$$

Thus an object F in $\mathsf{Map}_*(J,Z)$ is determined by the object $F_{\underline{1}}(1) \in Z\underline{1}$. Moreover, the naturality diagrams (9.29) for nonzero morphisms $f \in \mathcal{F}$ are determined by the naturality diagrams (9.36) for ν_j with $1 \le j \le n$. This is true by the computation (9.44) with θ replaced by F.

For objects $F, G \in \mathsf{Map}_*(J,Z)$, a *morphism* $\theta : F \longrightarrow G$ is determined by morphisms in $Z\underline{n}$

$$F_{\underline{n}}(j) \xrightarrow{\theta_{\underline{n},j}} G_{\underline{n}}(j) \quad \text{for} \quad 0 \le j \le n \tag{9.38}$$

such that

- $\theta_{\underline{n},0} = 1_*$ for each $n \ge 0$ and
- the modification axiom (9.30) is satisfied.

In fact, the morphisms $\theta_{\underline{n},j}$ in (9.38) are determined by the single morphism

$$F_{\underline{1}}(1) \xrightarrow{\theta_{\underline{1},1}} G_{\underline{1}}(1) \quad \text{in} \quad Z\underline{1} \tag{9.39}$$

subject to the modification axiom (9.30) for $f = \nu_j$ in (9.35) for $1 \le j \le n$. In more detail, the modification axiom for ν_j states that the two whiskered natural transformations below are equal.

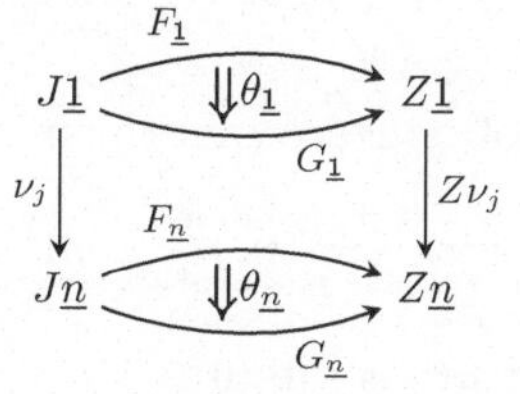

(9.40)

Evaluating at the object $1 \in \underline{1}$ yields the morphism equality

$$\theta_{\underline{n},j} = (Z\nu_j)\theta_{\underline{1},1} \quad \text{in} \quad Z\underline{n}. \tag{9.41}$$

Moreover, for any nonzero morphism $f : \underline{m} \longrightarrow \underline{n}$ in $\mathcal{F}$, the modification axiom (9.30)—which asserts the equality of the two whiskered natural transformations below—is automatically satisfied.

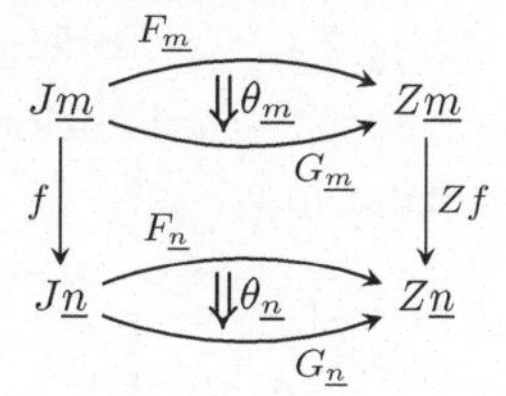

(9.42)

Indeed, for each $1 \le i \le m$ the following diagram in $\mathcal{F}$ is commutative.

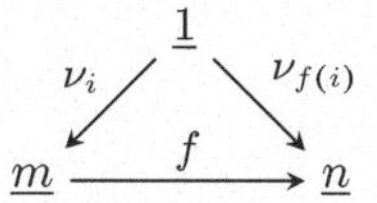

(9.43)

Thus there are morphism equalities in $Z\underline{n}$ as follows.

$$\begin{cases} (Zf)\theta_{\underline{m},i} = (Zf)\big((Z\nu_i)\theta_{\underline{1},1}\big) & \text{by (9.41)} \\ \qquad\qquad = (Z\nu_{f(i)})\theta_{\underline{1},1} & \text{by (9.43)} \\ \qquad\qquad = \theta_{\underline{n},f(i)} & \text{by (9.41)} \end{cases} \tag{9.44}$$

This proves the modification axiom (9.42) for $f : \underline{m} \longrightarrow \underline{n}$.

Multimorphism Categories in Higher Arity

For arity $n \geq 2$, we combine Explanations 9.26 and 9.28. An *object* F in the n-ary multimorphism category $\Gamma\text{-}\mathsf{Cat}\big(\langle X\rangle\,;\,Z\big)$ in (9.32) is a Γ-category morphism, that is, a pointed natural transformation

$$X_1 \wedge \cdots \wedge X_n \xrightarrow{\;F\;} Z. \tag{9.45}$$

Such a Γ-category morphism F consists of, for each n-tuple of objects $\langle \underline{m_j}\rangle_{j=1}^n$ in $\mathcal{F}$, a component pointed functor

$$\textstyle\bigwedge_{j=1}^n X_j\underline{m_j} \xrightarrow{\;F_{\langle \underline{m_j}\rangle}\;} Z\big(\wedge_{j=1}^n \underline{m_j}\big) = Z\big(\underline{m_1\cdots m_n}\big) \tag{9.46}$$

such that, for each n-tuple of nonzero morphisms

$$\big\langle h_j : \underline{m_j} \longrightarrow \underline{p_j}\big\rangle_{j=1}^n \quad \text{in} \quad \mathcal{F}, \tag{9.47}$$

the following *naturality diagram* commutes.

$$\begin{array}{ccc} \bigwedge_{j=1}^n X_j\underline{m_j} & \xrightarrow{\;F_{\langle \underline{m_j}\rangle}\;} & Z\big(\underline{m_1\cdots m_n}\big) \\ {\scriptstyle \wedge_j X_j h_j}\big\downarrow & & \big\downarrow{\scriptstyle Z(\wedge_j h_j)} \\ \bigwedge_{j=1}^n X_j\underline{p_j} & \xrightarrow{\;F_{\langle \underline{p_j}\rangle}\;} & Z\big(\underline{p_1\cdots p_n}\big) \end{array} \tag{9.48}$$

For objects $F, G \in \Gamma\text{-}\mathsf{Cat}\big(\langle X\rangle\,;\,Z\big)$, a *morphism* $\theta : F \longrightarrow G$ is a pointed modification. It consists of, for each n-tuple of objects $\langle \underline{m_j}\rangle_{j=1}^n$ in $\mathcal{F}$, a component pointed natural transformation

$$\textstyle\bigwedge_{j=1}^n X_j\underline{m_j} \underset{G_{\langle \underline{m_j}\rangle}}{\overset{F_{\langle \underline{m_j}\rangle}}{\rightrightarrows}} Z\big(\underline{m_1\cdots m_n}\big), \quad \Downarrow \theta_{\langle \underline{m_j}\rangle} \tag{9.49}$$

such that, for each n-tuple of nonzero morphisms $\langle h_j \rangle$ as in (9.47), the following two whiskered natural transformations are equal.

$$\begin{tikzcd}[column sep=large] \bigwedge_{j=1}^n X_j \underline{m_j} \arrow[r, bend left, "F_{\langle \underline{m_j} \rangle}"] \arrow[r, bend right, "G_{\langle \underline{m_j} \rangle}"'] \arrow[d, "\wedge_j X_j h_j"'] & Z(\underline{m_1 \cdots m_n}) \arrow[d, "Z(\wedge_j h_j)"] \\ \bigwedge_{j=1}^n X_j \underline{p_j} \arrow[r, bend left, "F_{\langle \underline{p_j} \rangle}"] \arrow[r, bend right, "G_{\langle \underline{p_j} \rangle}"'] & Z(\underline{p_1 \cdots p_n}) \end{tikzcd} \qquad \Downarrow\theta_{\langle \underline{m_j} \rangle},\ \Downarrow\theta_{\langle \underline{p_j} \rangle} \tag{9.50}$$

This is called the *modification axiom.*

Colored Units

For a Γ-category Z as in (9.31), the Z-colored unit (3.4)

$$\mathbf{1} \xrightarrow{1_Z} \Gamma\text{-}\mathsf{Cat}(Z\,;\,Z) = \mathsf{Map}_*(Z, Z) \tag{9.51}$$

is the identity natural transformation of the functor $Z : \mathcal{F} \longrightarrow \mathsf{Cat}$.

Symmetric Group Action

Suppose $\sigma \in \Sigma_n$ is a permutation. The right action functor (3.3)

$$\begin{array}{ccc} \Gamma\text{-}\mathsf{Cat}(\langle X \rangle\,;\, Z) & \xrightarrow[\cong]{\ \sigma\ } & \Gamma\text{-}\mathsf{Cat}(\langle X \rangle\sigma\,;\, Z) \\ \parallel & & \parallel \\ \mathsf{Map}_*\big(\bigwedge_{j=1}^n X_j\,,\, Z\big) & & \mathsf{Map}_*\big(\bigwedge_{j=1}^n X_{\sigma(j)}\,,\, Z\big) \end{array} \tag{9.52}$$

sends an object F as in (9.45) to the following vertical composite pointed natural transformation.

$$\begin{tikzcd} \bigwedge_{j=1}^n X_{\sigma(j)} \arrow[d, "\sigma"', "\cong"] \arrow[dr, "F^\sigma"] & \\ \bigwedge_{j=1}^n X_j \arrow[r, "F"'] & Z \end{tikzcd} \tag{9.53}$$

The arrow σ permutes the factors $\langle X_{\sigma(j)} \rangle_{j=1}^n$ using the braiding (9.21) for the pointed Day convolution. More explicitly, for each n-tuple of objects $\langle \underline{m_j} \rangle_{j=1}^n$ in $\mathcal{F}$, the component pointed functor of F^σ is the following composite.

$$\begin{tikzcd}[column sep=huge] \bigwedge_{j=1}^n X_{\sigma(j)} \underline{m_j} \arrow[r, "F^\sigma_{\langle \underline{m_j} \rangle}"] \arrow[d, "\sigma"', "\cong"] & Z(\underline{m_1 \cdots m_n}) \\ \bigwedge_{j=1}^n X_j \underline{m_{\sigma^{-1}(j)}} \arrow[r, "F_{\langle \underline{m_{\sigma^{-1}(j)}} \rangle}"'] & Z(\underline{m_{\sigma^{-1}(1)} \cdots m_{\sigma^{-1}(n)}}) \arrow[u, "\cong", "Z(\sigma^{-1})"'] \end{tikzcd} \tag{9.54}$$

The detail of (9.54) is as follows.

- The left vertical arrow σ permutes the factors $\langle X_{\sigma(j)}\underline{m_j}\rangle_{j=1}^{n}$ according to σ using the braiding for the smash product (9.5) in Cat_*. The latter is induced by the braiding for the Cartesian product in Cat.
- The bottom horizontal arrow is the component pointed functor of F as in (9.46) for the n-tuple of objects $\underline{\langle m_{\sigma^{-1}(j)}\rangle_{j=1}^{n}}$.
- The right vertical arrow is the image under Z of the isomorphism in $\mathcal{F}$

$$\begin{array}{ccc} \underline{\wedge_{j=1}^{n} m_{\sigma^{-1}(j)}} & \xrightarrow[\cong]{\sigma^{-1}} & \underline{\wedge_{j=1}^{n} m_j} \\ \| & & \| \\ \underline{m_{\sigma^{-1}(1)}\cdots m_{\sigma^{-1}(n)}} & & \underline{m_1\cdots m_n} \end{array} \tag{9.55}$$

  that permutes the factors $\underline{\langle m_{\sigma^{-1}(j)}\rangle_{j=1}^{n}}$ according to σ^{-1} using the multiplicative braiding $\xi^{\wedge}$ (2.60) in $\mathcal{F}$.

Suppose $\theta : F \longrightarrow G$ is a morphism in $\Gamma\text{-}\mathsf{Cat}(\langle X\rangle; Z)$ with component pointed natural transformations as in (9.49). The right action functor σ in (9.52) sends θ to the horizontal composite pointed modification

$$\theta^{\sigma} = \theta * 1_{\sigma} : F^{\sigma} \longrightarrow G^{\sigma}$$

as in (9.53) with F replaced by θ. Its component pointed natural transformations are given by the horizontal composites

$$\theta^{\sigma}_{\langle\underline{m_j}\rangle} = 1_{Z(\sigma^{-1})} * \theta_{\langle\underline{m_{\sigma^{-1}(j)}}\rangle} * 1_{\sigma} \tag{9.56}$$

as in (9.54) with F replaced by θ.

Multicategorical Composition

Suppose given, for each $j \in \{1,\ldots,n\}$, an ℓ_j-tuple of Γ-categories $\langle W_j\rangle = \langle W_{ji}\rangle_{i=1}^{\ell_j}$ along with

$$\ell = \ell_1 + \cdots + \ell_n \quad \text{and} \quad \langle W\rangle = \langle\langle W_j\rangle\rangle_{j=1}^{n} \tag{9.57}$$

the concatenated ℓ-tuple. The multicategorical composition (3.5)

$$(\Gamma\text{-}\mathsf{Cat})(\langle X\rangle; Z) \times \textstyle\prod_{j=1}^{n}(\Gamma\text{-}\mathsf{Cat})(\langle W_j\rangle; X_j) \xrightarrow{\gamma} (\Gamma\text{-}\mathsf{Cat})(\langle W\rangle; Z) \tag{9.58}$$

is the following composite functor.

$$\begin{array}{ccc} \mathsf{Map}_*\big(\wedge_{j=1}^{n} X_j, Z\big) \times \prod_{j=1}^{n} \mathsf{Map}_*\big(\wedge_{i=1}^{\ell_j} W_{ji}, X_j\big) & & \\ & \searrow^{\gamma} & \\ \big\downarrow{\scriptstyle 1\times\wedge} & & \mathsf{Map}_*\big(\wedge_{j=1}^{n}\wedge_{i=1}^{\ell_j} W_{ji}, Z\big) \\ & \nearrow_{m} & \\ \mathsf{Map}_*\big(\wedge_{j=1}^{n} X_j, Z\big) \times \mathsf{Map}_*\big(\wedge_{j=1}^{n}\wedge_{i=1}^{\ell_j} W_{ji}, \wedge_{j=1}^{n} X_j\big) & & \end{array} \tag{9.59}$$

The detail of (9.59) is as follows.

- The arrow labeled m is the composition in Γ-Cat as a Cat-category. It is given by
  - vertical composition of pointed natural transformations on objects and
  - horizontal composition of pointed modifications on morphisms.
- In the left vertical arrow, $\wedge$ is the pointed Day convolution (9.18), which extends to pointed natural transformations and pointed modifications componentwise.

The explicit formulas of γ in terms of component functors and component natural transformations are as in (5.25) and (5.27) with the smash product instead of the Cartesian product.

If some $\ell_j = 0$, then (9.59) needs to be slightly adjusted as in the discussion following (5.20) using (9.33) and the monoidal unit diagram J (9.19).

9.3 The Functor A

The purpose of this section is to define the functor

$$\Gamma\text{-Cat} \xrightarrow{A} \mathcal{A}\text{Cat}$$

that is one of the two constituent functors of the inverse K-theory functor $\mathcal{P}$ (Section 10.1).

- The object assignment of A is given in Definition 9.60.
- The morphism assignment of A is given in Definition 9.69.
- The functoriality of A is proved in Lemma 9.79.

Object Assignment of A

Recall the small tight bipermutative category

$$\left(\mathcal{A}, (\oplus, \langle\rangle, \beta^{\oplus}), (\otimes, (1), \beta^{\otimes}), \partial^l = (\delta^r)^{-1}, \partial^r = (\delta^l)^{-1} = 1\right)$$

in Example 2.63 with

- additive structure $\mathcal{A}^{\oplus} = (\mathcal{A}, \oplus, \langle\rangle, \beta^{\oplus})$ and
- multiplicative structure $\mathcal{A}^{\otimes} = (\mathcal{A}, \otimes, (1), \beta^{\otimes})$.

The inverse K-theory functor (Definition 10.2) only uses the additive structure $\mathcal{A}^{\oplus}$. The multiplicative structure $\mathcal{A}^{\otimes}$ is used when we consider the Cat-multifunctoriality of A; see (9.91).

For each integer $n \geq 0$, recall that

- $\overline{n} = \{1, \ldots, n\}$ denotes the unpointed finite set with $\overline{0} = \varnothing$, and
- $\underline{n} = \{0, 1, \ldots, n\}$ denotes the pointed finite set with basepoint 0.

First we define the object assignment of A.

Definition 9.60. Suppose X is a Γ-category (Definition 9.11). Define a functor

$$\mathcal{A} \xrightarrow{AX} \mathsf{Cat} \tag{9.61}$$

as follows.

Objects For an object $m = \langle m_i \rangle_{i=1}^p \in \mathcal{A}$ with each $m_i > 0$, define the category

$$(AX)(m) = \begin{cases} \prod_{i=1}^p X\underline{m_i} & \text{if } p > 0 \text{ and} \\ X\underline{0} = \mathbf{1} & \text{if } p = 0. \end{cases} \tag{9.62}$$

Morphisms For a morphism

$$\langle m_i \rangle_{i=1}^p = m \xrightarrow{\phi} n = \langle n_j \rangle_{j=1}^q \quad \text{in} \quad \mathcal{A}$$

as in (2.64), the functor

$$\prod_{i=1}^p X\underline{m_i} = (AX)(m) \xrightarrow{\phi_* = (AX)(\phi)} (AX)(n) = \prod_{j=1}^q X\underline{n_j} \tag{9.63}$$

is defined as follows, where $j \in \{1, \ldots, q\}$.

(i) If $q = 0$ then ϕ_* is the unique functor to the terminal category $\mathbf{1}$.

(ii) If $p = 0$ and $q > 0$, then the composite of ϕ_* with the j-th coordinate projection is defined as the composite

$$\begin{array}{ccc} (AX)\langle\rangle & \xrightarrow{\phi_*} & (AX)(n) = \prod_{j=1}^q X\underline{n_j} \\ \| & & \downarrow{\scriptstyle \text{project}} \\ X\underline{0} & \xrightarrow{X\iota_{n_j}} & X\underline{n_j} \end{array}$$

with $\iota_{n_j} : \underline{0} \longrightarrow \underline{n_j}$ the unique morphism of pointed finite sets.

(iii) Suppose $p, q > 0$ for this and the next cases. If $\phi^{-1}(\overline{n_j}) = \varnothing$, then the composite of ϕ_* with the j-th coordinate projection is defined as the following composite.

$$\begin{array}{ccc} (AX)(m) = \prod_{i=1}^p X\underline{m_i} & \xrightarrow{\phi_*} & (AX)(n) = \prod_{j=1}^q X\underline{n_j} \\ \downarrow & & \downarrow{\scriptstyle \text{project}} \\ \mathbf{1} = X\underline{0} & \xrightarrow{X\iota_{n_j}} & X\underline{n_j} \end{array}$$

(iv) If $\phi^{-1}(\overline{n_j}) \neq \varnothing$, then there is a unique index $i \in \{1, \ldots, p\}$ such that

$$\varnothing \neq \phi^{-1}(\overline{n_j}) \subset \overline{m_i}.$$

Define the morphism of pointed finite sets

$$\underline{m_i} \xrightarrow{\phi_{i,j}} \underline{n_j} \quad \text{in} \quad \mathcal{F} \tag{9.64}$$

by

$$\phi_{i,j}(x) = \begin{cases} \phi(x) & \text{if } x \in \phi^{-1}(\overline{n_j}) \subset \overline{m_i} \text{ and} \\ 0 & \text{if } x \in \underline{m_i} \smallsetminus \phi^{-1}(\overline{n_j}). \end{cases}$$

The composite of ϕ_* with the j-th coordinate projection is defined as the following composite.

$$\begin{array}{ccc} (AX)(m) = \prod_{i=1}^p X\underline{m_i} & \xrightarrow{\phi_*} & (AX)(n) = \prod_{j=1}^q X\underline{n_j} \\ \Big\downarrow{\scriptstyle \text{project}} & & \Big\downarrow{\scriptstyle \text{project}} \\ X\underline{m_i} & \xrightarrow{(\phi_{i,j})_* = X\phi_{i,j}} & X\underline{n_j} \end{array} \tag{9.65}$$

The functoriality of AX follows from the functoriality of X. This finishes the definition of the functor AX. ◇

Recall that the additive structure $\mathcal{A}^{\oplus} = (\mathcal{A}, \oplus, \langle\rangle, \beta^{\oplus})$ is given by concatenation $\oplus$ of sequences on objects, with the empty sequence $\langle\rangle$ as the strict monoidal unit (Example 2.63).

Lemma 9.66. *In the context of Definition 9.60, the triple*

$$\mathcal{A}^{\oplus} \xrightarrow{(AX,(AX)^2,(AX)^0)} \mathsf{Cat}$$

is a strictly unital strong symmetric monoidal functor with monoidal constraint $(AX)^2$ given by the associativity and unit isomorphisms for the Cartesian product.

Proof. The unit constraint

$$\mathbf{1} \xrightarrow{(AX)^0} (AX)\langle\rangle = X\underline{0} = \mathbf{1} \tag{9.67}$$

is the identity functor by definition. For objects $m = \langle m_i \rangle_{i=1}^p$ and $n = \langle n_j \rangle_{j=1}^p$ in $\mathcal{A}$, the monoidal constraint

$$\begin{array}{ccc} (AX)(m) \times (AX)(n) & \xrightarrow[\cong]{(AX)^2} & (AX)(m \oplus n) \\ \| & & \| \\ \left(\prod_{i=1}^p X\underline{m_i}\right) \times \left(\prod_{j=1}^q X\underline{n_j}\right) & & X\underline{m_1} \times \cdots \times X\underline{m_p} \times X\underline{n_1} \times \cdots \times X\underline{n_q} \end{array} \tag{9.68}$$

is given by

- the associativity isomorphism for the Cartesian product if $p, q > 0$,
- the left unit isomorphism for the Cartesian product if $p = 0$, and
- the right unit isomorphism for the Cartesian product if $q = 0$.

The naturality of $(AX)^2$ and the axioms of a symmetric monoidal functor (Definition 1.17) for $\big(AX, (AX)^2, (AX)^0\big)$ follow from the corresponding properties for the Cartesian product. □

For a Γ-category X, we always regard AX as a strictly unital strong symmetric monoidal functor as in Lemma 9.66.

Morphism Assignment of A

Next we define the assignment of A on morphisms.

Definition 9.69. For a Γ-category morphism $F : X \longrightarrow Y$ (Definition 9.11), define a natural transformation

$$\mathcal{A} \underset{AY}{\overset{AX}{\Downarrow AF}} \mathsf{Cat} \tag{9.70}$$

by, for each object $m = \langle m_i \rangle_{i=1}^p \in \mathcal{A}$, the component functor

$$(AF)_m = \begin{cases} \prod_{i=1}^p F_{\underline{m_i}} : (AX)(m) \longrightarrow (AY)(m) & \text{if } p > 0 \text{ and} \\ 1_1 : (AX)\langle\rangle \longrightarrow (AY)\langle\rangle & \text{if } p = 0. \end{cases} \tag{9.71}$$

The naturality of $AF : AX \longrightarrow AY$ follows from the pointed naturality of F. ◇

Explanation 9.72 (Injectivity)**.** The assignment $F \longmapsto AF$ in Definition 9.69 is injective. In other words, if $F, G : X \longrightarrow Y$ are two Γ-category morphisms such that $AF = AG$ as natural transformations, then $F = G$. This is true by the $p = 1$ case of (9.71). ◇

Recall from Definition 1.21 that a *monoidal natural transformation* between monoidal functors is a natural transformation that is compatible with the unit constraints and the monoidal constraints in the sense of (1.22).

Lemma 9.73. *For a Γ-category morphism $F : X \longrightarrow Y$, the natural transformation in Definition 9.69*

$$AX \xrightarrow{AF} AY$$

is a monoidal natural transformation, which is, furthermore, a natural isomorphism if and only if F is.

Proof. For the first assertion, we consider the two constraint diagrams in (1.22).

- The unit constraint diagram is commutative because each of $(AF)_{\langle\rangle}$, $(AX)^0$, and $(AY)^0$ is the identity functor $1_{\mathbf{1}}$.
- The commutativity of the monoidal constraint diagram for AF follows from the naturality of the associativity and unit isomorphisms for the Cartesian product.

The second assertion about natural isomorphism follows from the definition (9.71) of $(AF)_m$ as the product of the component functors $F_{\underline{m_i}}$. □

Functoriality of A

To express the functoriality of A in Definitions 9.60 and 9.69 and to define the inverse K-theory functor, we use the following concept.

Definition 9.74. For a (non-symmetric) Cat-multicategory M (Definition 3.2), its *underlying category* M_1 is the category defined as follows.

- Its objects are those in M.
- Its morphism sets are

$$\mathsf{M}_1(a,b) = \mathsf{Ob}\big(\mathsf{M}(a;b)\big)$$

  for objects $a, b \in \mathsf{M}$.
- Identity morphisms are the colored units in M.
- Composition of morphisms is the multicategorical composition in M restricted to the objects in the 1-ary multimorphism categories.

Definition 1.32, Proposition 1.35, and Example 3.18 imply that M_1 is a well-defined category. If the context is clear, then we omit the subscript and denote M_1 by M to simplify the notation. ◇

Example 9.75 (Permutative Opfibrations)**.** Suppose $(\mathcal{D}, \oplus, \otimes)$ is a small tight bipermutative category (Definition 2.41). The underlying category of the non-symmetric Cat-multicategory $\mathsf{PFib}^{\mathcal{D}}$ (Theorem 7.51) has

- permutative opfibrations over $\mathcal{D}$ (Definition 7.6) as objects;
- opcartesian 1-linear functors (Definition 7.11) as morphisms; and
- identities and composition (Definition 7.36) given by those of symmetric monoidal functors.

Permutative opfibrations over $\mathcal{D}$ only use the additive structure $\oplus$. Thus, by the description of opcartesian 1-linear functors in Example 7.24, only the additive structure $\oplus$, but *not* the multiplicative structure $\otimes$, is used in the underlying category of $\mathsf{PFib}^{\mathcal{D}}$. ◇

Example 9.76 (Γ-Categories)**.** The underlying category of the Cat-multicategory Γ-Cat (Theorem 9.23 and Section 9.2) is the category $[\mathcal{F}, \mathsf{Cat}]_*$ of Γ-categories and their morphisms (Definition 9.11). In Lemma 9.79 and Definition 10.2 we consider the underlying category of Γ-Cat, which is also denoted Γ-Cat. ◇

Example 9.77 (Bipermutative-Indexed Categories)**.** Suppose $(\mathcal{D}, \oplus, \otimes)$ is a small tight bipermutative category (Definition 2.41). The underlying category of the Cat-multicategory $\mathcal{D}$Cat (Theorem 5.71) has

- symmetric monoidal functors $\mathcal{D}^{\oplus} \longrightarrow \mathsf{Cat}$ as objects;
- monoidal natural transformations (Explanation 5.48 (4)) as morphisms;
- identity natural transformations as identities; and
- vertical composition of natural transformations (5.20) as composition.

Note that only the additive structure ⊕, but *not* the multiplicative structure ⊗, is used in the underlying category of $\mathcal{D}$Cat. In Lemma 9.79 and Definition 10.2 we consider the underlying category of $\mathcal{A}$Cat, which is also denoted $\mathcal{A}$Cat, for the small tight bipermutative category $\mathcal{A}$ in Example 2.63. ◇

Example 9.78 (Permutative Categories)**.** The underlying category of the Cat-multicategory $\mathsf{PermCat}^{\mathsf{sg}}$ (Theorem 6.31) has

- small permutative categories as objects;
- strictly unital strong symmetric monoidal functors (Example 6.13) as morphisms; and
- identities and composition given by those of monoidal functors.

In Definition 10.2 we consider the underlying category of $\mathsf{PermCat}^{\mathsf{sg}}$, which is also denoted $\mathsf{PermCat}^{\mathsf{sg}}$. ◇

The next result uses the underlying categories of Γ-Cat and $\mathcal{A}$Cat in, respectively, Examples 9.76 and 9.77.

Lemma 9.79. *The object and morphism assignments*

$$X \longmapsto AX \quad \text{and} \quad F \longmapsto AF$$

in, respectively, Definitions 9.60 and 9.69 define a functor

$$\Gamma\text{-}\mathsf{Cat} \xrightarrow{\;A\;} \mathcal{A}\mathsf{Cat}. \tag{9.80}$$

Proof. The object and morphism assignments are well defined by, respectively, Lemmas 9.66 and 9.73. The functoriality of A follows from the definition (9.71) of each component functor $(AF)_m$ and the functoriality of the Cartesian product. □

The functor A in (9.80) is extended to a Cat-multifunctor in Theorem 9.120.

9.4 Multimorphism Functors of A

Our next objective is to extend the functor

$$\Gamma\text{-}\mathsf{Cat} \xrightarrow{A} \mathcal{A}\mathsf{Cat}$$

in (9.80) to a Cat-multifunctor; see Theorem 9.120. In this section we first construct its multimorphism functors. For context recall the following.

- $\Gamma\text{-}\mathsf{Cat}$ (Definition 9.11) has the structure of a Cat-multicategory (Theorem 9.23) induced by a symmetric monoidal closed structure. The monoidal product is given by the *pointed* Day convolution (9.18).
- $\mathcal{D}\mathsf{Cat}$ is a Cat-multicategory for each small tight bipermutative category $\mathcal{D}$ (Theorem 5.71). In particular, $\mathcal{A}\mathsf{Cat}$ is a Cat-multicategory for the small tight bipermutative category $\mathcal{A}$ in Example 2.63. The Cat-multicategory structure on $\mathcal{A}\mathsf{Cat}$ is *not* induced by a monoidal structure (Explanation 5.64).

Here is an outline of this section.

- The multimorphism functors of A in arity 0 are given in Definition 9.81. They are verified to be isomorphisms in Lemma 9.85.
- The multimorphism functors of A in positive arity are constructed in
  - Definition 9.93 and Lemma 9.102 for objects and
  - Definition 9.103 and Lemma 9.109 for morphisms.
- Lemma 9.110 shows that these assignments define a functor.

Multimorphism Functors of A in Arity 0

The object assignment of A, $X \longmapsto AX$ for Γ-categories X, is in Definition 9.60 and Lemma 9.66. To define the multimorphism functors of A, first we consider the case where the input has arity 0.

Definition 9.81. Suppose Z is a Γ-category. Define a functor

$$\begin{array}{ccc} \Gamma\text{-}\mathsf{Cat}(\langle\rangle\,;Z) & \xrightarrow{A} & \mathcal{A}\mathsf{Cat}(\langle\rangle\,;AZ) \\ \| & & \| \\ \mathsf{Map}_*(J,Z) & & (AZ)(1) = Z\underline{1} \end{array} \tag{9.82}$$

as follows. In (9.82) the equality on the left is from (9.33), and the equalities on the right are from (5.62) and (9.62).

Objects By (9.37) an object F in $\mathsf{Map}_*(J, Z)$ is determined by the object $F_{\underline{1}}(1)$ in $Z\underline{1}$. We define the object

$$AF = F_{\underline{1}}(1) \quad \text{in} \quad Z\underline{1}. \tag{9.83}$$

Morphisms By (9.41) a morphism $\theta : F \longrightarrow G$ in $\mathsf{Map}_*(J, Z)$ is determined by the morphism $\theta_{\underline{1},1}$ in (9.39). We define the morphism

$$A\theta = \theta_{\underline{1},1} : AF = F_{\underline{1}}(1) \longrightarrow AG = G_{\underline{1}}(1) \quad \text{in} \quad Z\underline{1}. \tag{9.84}$$

This finishes the definition of A in (9.82). ◇

Lemma 9.85. *In the context of Definition 9.81,*

$$\Gamma\text{-}\mathsf{Cat}(\langle\rangle\,;Z) \xrightarrow[\cong]{A} \mathcal{A}\mathsf{Cat}(\langle\rangle\,;AZ)$$

is an isomorphism of categories.

Proof. The functoriality of A in (9.82) follows from the fact that identity morphisms and composition in $\mathsf{Map}_*(J, Z)$ are given componentwise by those of natural transformations (Explanation 9.28). The functor A is an isomorphism for the following reasons.

- By (9.37) and (9.41) A is injective on, respectively, objects and morphisms.
- Given an object $z \in Z\underline{1}$, we define an object F in $\mathsf{Map}_*(J, Z)$ using (9.34), (9.35), and (9.37):

$$F_{\underline{n}}(j) = \begin{cases} * \in Z\underline{n} & \text{if } j = 0 \text{ and} \\ (Z\nu_j)(z) \in Z\underline{n} & \text{if } 1 \le j \le n. \end{cases} \tag{9.86}$$

  The naturality diagram (9.36) for ν_j holds because

$$\begin{cases} F_{\underline{1}}(1) = (Z\nu_1)(z) & \text{by (9.86)} \\ \qquad\quad\;\, = z & \text{by functoriality of } Z. \end{cases} \tag{9.87}$$

  By (9.83) and (9.87)

$$AF = F_{\underline{1}}(1) = z,$$

  so A is surjective on objects.
- The surjectivity of A on morphism sets is proved similarly using (9.39), (9.41), and (9.84).

Therefore, A is an isomorphism of categories. □

Multimorphism Functors of A in Positive Arity: Object Assignment

Recall the following.

- With $\mathcal{D} = \mathcal{A}$ in Definition 5.39, the n-ary 1-cells in $\mathcal{A}\mathsf{Cat}$ are additive natural transformations. The latter are natural transformations as in (5.5) that satisfy the unity axiom (5.43) and the additivity axiom (5.45). In terms of objects and morphisms, the additivity axiom is (5.51).
- By Lemma 9.66 A sends each Γ-category to a strictly unital strong symmetric monoidal functor $\mathcal{A}^{\oplus} \longrightarrow \mathsf{Cat}$.

Next we define the object assignment of the multimorphism functor

$$\Gamma\text{-}\mathsf{Cat}(\langle X\rangle; Z) \xrightarrow{\quad A \quad} \mathcal{A}\mathsf{Cat}(\langle AX\rangle; AZ)$$

where the input $\langle X\rangle$ has arity $n > 0$. When the input has positive arity, we need to consider iterated products in $\mathcal{A}$ using its multiplicative structure $\otimes$. To simplify the presentation, we use the following notation.

Notation 9.88. For $1 \le j \le n$ suppose given an object

$$m^j = \langle m^j_{i_j}\rangle_{i_j=1}^{r_j} = \left(m^j_1, m^j_2, \ldots, m^j_{r_j}\right) \in \mathcal{A} \tag{9.89}$$

with length $r_j \ge 0$ and each $m^j_{i_j} > 0$. For each n-tuple $(i_1, \ldots, i_n)$ with each $1 \le i_j \le r_j$, define the positive integer

$$m^{1,\ldots,n}_{i_1,\ldots,i_n} = \textstyle\prod_{j=1}^n m^j_{i_j} = m^1_{i_1} m^2_{i_2} \cdots m^n_{i_n}. \tag{9.90}$$

Using these positive integers as entries, define the object

$$\begin{aligned} m^{1,\ldots,n} &= \textstyle\bigotimes_{j=1}^n m^j \in \mathcal{A} \\ &= \left\langle \cdots \left\langle \left\langle m^{1,\ldots,n}_{i_1,\ldots,i_n} \right\rangle_{i_1=1}^{r_1} \right\rangle_{i_2=1}^{r_2} \cdots \right\rangle_{i_n=1}^{r_n} \end{aligned} \tag{9.91}$$

with length $r_1 r_2 \cdots r_n$. Here $\otimes$ is the multiplicative structure (2.68) in the bipermutative category $\mathcal{A}$. ◇

Explanation 9.92. As we mentioned just after (2.68), the monoidal product

$$\langle m_i\rangle_{i=1}^p \otimes \langle n_j\rangle_{j=1}^q = \langle\langle m_i n_j\rangle_{i=1}^p\rangle_{j=1}^q \quad \text{in} \quad \mathcal{A}$$

is visualized as the $q \times p$ matrix with (j, i)-entry given by the positive integer $m_i n_j$, which is also regarded as the unpointed finite set $\overline{m_i n_j}$. Similarly, the n-fold monoidal product $m^{1,\ldots,n}$ in (9.91) may be visualized as an n-dimensional matrix. A typical entry has the form $m^{1,\ldots,n}_{i_1,\ldots,i_n}$ in (9.90), which is also regarded as the unpointed finite set

$$\overline{m^{1,\ldots,n}_{i_1,\ldots,i_n}} = \textstyle\prod_{j=1}^n \overline{m^j_{i_j}} = \overline{m^1_{i_1}} \times \overline{m^2_{i_2}} \times \cdots \times \overline{m^n_{i_n}}.$$

If any $r_j = 0$, then m^j is the empty sequence $\langle\rangle \in \mathcal{A}$, and so is $m^{1,\ldots,n}$. ◇

Definition 9.93. Suppose given

- Γ-categories $Z, X_1, \ldots, X_n$ with $n > 0$, $\langle X \rangle = \langle X_j \rangle_{j=1}^n$, and $\langle AX \rangle = \langle AX_j \rangle_{j=1}^n$, and
- an object F in the n-ary multimorphism category $\Gamma\text{-}\mathsf{Cat}\big(\langle X \rangle\,;\, Z\big)$, that is, a pointed natural transformation (9.45)

$$\textstyle\bigwedge_{j=1}^n X_j \xrightarrow{\;\;F\;\;} Z.$$

Define the data of an additive natural transformation (Definition 5.39)

$$\langle AX \rangle \xrightarrow{\;\;AF\;\;} AZ \tag{9.94}$$

as follows. For each n-tuple of objects

$$\langle m \rangle = \langle m^j \rangle_{j=1}^n \in \mathcal{A}^n \tag{9.95}$$

with each m^j as in (9.89), the component functor $(AF)_{\langle m \rangle}$ in the sense of (5.7) is defined by the following commutative diagrams for $1 \le i_j \le r_j$ and $1 \le j \le n$.

$$\begin{array}{ccccc}
\prod_{j=1}^n (AX_j) m^j & & \xrightarrow{\quad (AF)_{\langle m \rangle} \quad} & & (AZ) m^{1,\ldots,n} \\
\| & & & & \| \\
\prod_{j=1}^n \prod_{i_j=1}^{r_j} X_j \underline{m^j_{i_j}} & & & & \prod_{i_n=1}^{r_n} \cdots \prod_{i_1=1}^{r_1} Z \underline{m^{1,\ldots,n}_{i_1,\ldots,i_n}} \\
\Big\downarrow {\scriptstyle \prod_j \mathsf{pr}_{i_j}} & & & & \Big\downarrow {\scriptstyle \mathsf{pr}_{i_1,\ldots,i_n}} \\
\prod_{j=1}^n X_j \underline{m^j_{i_j}} & \xrightarrow{\;\mathsf{pr}\;} & \bigwedge_{j=1}^n X_j \underline{m^j_{i_j}} & \xrightarrow{\;F_{\langle m^j_{i_j} \rangle_{j=1}^n}\;} & Z \underline{m^{1,\ldots,n}_{i_1,\ldots,i_n}}
\end{array} \tag{9.96}$$

This finishes the definition of AF in (9.94). ◇

Explanation 9.97 (Components)**.** The detail of the diagram (9.96) is as follows.

- Each of AX_j and AZ is a strictly unital strong symmetric monoidal functor $\mathcal{A}^{\oplus} \longrightarrow \mathsf{Cat}$ (Lemma 9.66). The categories $(AX_j)m^j$ and $(AZ)m^{1,\ldots,n}$ are defined in (9.62).
- If any $r_j = 0$, then $m^{1,\ldots,n}$ in (9.91) is the empty sequence, and $(AF)_{\langle m \rangle}$ is the unique functor to the terminal category $(AZ)\langle\rangle = \mathbf{1}$.
- Each arrow $\mathsf{pr}_?$ is the ?-coordinate projection.
- The arrow pr is the universal arrow from the Cartesian product to the smash product (9.5). It is defined as the identity functor if $n = 1$.

- The arrow

$$F_{\langle m^j_{i_j}\rangle_{j=1}^n} = F_{\left(m^1_{i_1}, m^2_{i_2}, \ldots, m^n_{i_n}\right)}$$

  is the $\langle m^j_{i_j}\rangle_{j=1}^n$-component functor, in the sense of (9.46), of the given pointed natural transformation $F : \bigwedge_{j=1}^n X_j \longrightarrow Z$.

If $n = 1$, then $(AF)_{\langle m\rangle}$ agrees with the earlier definition in (9.71). ◇

Explanation 9.98 (Injectivity). The assignment

$$F \longmapsto AF$$

in Definition 9.93 is injective. This is true by

- the special case of (9.96) with each $r_j = 1$, which means that each $m^j \in \mathcal{A}$ has length 1, and
- the universal property of the smash product (9.5).

The special case with $n = 1$ is in Explanation 9.72. ◇

Explanation 9.99 (Formulas). The component functor $(AF)_{\langle m\rangle}$ in (9.96) can be described in terms of object and morphism assignments as follows. Suppose given, for each $1 \leq j \leq n$,

$$x^j = \left\langle x^j_{i_j}\right\rangle_{i_j=1}^{r_j} = \left(x^j_1, x^j_2, \ldots, x^j_{r_j}\right) \in \textstyle\prod_{i_j=1}^{r_j} X_j \underline{m^j_{i_j}} = (AX_j)m^j \tag{9.100}$$

with entries either all objects or all morphisms. The image of

$$\langle x^j\rangle_{j=1}^n \in \textstyle\prod_{j=1}^n (AX_j)m^j$$

under the component functor $(AF)_{\langle m\rangle}$ is given by

$$\begin{aligned}(AF)_{\langle m\rangle}\langle x^j\rangle_{j=1}^n &= \left\langle \cdots \left\langle \left\langle \left(F_{\langle m^j_{i_j}\rangle_{j=1}^n}\left\langle x^j_{i_j}\right\rangle_{j=1}^n\right)\right\rangle_{i_1=1}^{r_1}\right\rangle_{i_2=1}^{r_2} \cdots \right\rangle_{i_n=1}^{r_n} \\ &\in (AZ)m^{1,\ldots,n} = \textstyle\prod_{i_n=1}^{r_n} \cdots \prod_{i_1=1}^{r_1}\left(Z\underline{m^{1,\ldots,n}_{i_1,\ldots,i_n}}\right).\end{aligned} \tag{9.101}$$

Here

$$\left\langle x^j_{i_j}\right\rangle_{j=1}^n = \left(x^1_{i_1}, x^2_{i_2}, \ldots, x^n_{i_n}\right) \in \textstyle\prod_{j=1}^n X_j \underline{m^j_{i_j}}$$

also denotes its image in $\bigwedge_{j=1}^n X_j \underline{m^j_{i_j}}$ under the universal arrow pr. ◇

Next we show that AF in (9.94) defines an additive natural transformation (Definition 5.39).

Lemma 9.102. *In the context of Definition 9.93, AF is an additive natural transformation.*

Proof. For each n-tuple of morphisms in $\mathcal{A}$, the naturality diagram (5.9) for AF follows from

- the universal property of Cartesian products,
- the naturality of each $\mathsf{pr}_?$ and pr in (9.96), and
- the naturality diagram (9.48) for the given pointed natural transformation F.

Thus AF is a natural transformation in the sense of (5.5). It remains to check the two axioms in Definition 5.39.

Unity (5.43). The unit object (5.33) of

$$AZ : \mathcal{A}^{\oplus} \longrightarrow \mathsf{Cat}$$

belongs to the terminal category $(AZ)\langle\rangle = \mathbf{1}$. So the unity axiom holds.

Additivity (5.45). This axiom holds for the following reasons.

- Each component functor $(AF)_{\langle m \rangle}$ in (9.94) is defined by its coordinate projections as a component functor of F.
- By definition (9.68) the monoidal constraints $(AX_j)^2$ and $(AZ)^2$ are given by the associativity and unit isomorphisms for the Cartesian product.

Thus, in the desired diagram (5.45) in the current context, each composite followed by a coordinate projection to a factor of $(AZ)(?)$ is induced by the same component functor of F. □

Multimorphism Functors of A in Positive Arity: Morphism Assignment

Next we define the morphism assignment of the multimorphism functor

$$\Gamma\text{-}\mathsf{Cat}\big(\langle X \rangle ; Z\big) \xrightarrow{\;A\;} \mathcal{A}\mathsf{Cat}\big(\langle AX \rangle ; AZ\big)$$

where the input $\langle X \rangle$ has arity $n > 0$. Recall that a morphism in $\Gamma\text{-}\mathsf{Cat}\big(\langle X \rangle ; Z\big)$ is a pointed modification with component pointed natural transformations as in (9.49) that satisfy the modification axiom (9.50).

Definition 9.103. In the context of Definition 9.93, suppose given a morphism

$$\textstyle\bigwedge_{j=1}^{n} X_j \underset{G}{\overset{F}{\Downarrow \theta}} Z \quad \text{in} \quad \Gamma\text{-}\mathsf{Cat}\big(\langle X \rangle ; Z\big).$$

Define the data of an additive modification (Definition 5.52)

$$\langle AX \rangle \underset{AG}{\overset{AF}{\Downarrow A\theta}} AZ \tag{9.104}$$

by requiring that, for each n-tuple of objects $\langle m\rangle \in \mathcal{A}^n$ as in (9.95), the two whiskered natural transformations in the following diagram be equal.

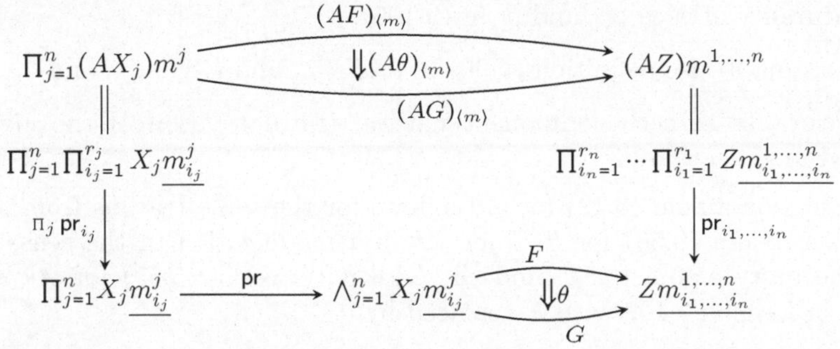

(9.105)

In this diagram, $\theta : F \longrightarrow G$ denotes the $\langle m_{i_j}^j\rangle_{j=1}^n$-component pointed natural transformation of θ. ◇

Explanation 9.106 (Components and Injectivity)**.** Consider Definition 9.103.

1. The diagram (9.105) defining the component $(A\theta)_{\langle m\rangle}$ is obtained from the diagram (9.96), which defines the component functor $(AF)_{\langle m\rangle}$, by replacing F with θ.
2. The assignment
$$\theta \longmapsto A\theta$$
is injective. This is true by
   - the special case of (9.105) with each $r_j = 1$ and
   - the fact that the universal arrow pr from the Cartesian product to the smash product (9.5) is surjective on objects.

   This extends the injectivity of the assignment $F \longmapsto AF$ in Explanation 9.98. ◇

Explanation 9.107 (Formulas)**.** Using the notation in Explanation 9.99, the component natural transformation $(A\theta)_{\langle m\rangle}$ in (9.105) is given by the morphism

$$\begin{aligned}\big((A\theta)_{\langle m\rangle}\big)_{\langle x^j\rangle_{j=1}^n} &= \Big\langle \cdots \Big\langle \Big\langle \theta_{\langle x_{i_j}^j\rangle_{j=1}^n} \Big\rangle_{i_1=1}^{r_1} \Big\rangle_{i_2=1}^{r_2} \cdots \Big\rangle_{i_n=1}^{r_n} \\ &\in (AZ)m^{1,\ldots,n} = \prod_{i_n=1}^{r_n} \cdots \prod_{i_1=1}^{r_1} \Big(Z\underline{m_{i_1,\ldots,i_n}^{1,\ldots,n}} \Big)\end{aligned} \tag{9.108}$$

for each object $\langle x^j\rangle_{j=1}^n$ in $\prod_{j=1}^n (AX_j)m^j$. ◇

Lemma 9.109. *In the context of Definition 9.103, $A\theta$ is an additive modification.*

Proof. The naturality of each component $(A\theta)_{\langle m\rangle}$ in (9.105) follows from

- the universal property of Cartesian products,
- the naturality of each $\mathsf{pr}_?$ and pr in (9.105),
- the coordinate-wise definition (9.96) of $(AF)_{\langle m\rangle}$ and $(AG)_{\langle m\rangle}$, and
- the naturality of each component natural transformation of the given pointed modification θ.

The modification axiom (5.12) for $A\theta$ follows for the same reasons from the modification axiom (9.50) for θ. Thus $A\theta$ is a modification in the sense of (5.10). The unity axiom (5.54) and the additivity axiom (5.55) for $A\theta$ are proved as in Lemma 9.102 with F replaced by θ. □

Lemma 9.110. *The object and morphism assignments*

$$F \longmapsto AF \quad \text{and} \quad \theta \longmapsto A\theta$$

in, respectively, Definitions 9.93 and 9.103 define a functor

$$\Gamma\text{-}\mathsf{Cat}(\langle X\rangle; Z) \xrightarrow{\;A\;} \mathcal{A}\mathsf{Cat}(\langle AX\rangle; AZ) \tag{9.111}$$

that is injective on objects and morphism sets.

Proof. The object and morphism assignments are well defined by, respective, Lemmas 9.102 and 9.109. By (5.4) and (9.32) and Explanation 9.28, in both the domain and the codomain of A, identity morphisms and composition are given componentwise by those of natural transformations. Thus the functoriality of A follows from the definition (9.105) of $A\theta$ as given componentwise by the corresponding component of θ. The injectivity of A on objects and morphism sets is discussed in Explanations 9.98 and 9.106. □

9.5 Cat-Multifunctoriality of A

The purpose of this section is to show that A with

- object assignment $X \longmapsto AX$ in Lemma 9.66 and
- multimorphism functors in Lemma 9.110

is a Cat-multifunctor in the sense of Definition 3.19. Here is an outline of this section.

- Lemma 9.112 shows that A preserves the symmetric group action of the Cat-multicategories Γ-Cat and $\mathcal{A}$Cat.
- Lemma 9.116 shows that A preserves the composition in Γ-Cat and $\mathcal{A}$Cat.
- The main result on the Cat-multifunctoriality of A is Theorem 9.120.

Preservation of Symmetric Group Action

Recall that the symmetric group action is discussed in

- (9.54) and (9.56) for the Cat-multicategory Γ-Cat and
- (5.15) and (5.18) for the Cat-multicategory $\mathcal{A}$Cat.

Lemma 9.112. *The functors A in* (9.111) *preserve the symmetric group action.*

Proof. We must show that, for each permutation $\sigma \in \Sigma_n$, the following diagram of functors commutes.

$$\begin{array}{ccc} \Gamma\text{-}\mathsf{Cat}\big(\langle X\rangle\,;\,Z\big) & \xrightarrow{\;A\;} & \mathcal{A}\mathsf{Cat}\big(\langle AX\rangle\,;\,AZ\big) \\ \sigma\big\downarrow\cong & & \cong\big\downarrow\sigma \\ \Gamma\text{-}\mathsf{Cat}\big(\langle X\rangle\sigma\,;\,Z\big) & \xrightarrow{\;A\;} & \mathcal{A}\mathsf{Cat}\big(\langle AX\rangle\sigma\,;\,AZ\big) \end{array} \tag{9.113}$$

Suppose F is an object in $\Gamma\text{-}\mathsf{Cat}\big(\langle X\rangle\,;\,Z\big)$, which means a pointed natural transformation $F : \bigwedge_{j=1}^{n} X_j \longrightarrow Z$ as in (9.45). With $(-)^{\sigma}$ denoting the right σ-action, to show that the additive natural transformations

$$\langle AX\rangle\sigma \overset{A(F^{\sigma})}{\underset{(AF)^{\sigma}}{\rightrightarrows}} AZ$$

are equal, it suffices to show that their corresponding component functors (5.7) are equal (Explanation 5.48 (1)). Moreover, by the definition (9.62) of AZ, it suffices to show that the corresponding component functors of $A(F^{\sigma})$ and $(AF)^{\sigma}$ followed by each coordinate projection are equal. Thus, using Notation 9.88 and the abbreviations

$$\tau = \sigma^{-1}, \qquad \widetilde{m} = m_{i_1,\ldots,i_n}^{1,\ldots,n}, \qquad \text{and} \qquad \widetilde{m}^{\tau} = m_{i_{\sigma^{-1}(1)},\ldots,i_{\sigma^{-1}(n)}}^{\sigma^{-1}(1),\ldots,\sigma^{-1}(n)} \tag{9.114}$$

it suffices to show that the following diagram, which will be explained further below, is commutative.

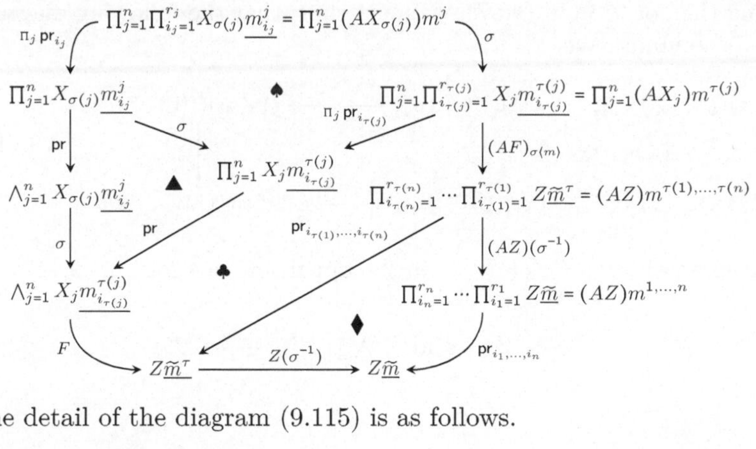

(9.115)

The detail of the diagram (9.115) is as follows.

- By (9.54) and (9.96) the left-bottom boundary composite is the composite functor
$$\mathsf{pr}_{i_1,\ldots,i_n} \circ A(F^\sigma)_{\langle m \rangle}.$$
- By (5.15) the right boundary composite is the composite functor
$$\mathsf{pr}_{i_1,\ldots,i_n} \circ (AF)^\sigma_{\langle m \rangle}.$$
- The region ♠ is commutative by the naturality of the braiding for the Cartesian product.
- The region ▲ is commutative by the universal property of the pushout (9.5) that defines the smash product.
- The region ♣ is commutative by (9.96) applied to the n-tuple of objects
$$\sigma\langle m \rangle = \big\langle m^{\sigma^{-1}(j)} \big\rangle_{j=1}^{n} \in \mathcal{A}^n.$$
- The region ♦ is commutative by (2.70) and (9.65).

This proves that the diagram (9.113) is commutative on objects.

To prove that the diagram (9.113) is commutative on morphisms, we reuse the proof for objects above by replacing (5.15), (9.54), and (9.96) with, respectively, (5.18), (9.56), and (9.105). □

Preservation of Composition

Next we show that A preserves the composition in the Cat-multicategories Γ-Cat and $\mathcal{A}$Cat, which are defined in, respectively, (9.59) and (5.20).

Lemma 9.116. *The functors A in* (9.111) *preserve the multicategorical composition.*

Proof. Using the notation in (9.57), we must show that the following diagram of functors is commutative.

$$\begin{array}{ccc} (\Gamma\text{-}\mathsf{Cat})(\langle X \rangle; Z) \times \prod_{j=1}^{n} (\Gamma\text{-}\mathsf{Cat})(\langle W_j \rangle; X_j) & \xrightarrow{\gamma} & (\Gamma\text{-}\mathsf{Cat})(\langle W \rangle; Z) \\ {\scriptstyle A \times \prod_j A} \downarrow & & \downarrow {\scriptstyle A} \\ (\mathcal{A}\mathsf{Cat})(\langle AX \rangle; AZ) \times \prod_{j=1}^{n} (\mathcal{A}\mathsf{Cat})(\langle AW_j \rangle; AX_j) & \xrightarrow{\gamma} & (\mathcal{A}\mathsf{Cat})(\langle AW \rangle; AZ) \end{array} \tag{9.117}$$

Suppose given an object $\big(F; \langle F_j \rangle_{j=1}^{n}\big)$ in the upper-left category in (9.117), so

$$\textstyle\bigwedge_{j=1}^{n} X_j \xrightarrow{F} Z \quad \text{and} \quad \bigwedge_{i=1}^{\ell_j} W_{ji} \xrightarrow{F_j} X_j$$

are pointed natural transformations as in (9.45). We must show that the two additive natural transformations below are equal.

$$\langle AW\rangle \underset{\gamma(AF;\langle AF_j\rangle)}{\overset{A\gamma(F;\langle F_j\rangle)}{\rightrightarrows}} AZ \tag{9.118}$$

As in the proof of Lemma 9.112, it suffices to show that in (9.118) the corresponding component functors (5.7) followed by each coordinate projection are equal. This is proved in the diagram (9.119) using the following notation.

Suppose given, for $1 \le j \le n$ and $1 \le i \le \ell_j$, objects

$$p^{ji} = \langle p^{ji}_{k_{ji}}\rangle_{k_{ji}=1}^{r_{ji}} = \left(p_1^{ji}, p_2^{ji}, \ldots, p_{r_{ji}}^{ji}\right) \in \mathcal{A}$$

with each $p^{ji}_{k_{ji}} > 0$ and $r_{ji} \ge 0$, along with the following notation, where $\ell = \ell_1 + \cdots + \ell_n$.

$$\begin{cases} p^{j\bullet} = \langle p^{ji}\rangle_{i=1}^{\ell_j} \in \mathcal{A}^{\ell_j} & p^j = \bigotimes_{i=1}^{\ell_j} p^{ji} \in \mathcal{A} & \langle p^j\rangle_j = \langle p^j\rangle_{j=1}^n \in \mathcal{A}^n \\ p^{\bullet\bullet} = \langle p^{j\bullet}\rangle_{j=1}^n \in \mathcal{A}^\ell & p = \bigotimes_{j=1}^n p^j \in \mathcal{A} & \\ (k_j) = \left(k_{j\ell_j}, \ldots, k_{j1}\right) & (k) = \left((k_n), \ldots, (k_1)\right) & \end{cases}$$

Moreover, in (smash) products the indices j, i, and k_{ji} run through, respectively,

$$1 \le j \le n, \qquad 1 \le i \le \ell_j, \quad \text{and} \quad 1 \le k_{ji} \le r_{ji}.$$

Here are the key examples that appear in the diagram (9.119).

$$\begin{cases} \prod_{j,i} = \prod_{j=1}^n \prod_{i=1}^{\ell_j} \qquad \prod_j \bigwedge_i = \prod_{j=1}^n \bigwedge_{i=1}^{\ell_j} \\ \prod_{j,i,k_{ji}} = \prod_{j=1}^n \prod_{i=1}^{\ell_j} \prod_{k_{ji}=1}^{r_{ji}} \\ \prod_{j,(k_j)} = \prod_{j=1}^n \prod_{k_{j\ell_j}=1}^{r_{j\ell_j}} \cdots \prod_{k_{j1}=1}^{r_{j1}} \\ \prod_{(k)} = \prod_{k_{n\ell_n}=1}^{r_{n\ell_n}} \cdots \prod_{k_{n1}=1}^{r_{n1}} \cdots \prod_{k_{1\ell_1}=1}^{r_{1\ell_1}} \cdots \prod_{k_{11}=1}^{r_{11}} \end{cases}$$

With the notation above, we consider the following diagram of functors.

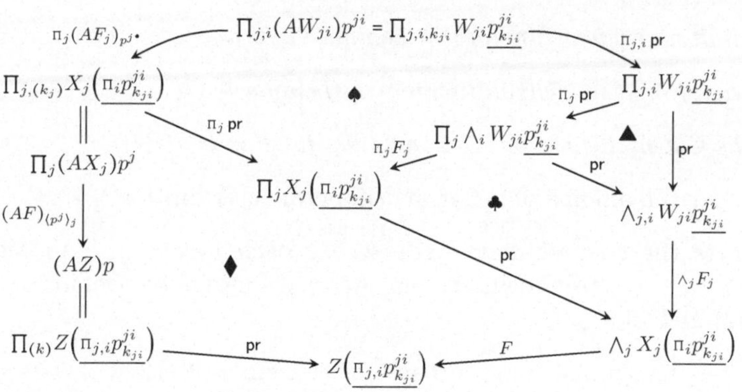

(9.119)

The detail of the diagram (9.119) is as follows.

- By (5.25) the top-left-bottom boundary composite is

$$\mathsf{pr}_{\langle\langle k_{ji}\rangle_{i=1}^{\ell_j}\rangle_{j=1}^{n}} \circ \gamma\big(AF\,;\langle AF_j\rangle\big)_{p^{\bullet\bullet}}.$$

- By (9.59) and (9.96) the top-right-bottom boundary composite is

$$\mathsf{pr}_{\langle\langle k_{ji}\rangle_{i=1}^{\ell_j}\rangle_{j=1}^{n}} \circ \big(A\gamma\big(F\,;\langle F_j\rangle\big)\big)_{p^{\bullet\bullet}}.$$

- The region ♠ is commutative by (9.96) for $(AF_j)_{p^{j\bullet}}$.
- The region ▲ is commutative by the universal property of the pushout (9.5) that defines the smash product.
- The region ♣ is commutative by the naturality of the pushout (9.5).
- The region ♦ is commutative by (9.96).

This proves that the diagram (9.117) is commutative on objects.

To prove that the diagram (9.117) is commutative on morphisms, we reuse the proof for objects above by replacing (5.25) and (9.96) with, respectively, (5.27) and (9.105). □

Next is the main result of this chapter.

Theorem 9.120. *There is a* Cat-*multifunctor*

$$\Gamma\text{-}\mathsf{Cat} \xrightarrow{\;A\;} \mathcal{A}\mathsf{Cat} \tag{9.121}$$

with

- Γ-Cat *the* Cat-*multicategory in Theorem 9.23;*
- $\mathcal{A}$Cat *the* Cat-*multicategory in Theorem 5.71 for $\mathcal{A}$ in Example 2.63;*
- *object assignment in Lemma 9.66;*
- *arity 0 multimorphism functors in Lemma 9.85; and*
- *positive arity multimorphism functors in Lemma 9.110.*

Moreover, the Cat-*multifunctor A extends the functor in* (9.80).

Proof. Consider the axioms of a Cat-multifunctor in Definition 3.19.

- A preserves the colored units because, for each Γ-category X, each of $A1_X$ and 1_{AX} is componentwise an identity functor by Definition 5.61 and (9.62) and (9.96).
- A preserves the symmetric group action by Lemma 9.112.
- A preserves the composition by Lemma 9.116.

Therefore, A is a Cat-multifunctor. The second assertion holds for the following reasons.

- The object assignment $X \longmapsto AX$ is the one in Lemma 9.66 in both cases.
- The $n = 1$ case of (9.96) agrees with (9.71).

Thus, A extends the functor in (9.80). □

10

Inverse K-Theory is a Pseudo Symmetric Cat-*Multifunctor*

The purpose of this chapter is to observe that the inverse K-theory functor

$$\overbrace{\Gamma\text{-}\mathsf{Cat} \xrightarrow{\ A\ } \mathcal{A}\mathsf{Cat} \xrightarrow{\ \int_{\mathcal{A}}\ } \mathsf{PermCat}^{\mathsf{sg}}}^{\mathcal{P}}$$

extends to a pseudo symmetric Cat-multifunctor; see Theorem 10.12. In fact, A is a Cat-multifunctor, and $\int_{\mathcal{A}}$ is a pseudo symmetric Cat-multifunctor. Moreover, there is a refinement (10.14)

$$\overbrace{\Gamma\text{-}\mathsf{Cat} \xrightarrow{\ A\ } \underbrace{\mathcal{A}\mathsf{Cat} \xrightarrow[\sim]{\ \int_{\mathcal{A}}\ } \mathsf{PFib}^{\mathcal{A}} \xrightarrow{\ \mathsf{U}\ } \mathsf{PermCat}^{\mathsf{sg}}}_{\int_{\mathcal{A}}}}^{\mathcal{P}}$$

that factors $\mathcal{P}$ through a non-symmetric Cat-*multiequivalence* $\int_{\mathcal{A}}$ that takes into account the opfibration structure of the Grothendieck construction. Below is a summary.

- $\int_{\mathcal{A}} : \mathcal{A}\mathsf{Cat} \longrightarrow \mathsf{PermCat}^{\mathsf{sg}}$ is a pseudo symmetric Cat-multifunctor (6.111).
- $\mathsf{U} : \mathsf{PFib}^{\mathcal{A}} \longrightarrow \mathsf{PermCat}^{\mathsf{sg}}$ is a non-symmetric Cat-multifunctor (7.53).
- $\int_{\mathcal{A}} : \mathcal{A}\mathsf{Cat} \xrightarrow{\sim} \mathsf{PFib}^{\mathcal{A}}$ is a non-symmetric Cat-multiequivalence (Theorem 8.76).
- $A : \Gamma\text{-}\mathsf{Cat} \longrightarrow \mathcal{A}\mathsf{Cat}$ is a Cat-multifunctor (9.121).
- $\mathcal{P} : \Gamma\text{-}\mathsf{Cat} \longrightarrow \mathsf{PermCat}^{\mathsf{sg}}$ is a pseudo symmetric Cat-multifunctor (Theorem 10.12).

Preservation of Algebraic Structures

As a pseudo symmetric Cat-multifunctor, $\mathcal{P}$ preserves algebraic structures parametrized by

DOI: 10.1201/9781003451235-10

- non-symmetric Cat-multifunctors (Corollary 10.15) and
- pseudo symmetric Cat-multifunctors from Cat-multicategories (Corollary 10.19).

In particular, inverse K-theory $\mathcal{P}$ preserves pseudo symmetric E_∞-algebras (Corollary 10.54). We emphasize that inverse K-theory $\mathcal{P}$ is *not* a Cat-multifunctor in the symmetric sense (Definition 3.19) because it does not strictly preserve the symmetric group action (Explanation 10.28). In general, $\mathcal{P}$ does not preserve algebraic structures parametrized by Cat-multicategories, such as E_∞-algebras (Definition 10.35). Instead, E_∞-algebras in Γ-Cat are sent by $\mathcal{P}$ to pseudo symmetric E_∞-algebras in $\mathsf{PermCat}^{\mathsf{sg}}$ (Example 10.55).

Some History

Inverse K-theory $\mathcal{P}$ is important because it is a homotopy inverse of Segal K-theory [Seg74]; see [JY∞, Chapter 8] for a detailed discussion of Segal K-theory. This homotopy inverse property of $\mathcal{P}$ is one of the main results of Mandell [Man10]. It is a sharpening of the result of Thomason [Tho95] that says that there is an equivalence between the stable homotopy categories of small permutative categories and of Γ-categories.

Although Segal K-theory provides an equivalence of stable homotopy categories, it is *not* a multifunctor, even in the non-symmetric sense. In particular, Segal K-theory does not, in general, preserve algebraic structures such as monoids. This fact is noted in [EM06] and is one of the main reasons for the development of Elmendorf-Mandell K-theory K^{EM}, which is Cat-multifunctorial. A detailed discussion of K^{EM} can be found in [JY∞, Chapters 10–13]. In view of the non-multifunctoriality of Segal K-theory, the pseudo symmetric Cat-multifunctoriality of inverse K-theory $\mathcal{P}$ (Theorem 10.12) is somewhat unexpected.

Organization

To prepare for the discussion of inverse K-theory as a pseudo symmetric Cat-multifunctor, in Section 10.1 we review the inverse K-theory functor $\mathcal{P} = \int_A \circ A$ from the underlying category of Γ-Cat to the underlying category of $\mathsf{PermCat}^{\mathsf{sg}}$.

In Section 10.2 we prove the main Theorem 10.12 on the pseudo symmetric Cat-multifunctoriality of inverse K-theory $\mathcal{P}$ by combining some of the results in earlier chapters. The associative operad, which parametrizes monoids, is a non-symmetric operad. Thus $\mathcal{P}$ preserves monoids (Example 10.17). This means that $\mathcal{P}$ sends each monoid in Γ-Cat to a small tight ring category.

In Section 10.3 we explain in detail the pseudo symmetric Cat-multifunctorial structure of $\mathcal{P}$ by describing its multimorphism functors and pseudo symmetry isomorphisms. In particular, the formulas (10.30) and (10.31) for the pseudo symmetry isomorphism $\mathcal{P}_\sigma$ show that $\mathcal{P}_\sigma$ is the identity if and only if σ is the identity permutation. In Explanations 10.33

and 10.34 we discuss related work in the literature [Elm∞, JY22b] about the multifunctoriality of $\mathcal{P}$.

To prepare for the discussion of $\mathcal{P}$ preserving pseudo symmetric E_∞-algebras, in Section 10.4 we review the categorical Barratt-Eccles E_∞-operad $E\Sigma_*$ and its Coherence Theorem 10.39. Small permutative categories and small (tight) bipermutative categories are E_∞-algebras in Cat, PermCat, and $\mathsf{PermCat}^{\mathsf{sg}}$ (Example 10.46). In Example 10.47 we explain the structure of an E_∞-algebra in Γ-Cat.

In Section 10.5 we define pseudo symmetric E_∞-algebras as pseudo symmetric Cat-multifunctors from the Barratt-Eccles operad $E\Sigma_*$. As a special case of the main Theorem 10.12, inverse K-theory $\mathcal{P}$ preserves pseudo symmetric E_∞-algebras. To better understand the structure involved, in Explanations 10.56, 10.70, and 10.82 we unpack the data and axioms of pseudo symmetric E_∞-algebras in a general Cat-multicategory, Γ-Cat, PermCat, and $\mathsf{PermCat}^{\mathsf{sg}}$.

We remind the reader of our left normalized bracketing Convention 1.24 for iterated monoidal product.

10.1 The Inverse K-Theory Functor

In this section we review the inverse K-theory functor $\mathcal{P}$ due to Mandell [Man10]; see also [JY22b, Section 6]. The inverse K-theory functor is a composite of two functors—A and $\int_{\mathcal{A}}$ with $\mathcal{A}$ in Example 2.63—from Γ-categories to small permutative categories. Here is an outline of this section.

- The inverse K-theory functor $\mathcal{P} = \int_{\mathcal{A}} \circ A$ is in Definition 10.2.
- Explanation 10.5 is a detailed description of the functor $\mathcal{P}$.
- Example 10.11 shows that $\mathcal{P}X$ recovers the permutative category $\mathcal{A}^\oplus$ when X is the terminal Γ-category.

Inverse K-Theory

For a small tight bipermutative category $(\mathcal{D}, \oplus, \otimes)$ (Definition 2.41), the Grothendieck construction $\int_{\mathcal{D}}$ has the following properties.

- It sends a symmetric monoidal functor $X : \mathcal{D}^\oplus \longrightarrow \mathsf{Cat}$ to a small permutative category $\int_{\mathcal{D}} X$ (Lemma 6.42).
- It sends a monoidal natural transformation (Explanation 5.48 (4))

$$\mathcal{D}^\oplus \underset{Z}{\overset{X}{\rightrightarrows}} \mathsf{Cat} \quad \Downarrow \phi$$

between symmetric monoidal functors to a strict symmetric monoidal functor $\int_{\mathcal{D}}\phi : \int_{\mathcal{D}} X \longrightarrow \int_{\mathcal{D}} Z$ (Remark 6.60).

Thus the Grothendieck construction (Theorem 6.110) restricts to a functor

$$\mathcal{D}\mathsf{Cat} \xrightarrow{\int_{\mathcal{D}}} \mathsf{PermCat}^{\mathsf{sg}} \tag{10.1}$$

between the underlying categories in Examples 9.77 and 9.78. The following definition is due to Mandell [Man10, 4.3 and 4.5].

Definition 10.2. The *inverse K-theory functor* $\mathcal{P}$ is defined as the composite functor

$$\begin{array}{ccc} \Gamma\text{-}\mathsf{Cat} & \xrightarrow{\mathcal{P}} & \mathsf{PermCat}^{\mathsf{sg}} \\ & {\scriptstyle A}\searrow \quad \nearrow{\scriptstyle \int_{\mathcal{A}}} & \\ & \mathcal{A}\mathsf{Cat} & \end{array} \tag{10.3}$$

with the following data.

- Γ-Cat, $\mathcal{A}\mathsf{Cat}$, and $\mathsf{PermCat}^{\mathsf{sg}}$ are the underlying categories of the respective Cat-multicategories in Examples 9.76 through 9.78.
- A is the functor in (9.80).
- $\int_{\mathcal{A}}$ is the functor in (10.1) with $\mathcal{D} = \mathcal{A}$ in Example 2.63.

This finishes the definition of the functor $\mathcal{P}$. ◇

Explanation 10.4 (Codomain of $\mathcal{P}$)**.** For the inverse K-theory functor (10.3)

$$\Gamma\text{-}\mathsf{Cat} \xrightarrow{\mathcal{P} = \int_{\mathcal{A}} A(-)} \mathsf{PermCat}^{\mathsf{sg}}$$

we can restrict the codomain to the subcategory of $\mathsf{PermCat}^{\mathsf{sg}}$ consisting of *strict* symmetric monoidal functors. The reason is that each monoidal natural transformation in $\mathcal{A}\mathsf{Cat}$ is sent by the Grothendieck construction

$$\mathcal{A}\mathsf{Cat} \xrightarrow{\int_{\mathcal{A}}} \mathsf{PermCat}^{\mathsf{sg}}$$

to a strict symmetric monoidal functor (Remark 6.60). However, when we consider the pseudo symmetric Cat-multifunctoriality of $\mathcal{P}$ in Section 10.2, the linearity constraints are natural isomorphisms but not identities in general (Explanation 10.25). ◇

Explanation 10.5 (Inverse K-Theory of Γ-Categories)**.** We unpack the inverse K-theory functor $\mathcal{P} = \int_{\mathcal{A}} A(-)$ in (10.3). Suppose X is a Γ-category.

Underlying Category The inverse K-theory

$$\mathcal{P}X = \int_{\mathcal{A}} AX \tag{10.6}$$

is the Grothendieck construction of the functor $AX : \mathcal{A} \longrightarrow \mathsf{Cat}$ in (9.61). Interpreting Definition 6.33 in this context, an *object* in the category $\mathcal{P}X$ is a pair (m, x) with

- $m = \langle m_i \rangle_{i=1}^p$ an object in $\mathcal{A}$ (Example 2.63) and
- $x = \langle x_i \rangle_{i=1}^p$ an object in $(AX)(m) = \prod_{i=1}^p X\underline{m_i}$ in (9.62).

A *morphism*

$$(m,x) \xrightarrow{(\phi,f)} (n,y) \quad \text{in} \quad \mathcal{P}X \tag{10.7}$$

consists of

- a morphism
$$m = \langle m_i \rangle_{i=1}^p \xrightarrow{\phi} n = \langle n_j \rangle_{j=1}^q \quad \text{in} \quad \mathcal{A}$$
as in (2.64) and
- a morphism
$$f = \langle f_j \rangle_{j=1}^q : \phi_*(x) \longrightarrow y \quad \text{in} \quad (AX)(n) = \textstyle\prod_{j=1}^q X\underline{n_j}$$
with
$$\phi_* = (AX)(\phi) : (AX)(m) \longrightarrow (AX)(n)$$
as in (9.63).

Composition is defined as

$$(\psi, g) \circ (\phi, f) = \big(\psi\phi, g \circ (\psi_* f)\big).$$

The *identity morphism* of an object (m,x) is the pair $(1_m, 1_x)$ of identity morphisms.

Permutative Structure Interpreting Definition 6.37 in this context, $\mathcal{P}X$ is a permutative category with strict *monoidal unit* given by the pair

$$\big(\langle\rangle, *\big) \in \mathcal{P}X \tag{10.8}$$

consisting of

- the empty tuple $\langle\rangle$ in $\mathcal{A}$ and
- the unique object $* \in (AX)\langle\rangle = X\underline{0} = \mathbf{1}$.

The *monoidal product* on objects is given by

$$(m,x) \,\square\, (n,y) = \begin{cases} \big((m,n),(x,y)\big) & \text{if } p,q > 0, \\ (m,x) & \text{if } q = 0, \text{ and} \\ (n,y) & \text{if } p = 0 \end{cases} \tag{10.9}$$

with

$$\begin{cases} (m,n) = m \oplus n = \big(\langle m_i \rangle_{i=1}^p, \langle n_j \rangle_{j=1}^q\big) \in \mathcal{A} \quad \text{and} \\ (x,y) = \big(\langle x_i \rangle_{i=1}^p, \langle y_j \rangle_{j=1}^q\big) \in (AX)(m,n). \end{cases}$$

The monoidal product on morphisms is defined similarly, using the fact that $\mathcal{A}^{\oplus}$ is a permutative category.

The *braiding* in $\mathcal{P}X$ is given componentwise by the isomorphism

$$\begin{array}{ccc} (m,x) \,\square\, (n,y) & \xrightarrow{\beta^{\square}_{(m,x),(n,y)}} & (n,y) \,\square\, (m,x) \\ \| & & \| \\ ((m,n),(x,y)) & \xrightarrow[\cong]{(\beta^{\oplus}_{m,n}\,,\,1)} & ((n,m),(y,x)) \end{array}$$

with $\beta^{\oplus}_{m,n}$ the additive braiding (2.66) in $\mathcal{A}$. This finishes the description of the permutative category $\mathcal{P}X = \int_{\mathcal{A}} AX$ for a Γ-category X.

Morphisms Interpreting Definition 6.45 in the current context, for a morphism $F : X \longrightarrow Y$ of Γ-categories, the strict symmetric monoidal functor

$$\mathcal{P}X = \textstyle\int_{\mathcal{A}} AX \xrightarrow{\mathcal{P}F = \int_{\mathcal{A}} AF} \mathcal{P}Y = \textstyle\int_{\mathcal{A}} AY$$

is given on objects by

$$\begin{aligned}(\mathcal{P}F)(m,x) &= \big(m, (AF)_m x\big) \\ &= \Big(m, \big\langle F_{\underline{m_i}}(x_i)\big\rangle_{i=1}^{p}\Big)\end{aligned} \tag{10.10}$$

and on morphisms by

$$\begin{aligned}(\mathcal{P}F)(\phi,f) &= \big(\phi, (AF)_n f\big) \\ &= \Big(\phi, \big\langle F_{\underline{n_j}}(f_j)\big\rangle_{j=1}^{q}\Big).\end{aligned}$$

The *unit constraint* is the identity morphism

$$(\mathcal{P}F)^0 = 1_{(\langle\rangle,*)} \quad \text{in} \quad \mathcal{P}Y.$$

The *monoidal constraint*

$$(\mathcal{P}F)(m,x) \,\square\, (\mathcal{P}F)(n,y) \xrightarrow{(\mathcal{P}F)^2} (\mathcal{P}F)\big((m,x) \,\square\, (n,y)\big)$$

is the identity morphism in $\mathcal{P}Y$ of the following object, using (10.9) to make the appropriate adjustment if either $p = 0$ or $q = 0$.

$$\begin{aligned}
&(\mathcal{P}F)\big((m,x) \,\square\, (n,y)\big) \\
&= (\mathcal{P}F)\big((m,n),(x,y)\big) && \text{by (10.9)} \\
&= \Big((m,n)\,,\, \big(\big\langle F_{\underline{m_i}}(x_i)\big\rangle_{i=1}^{p}\,,\, \big\langle F_{\underline{n_j}}(y_j)\big\rangle_{j=1}^{q}\big)\Big) && \text{by (10.10)} \\
&= \Big(m, \big\langle F_{\underline{m_i}}(x_i)\big\rangle_{i=1}^{p}\Big) \,\square\, \Big(n, \big\langle F_{\underline{n_j}}(y_j)\big\rangle_{j=1}^{q}\Big) && \text{by (10.9)} \\
&= (\mathcal{P}F)(m,x) \,\square\, (\mathcal{P}F)(n,y) && \text{by (10.10)}
\end{aligned}$$

This finishes the description of the inverse K-theory functor $\mathcal{P} = \int_{\mathcal{A}} A(-)$. ◇

Example 10.11 (Terminal Γ-Category). Suppose X is the Γ-category defined by

$$X\underline{n} = \mathbf{1} \quad \text{for each} \quad \underline{n} \in \mathcal{F}.$$

By definition (9.62) there is an isomorphism of categories

$$(AX)(m) \cong \mathbf{1} \quad \text{for each} \quad m \in \mathcal{A}.$$

By Explanation 10.5 there is a strict symmetric monoidal isomorphism

$$\begin{array}{ccc} \mathcal{P}X & \xrightarrow{\cong} & \mathcal{A}^{\oplus} \\ (m, *) & \longmapsto & m \\ (\phi, 1_*) & \longmapsto & \phi \end{array}$$

with $\mathcal{A}^{\oplus} = (\mathcal{A}, \oplus, \langle\rangle, \beta^{\oplus})$ the additive structure of $\mathcal{A}$ in Example 2.63. ◇

10.2 Pseudo Symmetric Cat-Multifunctorial Inverse K-Theory

The purpose of this section is to prove that the inverse K-theory functor $\mathcal{P}$ in (10.3) extends to a pseudo symmetric Cat-multifunctor; see Theorem 10.12. More specifically we obtain two factorizations of the pseudo symmetric Cat-multifunctor $\mathcal{P}$.

1. In (10.13) $\mathcal{P}$ factors as
   - a Cat-multifunctor A followed by
   - a pseudo symmetric Cat-multifunctor $\int_{\mathcal{A}}$ given by the Grothendieck construction on the small tight bipermutative category $\mathcal{A}$.
2. In the refinement (10.14), $\mathcal{P}$ factors as the composite of
   - a Cat-multifunctor A,
   - a non-symmetric Cat-*multiequivalence* $\int_{\mathcal{A}}$, and
   - a non-symmetric Cat-multifunctor U.

We emphasize that $\mathcal{P}$ is *not* a Cat-multifunctor because it does not strictly preserve the symmetric group action. See Explanations 10.33 and 10.34 for related work in the literature about the multifunctoriality of inverse K-theory.

Here is an outline of this section.

- Theorem 10.12 proves the pseudo symmetric Cat-multifunctoriality of inverse K-theory $\mathcal{P}$.

- In Corollary 10.15 we observe that $\mathcal{P}$ preserves any algebraic structure parametrized by a non-symmetric Cat-multifunctor, including the associative operad (Example 10.17).

- In Corollary 10.19 we observe that $\mathcal{P}$ preserves any algebraic structure parametrized by a pseudo symmetric Cat-multifunctor. Examples related to the Barratt-Eccles E_∞-operad are discussed in Sections 10.4 and 10.5.

Main Theorem

Recall from Definition 4.1 that a pseudo symmetric Cat-multifunctor

- strictly preserves the colored units (3.21) and composition (3.22) and

- is equipped with pseudo symmetry isomorphisms that satisfy the four coherence axioms (4.3) through (4.6).

Each Cat-multifunctor is a pseudo symmetric Cat-multifunctor with pseudo symmetry isomorphisms given by identities (Example 4.7). Composition of pseudo symmetric Cat-multifunctors is well defined, associative, and unital (Lemma 4.16).

Recall the Cat-multicategories

- Γ-Cat in Theorem 9.23,

- $\mathcal{A}$Cat in Theorem 5.71 with $\mathcal{A}$ in Example 2.63, and

- $\mathsf{PermCat}^{\mathsf{sg}}$ in Theorem 6.31.

Theorem 10.12. *The following statements hold for the inverse K-theory functor $\mathcal{P}$ in* (10.3).

1. *$\mathcal{P}$ extends to a pseudo symmetric* Cat*-multifunctor*

$$\begin{array}{ccc} \Gamma\text{-}\mathsf{Cat} & \xrightarrow{\ \mathcal{P}\ } & \mathsf{PermCat}^{\mathsf{sg}} \\ & {\scriptstyle A}\searrow \quad \nearrow{\scriptstyle \int_{\mathcal{A}}} & \\ & \mathcal{A}\mathsf{Cat} & \end{array} \tag{10.13}$$

   with

   - *A the* Cat*-multifunctor in* (9.121) *and*
   - *$\int_{\mathcal{A}}$ the pseudo symmetric* Cat*-multifunctor in* (6.111).

2. *The pseudo symmetric* Cat*-multifunctor $\mathcal{P}$ in* (10.13) *is not a* Cat*-multifunctor.*

3. As a non-symmetric Cat-*multifunctor,* $\mathcal{P}$ *in* (10.13) *factors as the boundary in*

$$\begin{array}{ccc} \Gamma\text{-}\mathsf{Cat} & \xrightarrow{\mathcal{P}} & \mathsf{PermCat}^{\mathsf{sg}} \\ {\scriptstyle A}\downarrow & \nearrow{\scriptstyle \int_{\mathcal{A}}} & \uparrow{\scriptstyle \mathsf{U}} \\ \mathcal{A}\mathsf{Cat} & \xrightarrow[\sim]{\int_{\mathcal{A}}} & \mathsf{PFib}^{\mathcal{A}} \end{array} \tag{10.14}$$

with the following data.

- $\mathsf{PFib}^{\mathcal{A}}$ *is the non-symmetric* Cat-*multicategory in Example 7.55.*
- *The bottom* $\int_{\mathcal{A}}$ *is the non-symmetric* Cat-*multiequivalence in Theorem 8.76.*
- U *is the non-symmetric* Cat-*multifunctor in* (7.53).

Proof. In (10.13) A is a Cat-multifunctor, and $\int_{\mathcal{A}}$ is a pseudo symmetric Cat-multifunctor. Thus the composite $\int_{\mathcal{A}} \circ A$ is a pseudo symmetric Cat-multifunctor by Example 4.7 and Lemma 4.16. Moreover, the following two statements hold.

- The Cat-multifunctor A extends the functor A in the inverse K-theory functor by Theorem 9.120.
- The pseudo symmetric Cat-multifunctor $\int_{\mathcal{A}}$ extends the functor $\int_{\mathcal{A}}$ in the inverse K-theory functor by definition (10.1).

Therefore, the composite pseudo symmetric Cat-multifunctor $\int_{\mathcal{A}} \circ A$ extends the inverse K-theory functor $\mathcal{P}$ in (10.3). This proves assertion (1).

To prove assertion (2), consider an object $F \in \Gamma\text{-}\mathsf{Cat}\big(\langle X \rangle ; Z\big)$ as in (9.45). The two strong n-linear functors

$$\textstyle\prod_{j=1}^{n} \big(\int_{\mathcal{A}} AX_{\sigma(j)}\big) \;\underset{(\mathcal{P}F)^\sigma \,=\, (\int_{\mathcal{A}} AF)^\sigma}{\overset{\mathcal{P}(F^\sigma) \,=\, \int_{\mathcal{A}} (AF)^\sigma}{\rightrightarrows}}\; \int_{\mathcal{A}} AZ$$

are different in general. Indeed, by (6.88) applied to $\mathcal{D} = \mathcal{A}$, the object assignments of the strong n-linear functors $\mathcal{P}(F^\sigma)$ and $(\mathcal{P}F)^\sigma$ are different in general because the multiplicative braiding in $\mathcal{A}$, which is defined in (2.68), is not the identity. Thus $\mathcal{P}$ does not strictly preserve the symmetric group action, so it is not a Cat-multifunctor.

Assertion (3) follows from Example 8.77 and (10.13). □

Preservation of Non-Symmetric Algebras

Each pseudo symmetric Cat-multifunctor (Definition 4.1) is, in particular, a non-symmetric Cat-multifunctor (Definition 3.19) after forgetting the pseudo symmetry isomorphisms (Example 4.8). Thus, Theorem 10.12 (1) yields the following result, which is first recorded in [JY22b, Section 10].

Corollary 10.15. *Suppose*

- Q *is a non-symmetric* Cat-*multicategory and*
- $F : \mathsf{Q} \longrightarrow \Gamma\text{-}\mathsf{Cat}$ *is a non-symmetric* Cat-*multifunctor.*

Then the composite

$$\mathsf{Q} \xrightarrow{F} \Gamma\text{-}\mathsf{Cat} \xrightarrow{\mathcal{P}} \mathsf{PermCat}^{\mathsf{sg}} \tag{10.16}$$

is a non-symmetric Cat-*multifunctor.*

Corollary 10.15 means that any algebraic structure in Γ-Cat parametrized by a non-symmetric Cat-multifunctor is preserved by inverse K-theory $\mathcal{P}$.

Example 10.17 (Monoids to Ring Categories)**.** Consider the *non-symmetric associative operad* $\mathsf{As}^{\mathsf{ns}}$ (Definition 3.2) with each

$$\mathsf{As}^{\mathsf{ns}}_n = \{*\},$$

a one-element set. It is regarded as a non-symmetric Cat-multicategory with one object and each $\mathsf{As}^{\mathsf{ns}}_n = \mathbf{1}$, the terminal category. See [Yau16, Section 14.2] and [JY∞, Section 11.1] for a detailed discussion of the free operad

$$\mathsf{As} = \left\{\Sigma_n\right\}_{n\geq 0} \tag{10.18}$$

generated by $\mathsf{As}^{\mathsf{ns}}$. In Γ-Cat and $\mathsf{PermCat}^{\mathsf{sg}}$, $\mathsf{As}^{\mathsf{ns}}$ parametrizes the following algebraic structures.

- By [Yau16, 14.2.18] or [JY∞, 11.1.15] a monoid in Γ-Cat is uniquely determined by a non-symmetric Cat-multifunctor

$$\mathsf{As}^{\mathsf{ns}} \longrightarrow \Gamma\text{-}\mathsf{Cat}.$$

- By [EM06, 3.4] or [JY∞, 11.2.16] a small tight ring category (Definition 2.29) is uniquely determined by a non-symmetric Cat-multifunctor

$$\mathsf{As}^{\mathsf{ns}} \longrightarrow \mathsf{PermCat}^{\mathsf{sg}}.$$

These facts and Corollary 10.15 imply that inverse K-theory $\mathcal{P}$ sends each monoid in Γ-Cat to a small tight ring category. This is first proved in [JY22b, 10.3]. ◇

Preservation of Pseudo Symmetric Algebras

By Lemma 4.16 pseudo symmetric Cat-multifunctors are closed under composition. Thus, Theorem 10.12 (1), which states that $\mathcal{P}$ is a pseudo symmetric Cat-multifunctor, yields the following result.

Corollary 10.19. *Suppose*

- Q *is a* Cat*-multicategory and*
- $F : \mathsf{Q} \longrightarrow \Gamma\text{-}\mathsf{Cat}$ *is a pseudo symmetric* Cat*-multifunctor.*

Then the composite

$$\mathsf{Q} \xrightarrow{F} \Gamma\text{-}\mathsf{Cat} \xrightarrow{\mathcal{P}} \mathsf{PermCat}^{\mathsf{sg}} \tag{10.20}$$

is a pseudo symmetric Cat*-multifunctor.*

Corollary 10.19 means that any algebraic structure in Γ-Cat parametrized by a pseudo symmetric Cat-multifunctor is preserved by inverse K-theory $\mathcal{P}$. The case where Q is the Barratt-Eccles E_∞-operad is discussed in Sections 10.4 and 10.5.

Example 10.21 (Cat-Multifunctors)**.** By Example 4.7 each Cat-multifunctor becomes a pseudo symmetric Cat-multifunctor when it is equipped with identity pseudo symmetry isomorphisms. In this case, Corollary 10.19 says that, if $F : \mathsf{Q} \longrightarrow \Gamma\text{-}\mathsf{Cat}$ is a Cat-multifunctor, then the composite

$$\mathsf{Q} \xrightarrow{\mathcal{P}\circ F} \mathsf{PermCat}^{\mathsf{sg}}$$

in (10.20) is a pseudo symmetric Cat-multifunctor. ◇

10.3 Unraveling Pseudo Symmetric Inverse K-Theory

In this section we explain in detail the pseudo symmetric Cat-multifunctor $\mathcal{P}$ in Theorem 10.12.

- The arity 0 multimorphism functors of $\mathcal{P}$ are discussed in Explanation 10.23.
- The positive arity multimorphism functors of $\mathcal{P}$ are discussed in Explanation 10.25.
- The pseudo symmetry isomorphisms $\mathcal{P}_\sigma$ of $\mathcal{P}$ are discussed in Explanation 10.28. The displays (10.30) and (10.31) show that $\mathcal{P}$ does *not* strictly preserve the symmetric group action because the multiplicative braiding in $\mathcal{A}$ is not the identity.
- Explanations 10.33 and 10.34 discuss related work in the literature about the multifunctoriality of $\mathcal{P}$.

For each Γ-category X, the inverse K-theory $\mathcal{P}X = \int_{\mathcal{A}} AX$ as a small permutative category is discussed in detail in Explanation 10.5. For Γ-categories

$$Z, X_1, \ldots, X_n : (\mathcal{F}, \underline{0}) \longrightarrow (\mathsf{Cat}, \mathbf{1})$$

with $\langle X \rangle = \langle X_j \rangle_{j=1}^n$ and $n \geq 0$, next we describe

- the multimorphism functor

$$\Gamma\text{-}\mathsf{Cat}\big(\langle X \rangle ; Z\big) \xrightarrow{\ \mathcal{P}\ } \mathsf{PermCat}^{\mathsf{sg}}\big(\langle \mathcal{P}X \rangle ; \mathcal{P}Z\big) \tag{10.22}$$

  in Explanations 10.23 and 10.25 and

- the pseudo symmetry isomorphisms in Explanation 10.28

for the pseudo symmetric Cat-multifunctor $\mathcal{P} = \int_{\mathcal{A}} \circ A$ in (10.13).

Explanation 10.23 (Inverse K-Theory in Arity 0)**.** If $n = 0$ then the multimorphism functor $\mathcal{P}$ in (10.22) is given by the following assignments for objects F and morphisms θ in $\Gamma\text{-}\mathsf{Cat}\big(\langle\rangle ; Z\big)$.

$$\begin{array}{ccccc} \Gamma\text{-}\mathsf{Cat}\big(\langle\rangle ; Z\big) & \xrightarrow[\cong]{A} & \mathcal{A}\mathsf{Cat}\big(\langle\rangle ; AZ\big) & \xrightarrow{\int_{\mathcal{A}}} & \mathsf{PermCat}^{\mathsf{sg}}\big(\langle\rangle ; \mathcal{P}Z\big) \\ \| & & \| & & \| \\ \mathsf{Map}_*(J, Z) & & (AZ)(1) = Z\underline{1} & & \mathcal{P}Z = \int_{\mathcal{A}} AZ \\ F & \longmapsto & AF = F_{\underline{1}}(1) & \longmapsto & \mathcal{P}F = \big((1)\,,\, F_{\underline{1}}(1)\big) \\ \theta & \longmapsto & A\theta = \theta_{\underline{1},1} & \longmapsto & \mathcal{P}\theta = \big(1_{(1)}\,,\, \theta_{\underline{1},1}\big) \end{array} \tag{10.24}$$

(The top arrow labelled $\mathcal{P}$ runs from $\Gamma\text{-}\mathsf{Cat}\big(\langle\rangle ; Z\big)$ to $\mathsf{PermCat}^{\mathsf{sg}}\big(\langle\rangle ; \mathcal{P}Z\big)$.)

The detail of (10.24) is as follows.

- The equality under $\Gamma\text{-}\mathsf{Cat}\big(\langle\rangle ; Z\big)$ is (9.33).
- The two equalities under $\mathcal{A}\mathsf{Cat}\big(\langle\rangle ; AZ\big)$ are from (5.62) and (9.62).
- The two equalities under $\mathsf{PermCat}^{\mathsf{sg}}\big(\langle\rangle ; \mathcal{P}Z\big)$ are from (6.23) and (10.3).
- The isomorphism A is from Lemma 9.85.
- The functor $\int_{\mathcal{A}}$ is from Definition 6.43.

The functor $\mathcal{P} = \int_{\mathcal{A}} \circ A$ in (10.24) is injective on objects and morphism sets. ◇

Notation 9.88 is used in Explanations 10.25 and 10.28 below.

Explanation 10.25 (Inverse K-Theory in Positive Arity)**.** If $n > 0$ then the multimorphism functor $\mathcal{P} = \int_{\mathcal{A}} \circ A$ in (10.22) is the composite

$$\begin{array}{ccccc} \Gamma\text{-}\mathsf{Cat}\big(\langle X \rangle ; Z\big) & \xrightarrow{A} & \mathcal{A}\mathsf{Cat}\big(\langle AX \rangle ; AZ\big) & \xrightarrow{\int_{\mathcal{A}}} & \mathsf{PermCat}^{\mathsf{sg}}\big(\langle \mathcal{P}X \rangle ; \mathcal{P}Z\big) \\ \| & & & & \\ \mathsf{Map}_*\big(\wedge_{j=1}^n X_j\,,\, Z\big) & & & & \end{array} \tag{10.26}$$

(The top arrow labelled $\mathcal{P}$ runs from $\Gamma\text{-}\mathsf{Cat}\big(\langle X \rangle ; Z\big)$ to $\mathsf{PermCat}^{\mathsf{sg}}\big(\langle \mathcal{P}X \rangle ; \mathcal{P}Z\big)$.)

with

- $\Gamma\text{-}\mathsf{Cat}\big(\langle X\rangle\,;\,Z\big) = \mathsf{Map}_*\big(\bigwedge_{j=1}^n X_j\,,\,Z\big)$ in (9.32),
- $\mathcal{A}\mathsf{Cat}\big(\langle AX\rangle\,;\,AZ\big)$ in Lemma 5.60 for $\mathcal{D} = \mathcal{A}$ in Example 2.63,
- $\mathsf{PermCat}^{\mathsf{sg}}\big(\langle \mathcal{P}X\rangle\,;\,\mathcal{P}Z\big)$ in (6.22),
- the functor A in Definitions 9.93 and 9.103, and
- the functor $\int_{\mathcal{A}}$ in Definitions 6.45 and 6.61.

We first describe the object assignment of $\mathcal{P}$ and then its morphism assignment.

Object Assignment. Suppose $F \in \Gamma\text{-}\mathsf{Cat}\big(\langle X\rangle\,;\,Z\big)$ is an object, which means a pointed natural transformation $F : \bigwedge_{j=1}^n X_j \longrightarrow Z$ as in (9.45). By Definitions 6.45 and 9.93 and Explanation 9.99, the strong n-linear functor (Definition 6.4)

$$\prod_{j=1}^n \mathcal{P}X_j = \prod_{j=1}^n \big(\textstyle\int_{\mathcal{A}} AX_j\big) \xrightarrow{\ \mathcal{P}F\ } \mathcal{P}Z = \textstyle\int_{\mathcal{A}} AZ$$

is given as follows.

Objects $\mathcal{P}F = \int_{\mathcal{A}} AF$ sends an object

$$\big\langle \big(m^j \in \mathcal{A}\,,\, x^j \in (AX_j)m^j\big)\big\rangle_{j=1}^n \in \prod_{j=1}^n \big(\textstyle\int_{\mathcal{A}} AX_j\big) \tag{10.27}$$

to the object

$$(\mathcal{P}F)\big\langle (m^j, x^j)\big\rangle_{j=1}^n = \big(m^{1,\ldots,n}\,,\, (AF)_{\langle m\rangle}\langle x^j\rangle_{j=1}^n\big) \in \textstyle\int_{\mathcal{A}} AZ$$

consisting of the object in (9.91)

$$m^{1,\ldots,n} = \textstyle\bigotimes_{j=1}^n m^j \in \mathcal{A}$$

and the object in (9.101)

$$(AF)_{\langle m\rangle}\langle x^j\rangle_{j=1}^n \in (AZ)m^{1,\ldots,n}.$$

Morphisms $\mathcal{P}F = \int_{\mathcal{A}} AF$ sends a morphism

$$\big\langle (f_j, p_j)\big\rangle_{j=1}^n : \big\langle (m^j, x^j)\big\rangle_{j=1}^n \longrightarrow \big\langle (k^j, w^j)\big\rangle_{j=1}^n \quad \text{in} \quad \prod_{j=1}^n \big(\textstyle\int_{\mathcal{A}} AX_j\big)$$

to the morphism

$$\begin{array}{c} (\mathcal{P}F)\big\langle (m^j, x^j)\big\rangle = \big(m^{1,\ldots,n}\,,\, (AF)_{\langle m\rangle}\langle x^j\rangle_{j=1}^n\big) \\ {\scriptstyle (\mathcal{P}F)\langle (f_j\,,\,p_j)\rangle}\Big\downarrow {\scriptstyle = \big(\bigotimes_{j=1}^n f_j\,,\,(AF)_{\langle k\rangle}\langle p_j\rangle_{j=1}^n\big)} \\ (\mathcal{P}F)\big\langle (k^j, w^j)\big\rangle = \big(k^{1,\ldots,n}\,,\, (AF)_{\langle k\rangle}\langle w^j\rangle_{j=1}^n\big) \end{array}$$

with the morphism in (9.101)

$$(AF)_{\langle k\rangle}\langle p_j\rangle_{j=1}^n \in (AZ)k^{1,\ldots,n}.$$

Linearity Constraints By (6.56), for each $j \in \{1, \ldots, n\}$, the j-th linearity constraint $(\mathcal{P}F)_j^2$ of $\mathcal{P}F$ is componentwise an isomorphism $(\mathsf{L}^{-1}, 1)$ with L the unique Laplaza coherence isomorphism in (5.46).

Morphism Assignment. Suppose $\theta : F \longrightarrow G$ is a morphism in $\Gamma\text{-}\mathsf{Cat}(\langle X \rangle ; Z)$, which means a pointed modification with component pointed natural transformations as in (9.49). By Definition 6.61 the n-linear natural transformation (Definition 6.16)

$$\prod_{j=1}^{n}\Big(\int_{\mathcal{A}} AX_j\Big) \;\underset{\mathcal{P}G \,=\, \int_{\mathcal{A}} AG}{\overset{\mathcal{P}F \,=\, \int_{\mathcal{A}} AF}{\xrightarrow[\qquad\qquad\qquad]{\Downarrow \mathcal{P}\theta \,=\, \int_{\mathcal{A}} A\theta}}}\; \int_{\mathcal{A}} AZ$$

has, for each object $\big\langle (m^j, x^j) \big\rangle_{j=1}^{n}$ as in (10.27), a component morphism

$$\begin{array}{c}
(\mathcal{P}F)\big\langle (m^j, x^j)\big\rangle = \big(m^{1,\ldots,n} \,,\, (AF)_{\langle m \rangle}\langle x^j \rangle_{j=1}^{n}\big) \\
{\scriptstyle (\mathcal{P}\theta)_{\langle (m^j, x^j)\rangle}}\Big\downarrow {\scriptstyle = \big(1_{m^{1,\ldots,n}} \,,\, \big((A\theta)_{\langle m \rangle}\big)_{\langle x^j \rangle_{j=1}^{n}}\big)} \\
(\mathcal{P}G)\big\langle (m^j, x^j)\big\rangle = \big(m^{1,\ldots,n} \,,\, (AG)_{\langle m \rangle}\langle x^j \rangle_{j=1}^{n}\big)
\end{array}$$

in $\int_{\mathcal{A}} AZ$. Here

$$\big((A\theta)_{\langle m \rangle}\big)_{\langle x^j \rangle_{j=1}^{n}} \in (AZ)m^{1,\ldots,n}$$

is the morphism in (9.108). This finishes our description of the multimorphism functor $\mathcal{P} = \int_{\mathcal{A}} \circ A$ in (10.26). ◇

Explanation 10.28 (Pseudo Symmetry Isomorphisms of Inverse K-Theory). Here we describe the pseudo symmetry isomorphisms (4.2) of the pseudo symmetric Cat-multifunctor $\mathcal{P} = \int_{\mathcal{A}} \circ A$ in (10.13). By definition (4.14), for each permutation $\sigma \in \Sigma_n$, the pseudo symmetry isomorphism $\mathcal{P}_\sigma$ is given by the following pasting of natural isomorphisms.

$$\begin{array}{ccccc}
 & & \xrightarrow{\qquad\qquad\qquad \mathcal{P} \qquad\qquad\qquad} & & \\
\Gamma\text{-}\mathsf{Cat}(\langle X \rangle ; Z) & \xrightarrow{A} & \mathcal{A}\mathsf{Cat}(\langle AX \rangle ; AZ) & \xrightarrow{\int_{\mathcal{A}}} & \mathsf{PermCat}^{\mathsf{sg}}(\langle \mathcal{P}X \rangle ; \mathcal{P}Z) \\
{\scriptstyle \sigma}\big\downarrow & = & \big\downarrow{\scriptstyle \sigma} & \overset{(\int_{\mathcal{A}})_\sigma}{\underset{\cong}{\Longrightarrow}} & \big\downarrow{\scriptstyle \sigma} \\
\Gamma\text{-}\mathsf{Cat}(\langle X \rangle\sigma ; Z) & \xrightarrow[A]{} & \mathcal{A}\mathsf{Cat}(\langle AX \rangle\sigma ; AZ) & \xrightarrow[\int_{\mathcal{A}}]{} & \mathsf{PermCat}^{\mathsf{sg}}(\langle \mathcal{P}X \rangle\sigma ; \mathcal{P}Z) \\
 & & \xrightarrow[\qquad\qquad\qquad \mathcal{P} \qquad\qquad\qquad]{} & &
\end{array} \tag{10.29}$$

The detail of (10.29) is as follows.

- By Theorem 9.120 A is a Cat-multifunctor, with the preservation of the symmetric group action proved in Lemma 9.112. As in Example 4.7, A is regarded as a pseudo symmetric Cat-multifunctor with identity pseudo symmetry isomorphisms. So the left square in (10.29) is strictly commutative.

- The pseudo symmetry isomorphism $(\int_{\mathcal{A}})_\sigma$ is from Definition 6.85 for $\mathcal{D} = \mathcal{A}$ in Example 2.63. By (6.89) the component isomorphisms of $(\int_{\mathcal{A}})_\sigma$ are always of the form $(\sigma, 1)$.

We unpack (10.29) further. Suppose $F \in \Gamma\text{-}\mathsf{Cat}(\langle X\rangle; Z)$ is an object, which means a pointed natural transformation $F : \bigwedge_{j=1}^n X_j \longrightarrow Z$ as in (9.45). The component of the pseudo symmetry isomorphism $(\int_{\mathcal{A}})_\sigma$ at the object

$$AF \in \mathcal{A}\mathsf{Cat}(\langle AX\rangle; AZ)$$

as in (9.94) is an n-linear natural isomorphism (Definition 6.16) as follows.

$$\prod_{j=1}^n \left(\textstyle\int_{\mathcal{A}} AX_{\sigma(j)}\right) \xrightarrow[\ (\mathcal{P}F)^\sigma = (\int_{\mathcal{A}} AF)^\sigma\ ]{\ \mathcal{P}(F^\sigma) = \int_{\mathcal{A}}(AF)^\sigma\ } \textstyle\int_{\mathcal{A}} AZ \qquad \Downarrow (\int_{\mathcal{A}})_{\sigma,AF}$$

To describe this natural isomorphism explicitly, consider an object

$$\langle (m^j, x^j)\rangle_{j=1}^n \in \prod_{j=1}^n \left(\textstyle\int_{\mathcal{A}} AX_{\sigma(j)}\right)$$

with, for each $j \in \{1, \ldots, n\}$, objects $m^j \in \mathcal{A}$ as in Notation 9.88 and

$$x^j = \langle x^j_{i_j}\rangle_{i_j=1}^{r_j} = \left(x^j_1, x^j_2, \ldots, x^j_{r_j}\right) \in \prod_{i_j=1}^{r_j} X_{\sigma(j)} \underline{m^j_{i_j}} = (AX_{\sigma(j)})m^j.$$

By (6.88) and (6.89) the component of $(\int_{\mathcal{A}})_{\sigma,AF}$ at the object $\langle (m^j, x^j)\rangle$ is the following isomorphism in $\int_{\mathcal{A}} AZ$.

$$\begin{array}{c} \mathcal{P}(F^\sigma)\langle (m^j,x^j)\rangle = \left(m^{1,\ldots,n}\,,\ \sigma_*^{-1}(AF)_{\sigma\langle m\rangle}\langle x^{\sigma^{-1}(j)}\rangle_{j=1}^n\right) \\ (\int_{\mathcal{A}})_{\sigma,AF,\langle (m^j,x^j)\rangle}\Big\downarrow = (\sigma,1) \\ (\mathcal{P}F)^\sigma\langle (m^j,x^j)\rangle = \left(m^{\sigma^{-1}(1),\ldots,\sigma^{-1}(n)}\,,\ (AF)_{\sigma\langle m\rangle}\langle x^{\sigma^{-1}(j)}\rangle_{j=1}^n\right) \end{array} \tag{10.30}$$

In (10.30) the object

$$(AF)_{\sigma\langle m\rangle}\langle x^{\sigma^{-1}(j)}\rangle_{j=1}^n = \left\langle \cdots \left\langle \left(F_{\langle m^{\sigma^{-1}(j)}_{i_{\sigma^{-1}(j)}}\rangle_{j=1}^n} \langle x^{\sigma^{-1}(j)}_{i_{\sigma^{-1}(j)}}\rangle_{j=1}^n\right)\right\rangle_{i_{\sigma^{-1}(1)}=1}^{r_{\sigma^{-1}(1)}} \cdots \right\rangle_{i_{\sigma^{-1}(n)}=1}^{r_{\sigma^{-1}(n)}}$$

belonging to

$$(AZ)m^{\sigma^{-1}(1),\ldots,\sigma^{-1}(n)} = \prod_{i_{\sigma^{-1}(n)}=1}^{r_{\sigma^{-1}(n)}} \cdots \prod_{i_{\sigma^{-1}(1)}=1}^{r_{\sigma^{-1}(1)}} \left(Z\underline{m^{\sigma^{-1}(1),\ldots,\sigma^{-1}(n)}_{i_{\sigma^{-1}(1)},\ldots,i_{\sigma^{-1}(n)}}}\right)$$

is as given in (9.101) for the objects

$$\begin{cases} \sigma\langle m\rangle = \langle m^{\sigma^{-1}(j)}\rangle_{j=1}^n \in \mathcal{A}^n \quad \text{and} \\ \langle x^{\sigma^{-1}(j)}\rangle_{j=1}^n \in \prod_{j=1}^n (AX_j)m^{\sigma^{-1}(j)}. \end{cases}$$

The isomorphism

$$m^{1,\ldots,n} = \bigotimes_{j=1}^{n} m^j \xrightarrow[\cong]{\sigma} \bigotimes_{j=1}^{n} m^{\sigma^{-1}(j)} = m^{\sigma^{-1}(1),\ldots,\sigma^{-1}(n)} \tag{10.31}$$

in (10.30) permutes the n factors $\langle m^j \rangle_{j=1}^n$ according to $\sigma \in \Sigma_n$ using the multiplicative braiding $\beta^{\otimes}$ in $\mathcal{A}$ (2.70).

The isomorphism $(\sigma, 1)$ in (10.30) is the identity morphism if and only if $\sigma \in \mathcal{A}$ in (10.31) is the identity morphism. Thus the pseudo symmetry isomorphism

$$\mathcal{P}_{\sigma} = (\textstyle\int_{\mathcal{A}})_{\sigma, A(-)} \tag{10.32}$$

for inverse K-theory $\mathcal{P}$ is the identity natural transformation if and only if $\sigma \in \Sigma_n$ is the identity permutation. ◇

Related Literature

Next we discuss the relationship between Theorem 10.12 and [Elm∞, JY22b] on the multifunctoriality of the inverse K-theory functor $\mathcal{P}$.

Explanation 10.33 (Related Work of Johnson-Yau)**.** The main result in [JY22b] says that inverse K-theory $\mathcal{P} = \int_{\mathcal{A}} \circ A$ is a non-symmetric Cat-multifunctor. Both [JY22b] and Theorem 10.12 consider Cat-enrichment. Theorem 10.12 is an extension and a refinement of the main result in [JY22b] in the following ways:

1. While $\mathcal{P}$ does not strictly preserve the symmetric group action, Theorem 10.12 shows that $\mathcal{P}$ is a pseudo symmetric Cat-multifunctor. In particular, $\mathcal{P}$ preserves the symmetric group action up to the pseudo symmetry isomorphisms $\mathcal{P}_{\sigma}$ in (10.29), which satisfy the four coherence axioms in Definition 4.1.

2. The proof in [JY22b] is a direct analysis of the functor $\mathcal{P}$ and does not consider the two functors A and $\int_{\mathcal{A}}$ separately. On the other hand, Theorem 10.12 shows that each of the two constituent functors of $\mathcal{P} = \int_{\mathcal{A}} \circ A$ is Cat-multifunctorial: in the symmetric sense for A and in the pseudo symmetric sense for $\int_{\mathcal{A}}$.

3. Theorem 10.12 has the further factorization (10.14) of $\mathcal{P}$ involving the non-symmetric Cat-*multiequivalence* $\int_{\mathcal{A}}$ that takes into account the permutative opfibration structure of the Grothendieck construction on $\mathcal{A}$. ◇

Explanation 10.34 (Related Work of Elmendorf)**.** The main claim in [Elm∞] is that inverse K-theory $\mathcal{P}$ is a composite of three multifunctors in the symmetric sense, and, therefore, $\mathcal{P}$ is also a multifunctor in the symmetric sense. That work differs from Theorem 10.12 and [JY22b] in the following ways:

1. Cat-enrichment is not considered in [Elm∞].
2. The constructions in [Elm∞] involve an indexing category $\mathbb{N}$ that is different from $\mathcal{A}$.
3. The three multifunctors in [Elm∞, Theorems 1–3] are not the ones in either (10.13) or (10.14).
4. The description of the monoidal unit in Γ-Cat in [Elm∞, Section 4], denoted B there, is not correct. The correct monoidal unit is J in (9.19) and Explanation 9.24.

Furthermore, the two main assertions in [Elm∞] are not correct, as we discuss next.

Contrary to the claim in [Elm∞, Corollary 5], $\mathcal{P}$ is *not* a multifunctor in the symmetric sense because it does not strictly preserve the symmetric group action. This is noted in [JY22b, Remark 8.1] and described in detail in Explanation 10.28. In particular, (10.30) shows how $\mathcal{P}(-)^{\sigma}$ and $(\mathcal{P}(-))^{\sigma}$ are different on objects.

Contrary to the claim in [Elm∞, Observation 4], the composite of the three constructions in [Elm∞, Theorems 1–3] does *not* yield $\mathcal{P}$. For a general Γ-category X, that composite contains many objects that are not in $\mathcal{P}X$ for the following reasons.

- The second construction in [Elm∞, Theorem 2] is a wreath product that creates a basepoint corresponding to the object $0 \in \mathbb{N}$ and the unique object $* \in X\underline{0} = \mathbf{1}$.
- The third construction in [Elm∞, Theorem 3] creates its own basepoint corresponding to the empty sequence, which should correspond to the monoidal unit in $\mathcal{P}X$ as in (10.8).

However, that composite construction also yields finite sequences $\langle * \rangle_{j=1}^{n}$ containing n copies of $* \in X\underline{0}$ created by the wreath product. The sequences $\langle * \rangle_{j=1}^{n}$ for $n \geq 1$ are not objects in $\mathcal{P}X$ as in Explanation 10.5. Similarly, finite sequences containing $*$ and other objects are also not supposed to be in $\mathcal{P}X$. For example, the composite construction in [Elm∞, Theorems 1–3] is already different from $\mathcal{P}X$ on objects for the terminal Γ-category X in Example 10.11. ◇

10.4 The Barratt-Eccles Operad and E_∞-Algebras

By Corollary 10.19 inverse K-theory $\mathcal{P}$ preserves any algebraic structure parametrized by a pseudo symmetric Cat-multifunctor from a Cat-multicategory. In particular, this holds for pseudo symmetric Cat-multifunctors from E_∞-operads. To explain in detail these pseudo symmetric E_∞-algebras in Γ-Cat and $\mathsf{PermCat}^{\mathsf{sg}}$, in this section we first recall the categorical Barratt-Eccles E_∞-operad.

- Theorem 10.39 is a coherence theorem for E_∞-algebras.
- Small permutative categories and small (tight) bipermutative categories are E_∞-algebras in, respectively, Cat, $\mathsf{PermCat}$, and $\mathsf{PermCat}^{\mathsf{sg}}$ (Example 10.46).
- In Example 10.47 we explain in detail E_∞-algebras in Γ-Cat.

In Section 10.5 we define the pseudo symmetric variant of E_∞-algebras using pseudo symmetric Cat-multifunctors from the Barratt-Eccles operad.

The following definition of the Barratt-Eccles operad is from [EM06, page 178]. For a detailed discussion, see [JY∞, Sections 11.4–11.6], which uses the notation $E\mathsf{As}$ instead of $E\Sigma_*$.

Definition 10.35. The *Barratt-Eccles operad*

$$E\Sigma_* = \{E\Sigma_n\}_{n\geq 0}$$

is the Cat-operad (Definition 3.2) defined by the following data. Its unique object is denoted by $*$.

- The category $E\Sigma_n$ has object set Σ_n. For each pair of permutations $\sigma, \tau \in \Sigma_n$, the morphism set $E\Sigma_n(\sigma,\tau)$ is the one-element set $\{*\}$.
- The $*$-*colored unit* is the object $\mathrm{id}_1 \in \Sigma_1$.
- The *symmetric group action* on objects is the group multiplication

$$\Sigma_n \times \Sigma_n \longrightarrow \Sigma_n.$$

- The *composition* on objects is defined by group multiplication

$$\gamma\big(\sigma ; \langle\sigma_j\rangle_{j=1}^n\big) = \widetilde{\sigma}\cdot\sigma^\times \in \Sigma_{k_1+\cdots+k_n} \tag{10.36}$$

  for $\sigma \in \Sigma_n$ and $\sigma_j \in \Sigma_{k_j}$ using the following notation.

  - The block permutation

  $$\widetilde{\sigma} = \sigma\langle k_1,\ldots,k_n\rangle \in \Sigma_{k_1+\cdots+k_n} \tag{10.37}$$

    permutes n consecutive blocks of lengths $k_1,\ldots,k_n$ according to σ (3.11).
  - The block sum

  $$\sigma^\times = \sigma_1 \times \cdots \times \sigma_n \in \Sigma_{k_1+\cdots+k_n} \tag{10.38}$$

    is given in (3.13).

- On morphisms the symmetric group action and composition are defined by the property that each morphism set in $E\Sigma_n$ is a one-element set.

This finishes the definition of $E\Sigma_*$.

Moreover, for a Cat-multicategory M, a Cat-multifunctor

$$E\Sigma_* \longrightarrow \mathsf{M}$$

is called an E_∞*-algebra* in M. ◇

The following coherence theorem for E_∞-algebras is from [Fre17, 6.3.3]. See also [JY∞, 11.4.14] for a detailed proof and [Yau22, 19.2.1] for an extension involving other types of equivariant structure. The transposition that permutes i and j is denoted (i, j), so $(1, 2)$ is the non-identity permutation in Σ_2.

Theorem 10.39 (E_∞-Coherence). *Suppose* $(\mathsf{M}, \gamma, 1)$ *is a* Cat-*multicategory. Then an* E_∞*-algebra in* M

$$E\Sigma_* \xrightarrow{\;F\;} \mathsf{M}$$

is uniquely determined by

- *the object* $x = F(*)$ *in* M*;*
- *the 0-ary 1-cell*
$$\mathit{1} = F(\mathrm{id}_0) \in \mathsf{M}(\langle\rangle\,;x);$$
- *the 2-ary 1-cell*
$$\mu = F(\mathrm{id}_2) \in \mathsf{M}(x, x\,; x); \quad \textit{and}$$
- *the 2-ary 2-cell isomorphism*
$$\mu \xrightarrow[\cong]{\xi = F(\tau)} \mu^{\mathsf{op}} = \mu(1,2) \quad \textit{in} \quad \mathsf{M}\big(x, x\,; x\big)$$

  with $\tau \in E\Sigma_2\big(\mathrm{id}_2, (1,2)\big)$ *the unique isomorphism.*

The above data are subject to the conditions (10.40) *through* (10.43) *below.*

1-Cell Unity and Associativity *The following object equalities hold.*

$$\begin{cases} \gamma\big(\mu\,; \mathit{1}, 1_x\big) = 1_x = \gamma\big(\mu\,; 1_x, \mathit{1}\big) \in \mathsf{M}(x\,; x) \\ \gamma\big(\mu\,; \mu, 1_x\big) = \gamma\big(\mu\,; 1_x, \mu\big) \in \mathsf{M}\big(x, x, x\,; x\big) \end{cases} \tag{10.40}$$

Symmetry *The following diagram in* $\mathsf{M}(x, x\,; x)$ *is commutative.*

$$\begin{array}{ccc} \mu & \xrightarrow{\;1_\mu\;} & \mu \\ & {\scriptstyle \xi}\searrow \quad \nearrow{\scriptstyle \xi(1,2)} & \\ & \mu^{\mathsf{op}} & \end{array} \tag{10.41}$$

Unity *The following diagram in $\mathsf{M}(x\,;x)$ is commutative.*

$$1_x = \gamma(\mu\,; 1_x, 1) \xrightarrow[\gamma(\xi\,;1_{1_x},1_1)]{1_{1_x}} \gamma(\mu\,; 1, 1_x) = \gamma(\mu^{\mathsf{op}}\,; 1_x, 1) = 1_x \tag{10.42}$$

Hexagon *The following diagram in $\mathsf{M}(x,x,x\,;x)$ is commutative.*

$$\begin{array}{ccc} \gamma(\mu\,;\mu,1_x) & \xrightarrow{\gamma(\xi\,;1_\mu,1_{1_x})} & \gamma(\mu^{\mathsf{op}}\,;\mu,1_x) \\ \| & & \| \\ \gamma(\mu\,;1_x,\mu) & & \gamma(\mu\,;\mu^{\mathsf{op}},1_x)(2,3) \\ {\scriptstyle\gamma(1_\mu\,;1_{1_x},\xi)}\searrow & & \nearrow{\scriptstyle\gamma(1_\mu\,;\xi,1_{1_x})(2,3)} \\ \gamma(\mu\,;1_x,\mu^{\mathsf{op}}) & = & \gamma(\mu\,;\mu,1_x)(2,3) \end{array} \tag{10.43}$$

Explanation 10.44 (E_∞-Coherence)**.** A key aspect of the proof of Theorem 10.39 is that, in addition to strictly preserving the unit and composition, a Cat-multifunctor $F : E\Sigma_* \longrightarrow \mathsf{M}$ also strictly preserves the symmetric group action. The last property is not available if F is merely a pseudo symmetric Cat-multifunctor (Definition 4.1). Therefore, the pseudo symmetric variant of Theorem 10.39 in Explanation 10.56 below has more data and more axioms. ◇

Explanation 10.45 (Underlying Monoid of an E_∞-Algebra)**.** In Theorem 10.39 the triple $(x,\mu,1)$ is a monoid in M in the sense that the unity and associativity axioms (10.40) are satisfied. This is a consequence of the Cat-multifunctor

$$\mathsf{As} \longrightarrow E\Sigma_*$$

from the associative operad (10.18) to the Barratt-Eccles operad given by the identity function on Σ_n for n-ary 1-cells. The unit, symmetric group action, and composition in As are as given in Definition 10.35. ◇

Example 10.46 ((Bi)permutative Categories as E_∞-Algebras)**.** The following examples are applications of Theorem 10.39 to the Cat-multicategories Cat (Example 1.16), $\mathsf{PermCat}$, and $\mathsf{PermCat}^{\mathsf{sg}}$ (Theorem 6.31).

1. By [JY∞, 11.4.26] an E_∞-algebra in Cat is precisely a small permutative category (Definition 1.11).
2. By [EM06, 3.8] or [JY∞, 11.5.5] an E_∞-algebra in $\mathsf{PermCat}$ is precisely a small bipermutative category (Definition 2.41).
3. Restricting the proof of [JY∞, 11.5.5] to $\mathsf{PermCat}^{\mathsf{sg}}$ shows that an E_∞-algebra in $\mathsf{PermCat}^{\mathsf{sg}}$ is precisely a small *tight* bipermutative category. ◇

Example 10.47 (E_∞-Algebras in Γ-Cat)**.** Recall the Cat-multicategory structure on Γ-Cat (Section 9.2). Applying Theorem 10.39, an E_∞-algebra in Γ-Cat is a quadruple

$$(X, 1, \mu, \xi)$$

consisting of the data (10.48) through (10.51) below.

Underlying Object It consists of a Γ-category (Definition 9.11)

$$(\mathcal{F}, \underline{0}) \xrightarrow{X} (\mathsf{Cat}, \mathbf{1}). \tag{10.48}$$

Unit It is equipped with a unit object

$$1 \in \mathsf{Map}_*(J, X).$$

This means a pointed natural transformation $1 : J \longrightarrow X$ with J the monoidal unit diagram (9.19). As discussed in (9.33) through (9.37), such an object 1 is uniquely determined by the object

$$1_{\underline{1}}(1) \in X\underline{1}. \tag{10.49}$$

Multiplication It is equipped with a 2-ary 1-cell

$$\mu \in \Gamma\text{-}\mathsf{Cat}\big(X, X\,;\, X\big) = \mathsf{Map}_*\big(X \wedge X\,,\, X\big).$$

This means a pointed natural transformation (Explanation 9.26)

$$X \wedge X \xrightarrow{\mu} X$$

with $\wedge$ the pointed Day convolution (9.18). It has component pointed functors

$$X\underline{m} \wedge X\underline{n} \xrightarrow{\mu_{\underline{m},\underline{n}}} X\underline{mn} \tag{10.50}$$

that are natural in $\underline{m}, \underline{n} \in \mathcal{F}$ in the sense of the naturality diagram (9.27).

Braiding It is equipped with a 2-ary 2-cell isomorphism

$$\mu \xrightarrow[\cong]{\xi} \mu \circ \xi^{\mathsf{Day}} \quad \text{in} \quad \Gamma\text{-}\mathsf{Cat}\big(X, X\,;\, X\big)$$

as follows, with ξ^{Day} the braiding for the pointed Day convolution (9.21).

$$\begin{array}{ccc} X \wedge X & \xrightarrow{\xi^{\mathsf{Day}}} & X \wedge X \\ & \overset{\xi}{\Longrightarrow} & \\ {\scriptstyle\mu}\searrow & X & \swarrow{\scriptstyle\mu} \end{array}$$

As discussed in (9.49), this means that ξ is a pointed modification with component pointed natural isomorphisms for $\underline{m}, \underline{n} \in \mathcal{F}$ as follows.

$$\begin{array}{ccc} X\underline{m} \wedge X\underline{n} & \xrightarrow{\xi^{\wedge}} & X\underline{n} \wedge X\underline{m} \\ {\scriptstyle\mu_{\underline{m},\underline{n}}}\big\downarrow & \overset{\xi_{\underline{m},\underline{n}}}{\underset{\cong}{\Longrightarrow}} & \big\downarrow{\scriptstyle\mu_{\underline{n},\underline{m}}} \\ X\underline{mn} & \xleftarrow[X(\xi^{\wedge}_{\underline{m},\underline{n}})^{-1}]{} & X\underline{nm} \end{array} \tag{10.51}$$

- The top $\xi^\wedge$ in (10.51) is the braiding for the smash product (9.5), which is induced by the braiding for the Cartesian product.
- The bottom

$$\underline{mn} \xrightarrow[\cong]{\xi^\wedge_{\underline{m},\underline{n}}} \underline{nm}$$

in (10.51) is the multiplicative braiding in $\mathcal{F}$ (2.60).

These component pointed natural isomorphisms satisfy the modification axiom (9.50), which means naturality in $\underline{m}, \underline{n} \in \mathcal{F}$.

In terms of the above data (10.48) through (10.51), the conditions (10.40) through (10.43) in Theorem 10.39 are given as follows for $m, n, p \geq 0$.

1-Cell Unity and Associativity The 1-cell unity condition in (10.40) means the equalities of functors

$$\mu_{\underline{1},\underline{n}}\big(1_{\underline{1}}(1), -\big) = 1_{X\underline{n}} = \mu_{\underline{n},\underline{1}}\big(-, 1_{\underline{1}}(1)\big).$$

The 1-cell associativity condition in (10.40) means that the following diagram of functors is commutative.

$$\begin{array}{ccc}
X\underline{m} \wedge X\underline{n} \wedge X\underline{p} & \xrightarrow{1 \wedge \mu_{\underline{n},\underline{p}}} & X\underline{m} \wedge X\underline{np} \\
{\scriptstyle \mu_{\underline{m},\underline{n}} \wedge 1}\downarrow & & \downarrow{\scriptstyle \mu_{\underline{m},\underline{np}}} \\
X\underline{mn} \wedge X\underline{p} & \xrightarrow{\mu_{\underline{mn},\underline{p}}} & X\underline{mnp}
\end{array} \tag{10.52}$$

Symmetry The condition (10.41) means that the following pasting is equal to the identity natural transformation on $\mu_{\underline{m},\underline{n}}$.

$$\begin{array}{ccccc}
& & \xrightarrow{\quad 1 \quad} & & \\
X\underline{m} \wedge X\underline{n} & \xrightarrow{\xi^\wedge} & X\underline{n} \wedge X\underline{m} & \xrightarrow{\xi^\wedge} & X\underline{m} \wedge X\underline{n} \\
{\scriptstyle \mu_{\underline{m},\underline{n}}}\downarrow & \overset{\xi_{\underline{m},\underline{n}}}{\underset{\cong}{\Longrightarrow}} & \downarrow{\scriptstyle \mu_{\underline{n},\underline{m}}} & \overset{\xi_{\underline{n},\underline{m}}}{\underset{\cong}{\Longrightarrow}} & \downarrow{\scriptstyle \mu_{\underline{m},\underline{n}}} \\
X\underline{mn} & \xleftarrow[X(\xi^\wedge_{\underline{m},\underline{n}})^{-1}]{} & X\underline{nm} & \xleftarrow[X(\xi^\wedge_{\underline{n},\underline{m}})^{-1}]{} & X\underline{mn} \\
& & \xleftarrow{\quad 1 \quad} & &
\end{array}$$

The top and bottom regions are commutative by the symmetry axiom (1.12) in, respectively, the symmetric monoidal categories

- $(\mathsf{Cat}_*, \wedge, \xi^\wedge)$ in Theorem 9.7 and
- $(\mathcal{F}, \wedge, \xi^\wedge)$ in Example 2.58.

Unity The condition (10.42) means that the natural transformation

$$X\underline{n} \underset{\mu_{\underline{1},\underline{n}}(1_{\underline{1}}(1),-) = 1}{\overset{\mu_{\underline{n},\underline{1}}(-,1_{\underline{1}}(1)) = 1}{\overset{\Downarrow (\xi_{\underline{n},\underline{1}})_{(-,1_{\underline{1}}(1))}}{\rightrightarrows}}} X\underline{n}$$

is equal to the identity on the identity functor $1_{X\underline{n}}$.

Hexagon In the following diagram we use the notation

$$\xi^{-\wedge} = (\xi^{\wedge})^{-1} \quad \text{and} \qquad (-)' = X(-)$$

and abbreviate $\wedge$ to concatenation. For example,

$$\underline{mn}' = X\underline{mn}, \quad \underline{m}'\underline{n}'\underline{p}' = X\underline{m} \wedge X\underline{n} \wedge X\underline{p}, \quad \text{and} \quad (\xi^{-\wedge}_{\underline{mn},\underline{p}})' = X(\xi^{\wedge}_{\underline{mn},\underline{p}})^{-1}.$$

The condition (10.43) means the following pasting diagram equality.

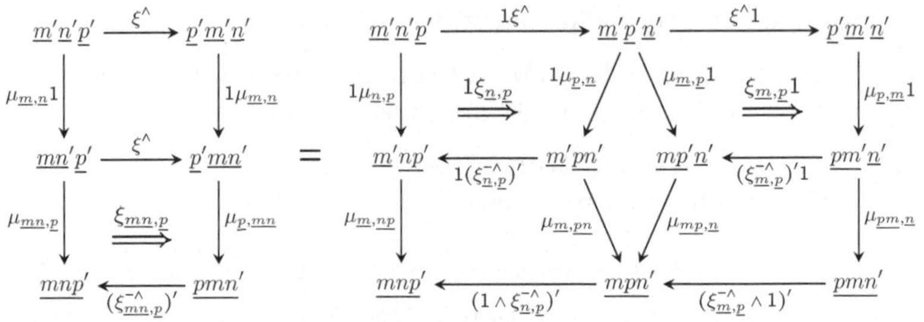

The detail of this pasting diagram equality is as follows.

- The top square on the left-hand side is commutative by the naturality of the braiding $\xi^\wedge$ for the smash product (9.5), which is induced by the braiding for the Cartesian product.
- Consider the right-hand side.
  - The center diamond is commutative by the 1-cell associativity condition (10.52).
  - The two unlabeled trapezoids are commutative by the naturality of $\mu : X \wedge X \longrightarrow X$ (9.27).
- The left, respectively, right, vertical boundaries of the two sides are equal by the 1-cell associativity condition (10.52).
- The top, respectively, bottom, horizontal boundaries of the two sides are equal by the hexagon axiom (1.12) in, respectively, the symmetric monoidal categories
  - $(\mathsf{Cat}_*, \wedge, \xi^\wedge)$ in Theorem 9.7 and
  - $(\mathcal{F}, \wedge, \xi^\wedge)$ in Example 2.58.

Equivalently, in terms of objects

$$a \in X\underline{m}, \quad b \in X\underline{n}, \quad \text{and} \quad c \in X\underline{p},$$

the condition (10.43) means the morphism equality

$$\begin{aligned}&\big(\xi_{\underline{mn},\underline{p}}\big)_{(\mu_{\underline{m},\underline{n}}(a,b)\,,\,c)}\\ &= X\Big(1_{\underline{m}} \wedge \xi^{\wedge}_{\underline{n},\underline{p}}\Big)^{-1}\Big(\mu_{\underline{mp},\underline{n}}\big((\xi_{\underline{m},\underline{p}})_{(a,c)}\,,\,1_b\big)\Big) \circ \mu_{\underline{m},\underline{np}}\Big(1_a\,,\,(\xi_{\underline{n},\underline{p}})_{(b,c)}\Big)\end{aligned}$$

in $X\underline{mnp}$.

This finishes our description of an E_∞-algebra in Γ-Cat. ◇

10.5 Inverse K-Theory Preserves Pseudo Symmetric E_∞-Algebras

Inverse K-theory $\mathcal{P}$ does *not* preserve E_∞-algebras in general, since E_∞-algebras are parametrized by Cat-multifunctors from the Barratt-Eccles operad $E\Sigma_*$ (Definition 10.35). Composing such Cat-multifunctors with the pseudo symmetric Cat-multifunctor $\mathcal{P}$ (10.13) yields pseudo symmetric Cat-multifunctors (Example 10.21), not Cat-multifunctors in general. In this section we

- define pseudo symmetric E_∞-algebras (Definition 10.53) and
- observe that inverse K-theory $\mathcal{P}$ preserves them (Corollary 10.54).

Then we completely unpack the structure of a pseudo symmetric E_∞-algebra in a general Cat-multicategory, Γ-Cat, PermCat, and $\mathsf{PermCat}^{\mathsf{sg}}$ (Explanations 10.56, 10.70, and 10.82).

Definition 10.53. Suppose M is a Cat-multicategory. A pseudo symmetric Cat-multifunctor (Definition 4.1) from the Barratt-Eccles operad (Definition 10.35)

$$E\Sigma_* \xrightarrow{(F,\{F_\sigma\})} \mathsf{M}$$

is called a *pseudo symmetric E_∞-algebra* in M. ◇

Applied to the Barratt-Eccles operad $E\Sigma_*$, Corollary 10.19 has the following special case.

Corollary 10.54. *The inverse K-theory pseudo symmetric* Cat-*multifunctor in* (10.13)

$$\Gamma\text{-}\mathsf{Cat} \xrightarrow{\mathcal{P}} \mathsf{PermCat}^{\mathsf{sg}}$$

preserves pseudo symmetric E_∞-algebras.

Example 10.55 (E_∞-Algebras)**.** By Example 4.7 each E_∞-algebra (Definition 10.35) is a pseudo symmetric E_∞-algebra when it is equipped with identity pseudo symmetry isomorphisms. By Corollary 10.54 inverse K-theory $\mathcal{P}$ sends each E_∞-algebra in Γ-Cat (Example 10.47) to a pseudo symmetric E_∞-algebra in $\mathsf{PermCat}^{\mathsf{sg}}$. ◇

Next we explain in detail the structure of a pseudo symmetric E_∞-algebra in

- a general Cat-multicategory (Explanation 10.56),
- Γ-Cat (Explanation 10.70), and
- PermCat, and $\mathsf{PermCat}^{\mathsf{sg}}$ (Explanation 10.82).

Explanation 10.56 below should be compared with the Coherence Theorem 10.39 for E_∞-algebras.

Explanation 10.56 (Pseudo Symmetric E_∞-Algebras)**.** Interpreting Definition 4.1 with domain the Barratt-Eccles operad $E\Sigma_*$ (Definition 10.35), a pseudo symmetric E_∞-algebra in a Cat-multicategory $(\mathsf{M}, \gamma, 1)$

$$E\Sigma_* \xrightarrow{(F,\{F_\sigma\})} \mathsf{M}$$

consists of the data (10.57) through (10.60) below for $n \geq 0$ and $\sigma, \tau \in \Sigma_n$.

Object The object assignment of F yields an object

$$x = F(*) \in \mathsf{M}. \tag{10.57}$$

1-Cells in Arity n The n-ary 1-cell assignment of F yields an object

$$\mu_\sigma = F(\sigma) \in \mathsf{M}(x^n\,; x) \tag{10.58}$$

with $x^n = \langle x \rangle_{j=1}^n$ the n-tuple consisting of n copies of x.

2-Cells in Arity n The n-ary 2-cell assignment of F yields an isomorphism

$$\mu_\sigma \xrightarrow[\cong]{\mu_\sigma^\tau} \mu_\tau \quad \text{in} \quad \mathsf{M}(x^n\,; x) \tag{10.59}$$

corresponding to the unique isomorphism in $E\Sigma_n(\sigma, \tau) = \{*\}$.

Pseudo Symmetry Isomorphisms The τ-component of the pseudo symmetry isomorphism F_σ (4.2) is an n-ary 2-cell isomorphism

$$\mu_{\tau\sigma} \xrightarrow[\cong]{F_{\sigma;\tau}} \mu_\tau \cdot \sigma \quad \text{in} \quad \mathsf{M}(x^n\,; x) \tag{10.60}$$

with $\mu_\tau \cdot \sigma$ the right σ-action on μ_τ.

The above data are subject to the conditions (10.61) through (10.69) below for $n \geq 0$ and $\sigma, \tau, \pi \in \Sigma_n$.

Component Functoriality The functoriality of each functor

$$E\Sigma_n \xrightarrow{F} \mathsf{M}(x^n\,; x)$$

means the following morphism equalities in $\mathsf{M}(x^n\,; x)$.

$$\begin{cases} \mu_\sigma^\sigma = 1_{\mu_\sigma} \\ \mu_\tau^\pi \circ \mu_\sigma^\tau = \mu_\sigma^\pi \end{cases} \tag{10.61}$$

Pseudo Symmetry Naturality The naturality of F_σ in (10.60) means that the following diagram in $\mathsf{M}(x^n\,;x)$ is commutative.

$$\begin{array}{ccc} \mu_{\tau\sigma} & \xrightarrow{F_{\sigma;\tau}} & \mu_\tau\cdot\sigma \\ {\scriptstyle \mu_{\tau\sigma}^{\pi\sigma}}\big\downarrow & & \big\downarrow{\scriptstyle \mu_\tau^\pi\cdot\sigma} \\ \mu_{\pi\sigma} & \xrightarrow{F_{\sigma;\pi}} & \mu_\pi\cdot\sigma \end{array} \tag{10.62}$$

Unit The unit axiom (3.21) is the object equality

$$\mu_{\mathrm{id}_1} = 1_x \in \mathsf{M}(x\,;x). \tag{10.63}$$

Composition Using the notation in (10.36) through (10.38), the composition axiom (3.22) on objects is the following object equality in $\mathsf{M}(x^{k_1+\cdots+k_n}\,;x)$.

$$\mu_{\widetilde{\sigma}\cdot\sigma^\times} = \gamma\big(\mu_\sigma\,;\langle\mu_{\sigma_j}\rangle_{j=1}^n\big) \tag{10.64}$$

The composition axiom (3.22) on morphisms is the following morphism equality for $\sigma,\tau\in\Sigma_n$ and $\sigma_j,\tau_j\in\Sigma_{k_j}$.

$$\mu_{\widetilde{\sigma}\cdot\sigma^\times}^{\widetilde{\tau}\cdot\tau^\times} = \gamma\big(\mu_\sigma^\tau\,;\langle\mu_{\sigma_j}^{\tau_j}\rangle_{j=1}^n\big) \tag{10.65}$$

The axioms (4.3) through (4.6) in Definition 4.1 are given as follows.

Unit Permutation The axiom (4.3) is the following morphism equality in $\mathsf{M}(x^n\,;x)$.

$$F_{\mathrm{id}_n;\tau} = 1_{\mu_\tau} : \mu_\tau \longrightarrow \mu_\tau \tag{10.66}$$

Product Permutation The axiom (4.4) means that the following diagram in $\mathsf{M}(x^n\,;x)$ is commutative, with the equality given by the symmetric group action axiom (3.6) in M.

$$\begin{array}{ccc} \mu_{\pi\sigma\tau} & \xrightarrow{F_{\sigma\tau;\pi}} & \mu_\pi\cdot\sigma\tau \\ {\scriptstyle F_{\tau;\pi\sigma}}\big\downarrow & & \big\| \\ \mu_{\pi\sigma}\cdot\tau & \xrightarrow{F_{\sigma;\pi}\cdot\tau} & (\mu_\pi\cdot\sigma)\cdot\tau \end{array} \tag{10.67}$$

Top Equivariance The axiom (4.5) means that, for $\sigma,\tau\in\Sigma_n$ and $\tau_j\in\Sigma_{k_j}$ with $j\in\{1,\ldots,n\}$, the following diagram in $\mathsf{M}(x^{k_1+\cdots+k_n}\,;x)$ is commutative.

$$\begin{array}{ccc} \mu_{\widetilde{\tau}\cdot\tau^\times\cdot\overline{\sigma}} & \xrightarrow{F_{\overline{\sigma};\,\widetilde{\tau}\cdot\tau^\times}} & \mu_{\widetilde{\tau}\cdot\tau^\times}\cdot\overline{\sigma} \\ \spadesuit\big\| & & \big\|\clubsuit \\ \mu_{\widetilde{\tau\sigma}\cdot(\prod_j\tau_{\sigma(j)})} & & \gamma\big(\mu_\tau\,;\langle\mu_{\tau_j}\rangle_{j=1}^n\big)\cdot\overline{\sigma} \\ \clubsuit\big\| & & \big\|\spadesuit \\ \gamma\big(\mu_{\tau\sigma}\,;\langle\mu_{\tau_{\sigma(j)}}\rangle_{j=1}^n\big) & \xrightarrow{\gamma(F_{\sigma;\tau}\,;\langle 1_{\mu_{\tau_{\sigma(j)}}}\rangle_{j=1}^n)} & \gamma\big(\mu_\tau\cdot\sigma\,;\langle\mu_{\tau_{\sigma(j)}}\rangle_{j=1}^n\big) \end{array} \tag{10.68}$$

- The block permutation

$$\overline{\sigma} = \sigma\langle k_{\sigma(1)}, \ldots, k_{\sigma(n)} \rangle \in \Sigma_{k_1+\cdots+k_n}$$

  is induced by σ as in (3.11).
- The two equalities labeled ♠ use the top equivariance axiom (3.10) in, respectively, $E\Sigma_*$ and M.
- The two equalities labeled ♣ use the composition axiom (10.64).

Bottom Equivariance The axiom (4.6) means that the following diagram in $\mathsf{M}(x^{k_1+\cdots+k_n}\,;x)$ is commutative for $\sigma \in \Sigma_n$ and $\sigma_j, \tau_j \in \Sigma_{k_j}$ with $j \in \{1, \ldots, n\}$.

$$\begin{array}{ccc}
\mu_{\widetilde{\sigma}\cdot\sigma^\times\cdot\tau^\times} & \xrightarrow{F_{\tau^\times;\,\widetilde{\sigma}\cdot\sigma^\times}} & \mu_{\widetilde{\sigma}\cdot\sigma^\times}\cdot\tau^\times \\
{\scriptstyle\heartsuit}\Vert & & \Vert{\scriptstyle\Diamond} \\
\mu_{\widetilde{\sigma}\cdot(\sqcap_j \sigma_j\tau_j)} & & \gamma\big(\mu_\sigma\,; \langle \mu_{\sigma_j}\rangle_{j=1}^n\big)\cdot\tau^\times \\
{\scriptstyle\Diamond}\Vert & & \Vert{\scriptstyle\heartsuit} \\
\gamma\big(\mu_\sigma\,; \langle \mu_{\sigma_j\tau_j}\rangle_{j=1}^n\big) & \xrightarrow{\gamma(1_{\mu_\sigma}\,;\, \langle F_{\tau_j;\sigma_j}\rangle_{j=1}^n)} & \gamma\big(\mu_\sigma\,; \langle \mu_{\sigma_j}\cdot\tau_j\rangle_{j=1}^n\big)
\end{array} \tag{10.69}$$

- The two equalities labeled ♡ use the bottom equivariance axiom (3.12) in, respectively, $E\Sigma_*$ and M.
- The two equalities labeled ◊ use the composition axiom (10.64).

This finishes the explicit description of a pseudo symmetric E_∞-algebra in a Cat-multicategory $(\mathsf{M}, \gamma, \mathbb{1})$. ◇

Explanation 10.70 below should be compared with the description of an E_∞-algebra in Γ-Cat (Example 10.47).

Explanation 10.70 (Pseudo Symmetric E_∞-Algebras in Γ-Cat)**.** Applying Explanation 10.56 to the Cat-multicategory M = Γ-Cat (Section 9.2), a pseudo symmetric E_∞-algebra in Γ-Cat

$$\big(X, \{\mu_\sigma\}, \{\mu_\sigma^\tau\}, \{F_\sigma\}\big) : E\Sigma_* \longrightarrow \Gamma\text{-}\mathsf{Cat}$$

consists of the data (10.71), (10.72), (10.74), (10.75), and (10.77) below for $n \geq 0$ and $\sigma, \tau \in \Sigma_n$.

Underlying Γ-Category It consists of a Γ-category (Definition 9.11)

$$(\mathcal{F}, \underline{0}) \xrightarrow{\;X\;} (\mathsf{Cat}, \mathbf{1}). \tag{10.71}$$

1-Cells in Arity n It has a pointed natural transformation (9.45)

$$\textstyle\bigwedge_{j=1}^n X \xrightarrow{\;\mu_\sigma\;} X \quad \text{in} \quad \mathsf{Map}_*\big(\bigwedge_{j=1}^n X\,, X\big). \tag{10.72}$$

For each n-tuple of objects $\langle m\rangle = \langle \underline{m_j}\rangle_{j=1}^n$ in $\mathcal{F}$, μ_σ consists of a component pointed functor (9.46)

$$\bigwedge_{j=1}^n X\underline{m_j} \xrightarrow{\mu_{\sigma;\langle m\rangle}} X(\underline{m_1 \cdots m_n}) \tag{10.73}$$

that satisfies the naturality diagram (9.48) with respect to $\langle \underline{m_j}\rangle_{j=1}^n$.

If $n = 0$, then $\bigwedge_{j=1}^n X$ is interpreted as the monoidal unit diagram J in (9.19). As discussed in the paragraph containing (9.37), the pointed natural transformation

$$J \xrightarrow{\mu_{\mathrm{id}_0}} X \quad \text{in} \quad \mathsf{Map}_*(J, X)$$

is determined by the object

$$(\mu_{\mathrm{id}_0})_{\underline{1}}(1) \in X\underline{1}. \tag{10.74}$$

2-Cells in Arity n It has an invertible pointed modification

$$\bigwedge_{j=1}^n X \underset{\mu_\tau}{\overset{\mu_\sigma}{\Downarrow \mu_\sigma^\tau}} X \quad \text{in} \quad \mathsf{Map}_*\big(\bigwedge_{j=1}^n X, X\big). \tag{10.75}$$

For each n-tuple of objects $\langle m\rangle = \langle \underline{m_j}\rangle_{j=1}^n$ in $\mathcal{F}$, μ_σ^τ consists of a component pointed natural isomorphism (9.49)

$$\bigwedge_{j=1}^n X\underline{m_j} \underset{\mu_{\tau;\langle m\rangle}}{\overset{\mu_{\sigma;\langle m\rangle}}{\Downarrow (\mu_\sigma^\tau)_{\langle m\rangle}}} X(\underline{m_1 \cdots m_n}) \tag{10.76}$$

that satisfies the modification axiom (9.50).

If $n = 0$, then by (10.61) $\mu_{\mathrm{id}_0}^{\mathrm{id}_0}$ consists of the identity morphism on the object determined by μ_{id_0} in (10.74).

Pseudo Symmetry Isomorphisms It has an invertible pointed modification

$$\bigwedge_{j=1}^n X \underset{\mu_\tau \cdot \sigma}{\overset{\mu_{\tau\sigma}}{\Downarrow F_{\sigma;\tau}}} X \quad \text{in} \quad \mathsf{Map}_*\big(\bigwedge_{j=1}^n X, X\big). \tag{10.77}$$

For each n-tuple of objects $\langle m\rangle = \langle \underline{m_j}\rangle_{j=1}^n$ in $\mathcal{F}$, $F_{\sigma;\tau}$ consists of a component pointed natural isomorphism (9.49)

$$\bigwedge_{j=1}^n X\underline{m_j} \underset{(\mu_\tau \cdot \sigma)_{\langle m\rangle}}{\overset{\mu_{\tau\sigma;\langle m\rangle}}{\Downarrow F_{\sigma;\tau,\langle m\rangle}}} X(\underline{m_1 \cdots m_n}) \tag{10.78}$$

that satisfies the modification axiom (9.50). By (9.54) the pointed functor $(\mu_\tau \cdot \sigma)_{\langle m\rangle}$ in (10.78) is the following composite.

$$\begin{tikzcd} \bigwedge_{j=1}^{n} X\underline{m_j} \ar[r, "(\mu_\tau\cdot\sigma)_{\langle m\rangle}"] \ar[d, "\sigma"', "\cong"] & X(\underline{m_1\cdots m_n}) \\ \bigwedge_{j=1}^{n} X\underline{m_{\sigma^{-1}(j)}} \ar[r, "\mu_{\tau;\sigma\langle m\rangle}"] & X(\underline{m_{\sigma^{-1}(1)}\cdots m_{\sigma^{-1}(n)}}) \ar[u, "\cong", "X(\sigma^{-1})"'] \end{tikzcd} \tag{10.79}$$

If $n = 0$, then by (10.66) $F_{\mathrm{id}_0;\mathrm{id}_0}$ is the identity morphism on the object determined by μ_{id_0} in (10.74).

This finishes the description of the list of data of a pseudo symmetric E_∞-algebra in Γ-Cat.

The conditions (10.61) through (10.69) for $n \geq 0$ and $\sigma, \tau, \pi \in \Sigma_n$ are interpreted componentwise using the data (10.73), (10.76), and (10.78), with appropriate adjustment using (10.74) for 0-ary inputs. Below we state explicitly the three most nontrivial conditions from that list.

Composition Consider objects

$$\underline{m_{ji}} \in \mathcal{F} \qquad \text{for } 1 \leq j \leq n \text{ and } 1 \leq i \leq k_j$$

along with the following notation.

$$\begin{cases} \langle \underline{m_j} \rangle = \langle \underline{m_{ji}} \rangle_{i=1}^{k_j} \in \mathcal{F}^{k_j} & m_{j\bullet} = \prod_{i=1}^{k_j} m_{ji} \\ \langle m \rangle = \langle \langle \underline{m_j} \rangle \rangle_{j=1}^{n} \in \mathcal{F}^{k_1+\cdots+k_n} & \end{cases} \tag{10.80}$$

Then the composition axiom on objects (10.64) is the following commutative diagram of functors.

$$\begin{tikzcd} \bigwedge_{j=1}^{n} \bigwedge_{i=1}^{k_j} X\underline{m_{ji}} \ar[rr, "\mu_{\overline{\sigma}\cdot\sigma^\times;\langle m\rangle}"] \ar[dr, "\wedge_j \mu_{\sigma_j;\langle \underline{m_j}\rangle}"'] & & X(\underline{\prod_{j=1}^{n} m_{j\bullet}}) \\ & \bigwedge_{j=1}^{n} X\underline{m_{j\bullet}} \ar[ur, "\mu_{\sigma;\langle \underline{m_{j\bullet}}\rangle_{j=1}^{n}}"'] & \end{tikzcd} \tag{10.81}$$

The composition axiom on morphisms (10.65) is the analogue of (10.81) with each arrow replaced by its 2-cell variant in (10.76).

Top Equivariance We use the notation in (10.80) together with

$$\langle m\rangle\overline{\sigma} = \langle \langle \underline{m_{\sigma(j)}} \rangle \rangle_{j=1}^{n} \in \mathcal{F}^{k_{\sigma(1)}+\cdots+k_{\sigma(n)}}.$$

The top equivariance axiom (10.68) means that the component natural isomorphism

$$\begin{tikzcd}[column sep=huge] \bigwedge_{j=1}^{n} \bigwedge_{i=1}^{k_{\sigma(j)}} X\underline{m_{\sigma(j),i}} \ar[r, bend left, "\mu_{\overline{\tau}\cdot\tau^\times\cdot\overline{\sigma};\langle m\rangle\overline{\sigma}}", ""{name=U, below}] \ar[r, bend right, "(\mu_{\overline{\tau}\cdot\tau^\times}\cdot\overline{\sigma})_{\langle m\rangle\overline{\sigma}}"', ""{name=D, above}] & X(\underline{\prod_{j=1}^{n} m_{\sigma(j),\bullet}}) \ar[Rightarrow, from=U, to=D, "F_{\overline{\sigma};\overline{\tau}\cdot\tau^\times,\langle m\rangle\overline{\sigma}}"] \end{tikzcd}$$

is equal to the following whiskered natural isomorphism.

$$\begin{array}{ccccc} & \overset{\wedge_j \mu_{\tau_{\sigma(j)};\langle \underline{m_{\sigma(j)}}\rangle}}{\nearrow} & \bigwedge_{j=1}^n X\underline{m_{\sigma(j),\bullet}} & \xrightarrow{\mu_{\tau\sigma;\langle \underline{m_{\sigma(j),\bullet}}\rangle_{j=1}^n}} & X\big(\underline{\prod_{j=1}^n m_{\sigma(j),\bullet}}\big) \\ \bigwedge_{j=1}^n \bigwedge_{i=1}^{k_{\sigma(j)}} X\underline{m_{\sigma(j),i}} & & \sigma \big\downarrow & \Downarrow F_{\sigma;\tau,\langle \underline{m_{\sigma(j),\bullet}}\rangle_{j=1}^n} & \big\uparrow X(\sigma^{-1}) \\ & & \bigwedge_{j=1}^n X\underline{m_{j\bullet}} & \xrightarrow[\mu_{\tau;\langle \underline{m_{j\bullet}}\rangle_{j=1}^n}]{} & X\big(\underline{\prod_{j=1}^n m_{j\bullet}}\big) \end{array}$$

In the previous two diagrams, the top, respectively, bottom, boundary composites are equal by the left, respectively, right, vertical equalities in (10.68), along with (10.79).

Bottom Equivariance We use (10.80) along with the following notation.

$$\begin{cases} \langle \underline{m_j}\rangle \tau_j = \langle \underline{m_{j,\tau_j(i)}}\rangle_{i=1}^{k_j} \in \mathcal{F}^{k_j} & m_{j,\tau_j}\bullet = \prod_{i=1}^{k_j} m_{j,\tau_j(i)} \\ \langle m\rangle \tau^\times = \big\langle \langle \underline{m_j}\rangle \tau_j\big\rangle_{j=1}^n \in \mathcal{F}^{k_1+\cdots+k_n} & \end{cases}$$

The bottom equivariance axiom (10.69) means that the component natural isomorphism

$$\bigwedge_{j=1}^n \bigwedge_{i=1}^{k_j} X\underline{m_{j,\tau_j(i)}} \;\overset{\mu_{\overline{\sigma}\cdot\sigma^\times\cdot\tau^\times;\langle m\rangle\tau^\times}}{\underset{(\mu_{\overline{\sigma}\cdot\sigma^\times}\cdot\tau^\times)_{\langle m\rangle\tau^\times}}{\Downarrow F_{\tau^\times;\overline{\sigma}\cdot\sigma^\times,\langle m\rangle\tau^\times}}}\; X\big(\underline{\prod_{j=1}^n m_{j,\tau_j}\bullet}\big)$$

is equal to the following whiskered natural isomorphism.

$$\begin{array}{ccccc} \bigwedge_{j=1}^n \bigwedge_{i=1}^{k_j} X\underline{m_{j,\tau_j(i)}} & \xrightarrow{\wedge_j \mu_{\sigma_j\tau_j;\langle \underline{m_j}\rangle\tau_j}} & \bigwedge_{j=1}^n X\underline{m_{j,\tau_j}\bullet} & \overset{\mu_{\sigma;\langle \underline{m_{j,\tau_j}\bullet}\rangle_{j=1}^n}}{\searrow} & \\ \wedge_j \tau_j \big\downarrow & \Downarrow \wedge_j F_{\tau_j;\sigma_j,\langle \underline{m_j}\rangle\tau_j} & \big\uparrow \wedge_j X(\tau_j^{-1}) & & X\big(\underline{\prod_{j=1}^n m_{j,\tau_j}\bullet}\big) \\ \bigwedge_{j=1}^n \bigwedge_{i=1}^{k_j} X\underline{m_{ji}} & \xrightarrow[\wedge_j \mu_{\sigma_j;\langle \underline{m_j}\rangle}]{} & \bigwedge_{j=1}^n X\underline{m_{j\bullet}} & & \end{array}$$

In the previous two diagrams, the top, respectively, bottom, boundary composites are equal by the left, respectively, right, vertical equalities in (10.69), along with (10.79).

This finishes the description of a pseudo symmetric E_∞-algebra in Γ-Cat. ◇

Recall from Example 10.46 that E_∞-algebras in PermCat and $\mathsf{PermCat}^{\mathsf{sg}}$ are, respectively, small bipermutative categories and their tight variants (Definition 2.41). Explanation 10.82 below also applies to $\mathsf{PermCat}^{\mathsf{sg}}$ (6.22) by restricting to *strong* n-linear functors, which have invertible linearity constraints (10.85).

Explanation 10.82 (Pseudo Symmetric E_∞-Algebras in PermCat)**.** Applying Explanation 10.56 to the Cat-multicategory M = PermCat (Theorem 6.31), a pseudo symmetric E_∞-algebra in PermCat

$$\big(\mathsf{C}, \{\mu_\sigma\}, \{\mu_\sigma^\tau\}, \{F_\sigma\}\big) : E\Sigma_* \longrightarrow \mathsf{PermCat}$$

consists of the data (10.83) through (10.87) below for $n \geq 0$ and $\sigma, \tau \in \Sigma_n$.

Underlying Permutative Category It consists of a small permutative category (Definition 1.11)

$$(\mathsf{C}, \oplus, e, \beta). \tag{10.83}$$

1-Cells in Arity n It has an n-linear functor (Definition 6.4)

$$\textstyle\prod_{j=1}^{n} \mathsf{C} = \mathsf{C}^n \xrightarrow{\left(\mu_\sigma\, ,\, \langle \mu^2_{\sigma;j} \rangle_{j=1}^n\right)} \mathsf{C}. \tag{10.84}$$

The natural transformation

$$\mu_\sigma \langle X \circ_j X_j \rangle \oplus \mu_\sigma \langle X \circ_j X'_j \rangle \xrightarrow{\mu^2_{\sigma;j}} \mu_\sigma \langle X \circ_j (X_j \oplus X'_j) \rangle \tag{10.85}$$

for objects $\langle X \rangle \in \mathsf{C}^n$ and $X'_j \in \mathsf{C}$ is the j-th linearity constraint (6.6) of μ_σ. If $\mathsf{M} = \mathsf{PermCat}^{\mathsf{sg}}$, then μ_σ is a *strong* n-linear functor. This means that each linearity constraint $\mu^2_{\sigma;j}$ is a natural isomorphism.

2-Cells in Arity n It has an n-linear natural isomorphism (Definition 6.16)

$$\mathsf{C}^n \;\underset{\left(\mu_\tau,\, \langle \mu^2_{\tau;j} \rangle_{j=1}^n\right)}{\overset{\left(\mu_\sigma,\, \langle \mu^2_{\sigma;j} \rangle_{j=1}^n\right)}{\Downarrow \mu_\sigma^\tau}}\; \mathsf{C} \quad \text{in} \quad \mathsf{PermCat}\big(\langle \mathsf{C} \rangle_{j=1}^n\, ; \mathsf{C}\big). \tag{10.86}$$

Pseudo Symmetry Isomorphisms It has an n-linear natural isomorphism

$$\mathsf{C}^n \;\underset{\left(\mu_\tau,\, \langle \mu^2_{\tau;j} \rangle_{j=1}^n\right) \cdot \sigma}{\overset{\left(\mu_{\tau\sigma},\, \langle \mu^2_{\tau\sigma;j} \rangle_{j=1}^n\right)}{\Downarrow F_{\sigma;\tau}}}\; \mathsf{C} \quad \text{in} \quad \mathsf{PermCat}\big(\langle \mathsf{C} \rangle_{j=1}^n\, ; \mathsf{C}\big). \tag{10.87}$$

Here $\left(\mu_\tau, \langle \mu^2_{\tau;j} \rangle_{j=1}^n\right) \cdot \sigma$ is the right σ-action on $\left(\mu_\tau, \langle \mu^2_{\tau;j} \rangle_{j=1}^n\right)$ as defined in (6.25) and (6.26).

This finishes the description of the list of data of a pseudo symmetric E_∞-algebra in $\mathsf{PermCat}$.

The conditions (10.61) through (10.69) for $n \geq 0$ and $\sigma, \tau, \pi \in \Sigma_n$ are interpreted using the data (10.83) through (10.87). Below we state explicitly the three most nontrivial conditions from that list.

Composition The composition axiom on objects (10.64) means that the following diagram of functors is commutative.

$$\begin{array}{ccc} \mathsf{C}^{k_1+\cdots+k_n} & \xrightarrow{\mu_{\bar{\sigma}\cdot\sigma^\times}} & \mathsf{C} \\ & {\scriptstyle \prod_{j=1}^n \mu_{\sigma_j}} \searrow \quad \nearrow {\scriptstyle \mu_\sigma} & \\ & \mathsf{C}^n & \end{array} \tag{10.88}$$

Moreover, their corresponding linearity constraints are equal. The linearity constraints for the composite $\mu_\sigma \circ (\prod_{j=1}^n \mu_{\sigma_j})$ are defined in (6.30).

The composition axiom on morphisms (10.65) is the analogue of (10.88) with each functor replaced by its 2-cell variant in (10.86).

Top Equivariance With the notation

$$k = k_1 + \cdots + k_n \quad \text{and} \quad k_\sigma = k_{\sigma(1)} + \cdots + k_{\sigma(n)},$$

the top equivariance axiom (10.68) is the following equality of natural isomorphisms.

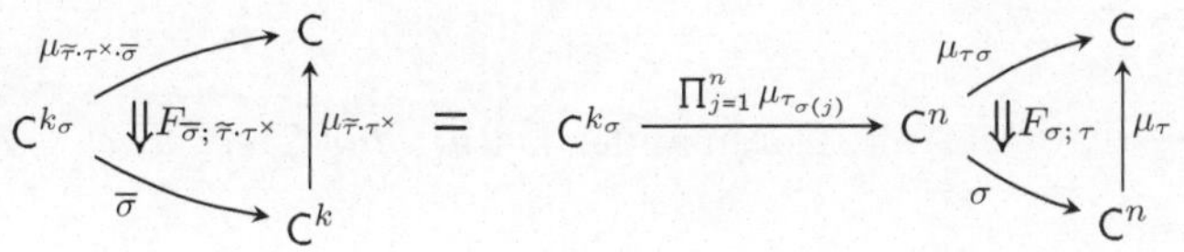

In the above equality, the top, respectively, bottom, boundary composites are equal by the left, respectively, right, vertical equalities in (10.68), along with (6.25) and (6.29).

Bottom Equivariance The bottom equivariance axiom (10.69) is the following equality of natural isomorphisms.

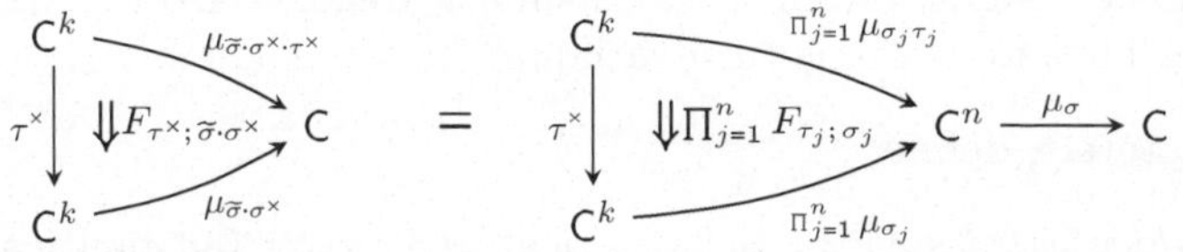

In the above equality, the top, respectively, bottom, boundary composites are equal by the left, respectively, right, vertical equalities in (10.69), along with (6.25) and (6.29).

This finishes the description of a pseudo symmetric E_∞-algebra in the Cat-multicategory PermCat. ◇

A

Open Questions

> *"Captain, the most elementary and valuable statement in science, the beginning of wisdom, is, 'I do not know'. I do not know what that is, sir."*
>
> – Lt. Commander Data, *Star Trek: The Next Generation*

In this Appendix, we discuss open questions related to the topics of this monograph: enriched multicategories, bipermutative categories, the Grothendieck construction, and inverse K-theory.

Question A.1 (Multi-Bicategories). Each Cat-multicategory has an underlying 1-ary 2-category (Example 3.18). A *bicategory* [JY21, 2.1.3] is a pseudo version of a 2-category.

- What is a pseudo variant of a Cat-multicategory? Do theorems for bicategories have multicategorical variants?

More concretely, define

- *multi-bicategories* as the bicategorical variants of Cat-multicategories and the multiple-input variants of bicategories,
- *lax* Cat-*multifunctors* between multi-bicategories, generalizing Definition 4.1 and Explanation 4.35, as the lax variants of Cat-multifunctors and the multiple-input variants of lax functors [JY21, 4.1.2],
- *lax* Cat-*multinatural transformations* between lax Cat-multifunctors as the lax variants of pseudo symmetric Cat-multinatural transformations (Definition 4.19) and the multiple-input variants of lax transformations [JY21, 4.2.1], and so on.

Then use these definitions to prove multi-bicategorical variants of

- the Bicategorical Pasting Theorem [JY21, 3.6.6],
- the Bicategorical Quillen Theorem A [JY21, 7.3.1],
- the Whitehead Theorem for Bicategories [JY21, 7.4.1], generalizing Theorems 3.69 and 4.64,

DOI: 10.1201/9781003451235-A

- the Bicategorical Yoneda Lemma [JY21, 8.3.16], and
- the Bicategorical Coherence Theorem [JY21, 8.4.1].

There may be a conceptual approach to some of these questions via 2-monads [JY21, 6.5].

- Is there a 2-monad with
  - Cat-multicategories with a given object set X as strict algebras and
  - multi-bicategories with object set X as pseudo algebras?

If the answer is yes, then the coherence theorems in [BKP89, Pow89] may apply to this 2-monad. ◇

Question A.2 (Pseudo Symmetry)**.** Recall the notion of a pseudo symmetric Cat-multifunctor $(F, \{F_\sigma\})$ in Definition 4.1.

- Is it possible for F—consisting of its object assignment and component functors—to admit several different pseudo symmetric Cat-multifunctor structures $\{F_\sigma\}$?

In Example 4.7 we equip a Cat-multifunctor F with identity natural transformations $F_\sigma = \mathrm{Id}$ to make $(F, \{\mathrm{Id}\})$ into a pseudo symmetric Cat-multifunctor. Here is a special case of the previous question.

- Is it possible for a Cat-multifunctor F to admit generally non-identity pseudo symmetry isomorphisms $\{F_\sigma\}$ such that $(F, \{F_\sigma\})$ is a pseudo symmetric Cat-multifunctor different from the one in Example 4.7? ◇

Question A.3 (Permutative Opfibrations)**.** By the Grothendieck Fibration Theorem [JY21, 9.5.6], there is a 2-monad on Cat/B whose pseudo, respectively, strict, algebras are in canonical bijections with cloven, respectively, split, opfibrations over B.

- Prove the analogous theorem for *permutative* opfibrations and the cloven variant in Definition 7.6.

In a permutative opfibration $P : \mathsf{E} \longrightarrow \mathcal{D}$, the functor P is a strict symmetric monoidal functor between permutative categories. Therefore, instead of Cat/B, one should consider the over-category of small permutative categories and strict symmetric monoidal functors over $\mathcal{D}$. ◇

Question A.4 (Pseudo Bipermutative-Indexed Categories)**.** For a small tight bipermutative category $\mathcal{D}$, the Cat-multicategory $\mathcal{D}$Cat (Theorem 5.71) has objects, n-ary 1-cells, and n-ary 2-cells given by, respectively, additive symmetric monoidal functors, additive natural transformations, and additive modifications (Definitions 5.31, 5.39, and 5.52).

- If the objects, n-ary 1-cells, and n-ary 2-cells are replaced by appropriate versions of, respectively, pseudo functors, strong transformations, and modifications [JY21, 4.1.2, 4.2.1, and 4.4.1], does $\mathcal{D}$Cat still form a Cat-multicategory or a multi-bicategory (Question A.1)?

One can simply replace an additive symmetric monoidal functor in Definition 5.31 by a pseudo functor. However, each of Definitions 5.39 and 5.52 has a unity axiom and an additivity axiom. Their pseudo variants are, therefore, more subtle to formulate. ◇

Question A.5 (Pseudo Permutative Opfibrations). For a small tight bipermutative category $\mathcal{D}$, the non-symmetric Cat-multicategory $\mathsf{PFib}^{\mathcal{D}}$ (Theorem 7.51) has objects, n-ary 1-cells, and n-ary 2-cells given by, respectively, small permutative opfibrations over $\mathcal{D}$, opcartesian strong n-linear functors, and opcartesian n-linear transformations (Definitions 7.6, 7.11, and 7.27). Similar to Question A.4, we ask the following.

- If the objects, n-ary 1-cells, and n-ary 2-cells are replaced by, respectively, cloven opfibrations and some pseudo versions of n-ary 1-cells and n-ary 2-cells, does $\mathsf{PFib}^{\mathcal{D}}$ still form a non-symmetric Cat-multicategory or a non-symmetric multi-bicategory (Question A.1)? ◇

Question A.6 (Pseudo Grothendieck Cat-Multiequivalences). For each small tight bipermutative category $\mathcal{D}$, there is a non-symmetric Cat-multiequivalence given by the Grothendieck construction (Theorem 8.76)

$$\mathcal{D}\mathsf{Cat} \xrightarrow[\sim]{\int_{\mathcal{D}}} \mathsf{PFib}^{\mathcal{D}}.$$

If Questions A.4 and A.5 can both be answered in the positive, then we ask the following.

- Can the non-symmetric Cat-multiequivalence $\int_{\mathcal{D}}$ be extended to the pseudo setting?

In fact, one can regard this question as a litmus test for the correctness of the definitions of the pseudo variants of $\mathcal{D}\mathsf{Cat}$ and $\mathsf{PFib}^{\mathcal{D}}$ in Questions A.4 and A.5. ◇

Question A.7 (Global Grothendieck Cat-Multiequivalences). In the construction $\int_{\mathcal{D}}$ the small tight bipermutative category $\mathcal{D}$ is fixed. So a natural question is whether $\mathcal{D}$ can be used as a variable.

- Is there a global version of the non-symmetric Cat-multiequivalence $\int_{\mathcal{D}}$ where $\mathcal{D}$ varies in tight bipermutative categories?

To answer this question, one would need

- a notion of functors between bipermutative categories, which should be analogous to symmetric bimonoidal functors in [Yau∞I, Section 5.1];
- a global version of the Cat-multicategory $\mathcal{D}\mathsf{Cat}$ (Theorem 5.71) where $\mathcal{D}$ varies in tight bipermutative categories; and
- a global version of the non-symmetric Cat-multicategory $\mathsf{PFib}^{\mathcal{D}}$ (Theorem 7.51) where $\mathcal{D}$ varies in tight bipermutative categories.

Moreover, if Question A.6 has a positive answer, then one can also ask whether there is a global variant where $\mathcal{D}$ varies. $\diamond$

Question A.8 (Operadic Grothendieck Cat-Multiequivalences)**.** Small tight bipermutative categories are E_∞-algebras in $\mathsf{PermCat}^{\mathsf{sg}}$ (Example 10.46 (3)) with respect to the Barratt-Eccles operad $E\Sigma_*$ (Definition 10.35).

- Is there a generalization of the non-symmetric Cat-multiequivalence $\int_{\mathcal{D}}$ where $\mathcal{D}$ is a fixed Q-algebra in $\mathsf{PermCat}^{\mathsf{sg}}$ for a categorical operad Q?

In addition to the Barratt-Eccles E_∞-operad $E\Sigma_*$, other categorical operads of interest include the E_n-operads in [JY∞, 13.2.1] and the G-monoidal category operads for an action operad G in [Yau22, 19.2.1]. If the answer to this question is yes, then we can extend Question A.7 to this setting.

- Is there a global version of the non-symmetric Cat-multiequivalence $\int_{\mathcal{D}}$ where $\mathcal{D}$ varies among Q-algebras in $\mathsf{PermCat}^{\mathsf{sg}}$?

One can extend this question further by

- varying the categorical operad Q and
- considering it in the context of Question A.6. $\diamond$

Question A.9 (Cat-Multifunctorial Inverse K-Theory)**.** Inverse K-theory $\mathcal{P}$ is a pseudo symmetric Cat-multifunctor but *not* a Cat-multifunctor in the symmetric sense (Theorem 10.12).

- Is there an equivalent variant of $\mathcal{P}$ that is a Cat-multifunctor in the symmetric sense?

More precisely, this question asks for the existence of

- a Cat-multifunctor

$$\Gamma\text{-}\mathsf{Cat} \xrightarrow{\mathcal{P}'} \mathsf{PermCat}^{\mathsf{sg}}$$

  together with

- a natural transformation $\theta : \mathcal{P} \longrightarrow \mathcal{P}'$ or $\mathcal{P}' \longrightarrow \mathcal{P}$ that is componentwise a stable equivalence in $\mathsf{PermCat}^{\mathsf{sg}}$.

A morphism in $\mathsf{PermCat}^{\mathsf{sg}}$ is a *stable equivalence* if its image under Segal K-theory [Seg74] is a stable equivalence; see [GJO17, JY22b, JY22c, BF78, Man10, Tho95] for more discussion of stable equivalences. Since $\mathcal{P}$ is a homotopy inverse of Segal K-theory [Man10], θ ensures that $\mathcal{P}'$ is also a homotopy inverse of Segal K-theory. Moreover, being a Cat-multifunctor, $\mathcal{P}'$ preserves any algebraic structure—such as E_∞-algebras and E_n-algebras—parametrized by a Cat-multifunctor, strengthening Corollaries 10.15 and 10.19. $\diamond$

B

List of Main Facts

Chapter 1. Preliminaries on Enriched Categories and 2-Categories

(1.5) Each monoidal category satisfies $\lambda_1 = \rho_1$.
(1.6) Each monoidal category satisfies the left and right unity axioms.
(1.10) Each braided monoidal category satisfies $\rho = \lambda\xi_{-,1}$ and $\lambda = \rho\xi_{1,-}$
(1.13) A symmetric monoidal category is precisely a braided monoidal category that satisfies the symmetry axiom.
(1.16) Cat is a symmetric monoidal closed category.
(1.33) Cat is a 2-category.
(1.34) V-Cat is a 2-category.
(1.35) A locally small 2-category is precisely a Cat-category.
(1.63) The diagram category $\mathcal{D}$-V is symmetric monoidal closed.
(1.65) $\mathcal{D}$-V is enriched, tensored, and cotensored over V.

Chapter 2. Symmetric Bimonoidal and Bipermutative Categories

(2.28) In a symmetric bimonoidal category, 12 of the 24 axioms follow from the other 12 axioms.
(2.45) There is a canonical bijection between tight bipermutative categories and tight symmetric bimonoidal categories with permutative additive structure, permutative multiplicative structure, $\lambda^\bullet = 1$, and $\rho^\bullet = 1$
(2.48) There is a small tight bipermutative category $\mathsf{Fin}^{\mathsf{sk}}$ with unpointed finite sets $\overline{n}$ for $n \geq 0$ as objects and functions as morphisms.
(2.57) There is a small tight bipermutative category Σ with the same objects as $\mathsf{Fin}^{\mathsf{sk}}$ and permutations as morphisms.
(2.58) There is a small tight bipermutative category $\mathcal{F}$ with pointed finite sets $\underline{n}$ for $n \geq 0$ as objects and pointed functions as morphisms.
(2.63) Mandell's category $\mathcal{A}$ for inverse K-theory is a small tight bipermutative category.
(2.90) In a symmetric bimonoidal category C in which the value of each δ-prime edge is a monomorphism, any two parallel paths in $\mathsf{Gr}(X)$ with a regular domain have the same value in C.
(2.91) Laplaza's Coherence Theorem 2.90 applies to tight bipermutative categories.

DOI: 10.1201/9781003451235-B

Chapter 3. Enriched Multicategories and Multiequivalences

(3.15) The terminal multicategory T consists of a single object and a single n-ary operation for each n.
(3.16) Each object in a V-multicategory has an endomorphism V-operad.
(3.17) Each permutative category has an endomorphism multicategory.
(3.18) Each V-multicategory has an underlying V-category.
(3.28) There is a 2-category with small V-multicategories as objects.
(3.29) V-Multicat and V-Multicat^{ns} are complete and cocomplete.
(3.30) The initial V-multicategory has an empty set of objects. The terminal V-multicategory has one object and each multimorphism object given by the terminal object in V.
(3.33) A Cat-multinatural transformation consists of component 1-ary 1-cells that satisfy two Cat-naturality conditions, one for objects and one for morphisms.
(3.45) Each Cat-multiequivalence is essentially surjective on objects and an isomorphism on each multimorphism category.
(3.69) A Cat-multifunctor is a Cat-multiequivalence if and only if it is essentially surjective on objects and an isomorphism on each multimorphism category.
(3.70) A 2-functor between 2-categories is a 2-equivalence if and only if it is essentially surjective on objects and an isomorphism on each hom category.

Chapter 4. Pseudo Symmetry

(4.1) A pseudo symmetric Cat-multifunctor strictly preserves the colored units and composition and has pseudo symmetry isomorphisms that satisfy four coherence axioms.
(4.7) There is a canonical bijection between Cat-multifunctors and pseudo symmetric Cat-multifunctors with identity pseudo symmetry isomorphisms.
(4.8) A pseudo symmetric Cat-multifunctor yields a non-symmetric Cat-multifunctor.
(4.16) Pseudo symmetric Cat-multifunctors are closed under composition, which is associative and unital.
(4.19) A pseudo symmetric Cat-multinatural transformation satisfies two Cat-naturality conditions and the pseudo symmetry preservation axiom.
(4.28) Identities and vertical composites are well defined for pseudo symmetric Cat-multinatural transformations.
(4.29) Pseudo symmetric Cat-multinatural transformations are closed under horizontal composition.
(4.34) There is a 2-category Cat-Multicat^{ps} with small Cat-multicategories as objects, pseudo symmetric Cat-multifunctors as 1-cells, and pseudo symmetric Cat-multinatural transformations as 2-cells.
(4.42) Each pseudo symmetric Cat-multiequivalence is essentially surjective on objects and an isomorphism on each multimorphism category.

(4.64) A pseudo symmetric Cat-multifunctor is a pseudo symmetric Cat-multiequivalence if and only if it is essentially surjective on objects and an isomorphism on each multimorphism category.

Chapter 5. Enriched Multicategories of Indexed Categories

(5.1) $\mathcal{D}$-V is a V-multicategory.
(5.39) An additive natural transformation is a natural transformation that satisfies the unity axiom and the additivity axiom.
(5.52) An additive modification is a modification that satisfies the unity axiom and the additivity axiom.
(5.60) $\mathcal{D}\mathsf{Cat}(\langle X\rangle ; Z)$ is a subcategory of $\mathcal{D}^{\otimes}\text{-}\mathsf{Cat}(\langle X\rangle ; Z)$.
(5.64) $\mathcal{D}\mathsf{Cat}$ is not a symmetric monoidal category.
(5.71) $\mathcal{D}\mathsf{Cat}$ is a Cat-multicategory.
(5.72) There is a forgetful Cat-multifunctor $\mathcal{D}\mathsf{Cat} \longrightarrow \mathcal{D}^{\otimes}\text{-}\mathsf{Cat}$.

Chapter 6. The Grothendieck Construction is a Pseudo Symmetric Cat-Multifunctor

(6.13) A 1-linear functor is precisely a strictly unital symmetric monoidal functor.
(6.14) The multiplicative structure in a tight bipermutative category is a strong n-linear functor relative to the additive structure.
(6.31) PermCat and $\mathsf{PermCat}^{\mathsf{sg}}$ are Cat-multicategories.
(6.36) The Grothendieck construction of an indexed category is a category.
(6.42) The Grothendieck construction of a symmetric monoidal functor to Cat is a permutative category.
(6.57) The Grothendieck construction of an additive natural transformation is a strong n-linear functor.
(6.67) The Grothendieck construction of an additive modification is an n-linear natural transformation.
(6.69) The Grothendieck construction defines a functor from $\mathcal{D}\mathsf{Cat}(\langle X\rangle ; Z)$ to $\mathsf{PermCat}^{\mathsf{sg}}(\langle \int_{\mathcal{D}} X\rangle ; \int_{\mathcal{D}} Z)$.
(6.73) The Grothendieck construction preserves colored units.
(6.74) The Grothendieck construction preserves composition.
(6.100) The Grothendieck construction has pseudo symmetry isomorphisms.
(6.110) The Grothendieck construction $\int_{\mathcal{D}}$ is a pseudo symmetric Cat-multifunctor from $\mathcal{D}\mathsf{Cat}$ to $\mathsf{PermCat}^{\mathsf{sg}}$. It is a Cat-multifunctor if the multiplicative braiding in $\mathcal{D}$ is the identity. Conversely, if $\int_{\mathcal{D}}$ is a Cat-multifunctor, then $x \otimes y = y \otimes x$ for all objects $x, y \in \mathcal{D}$.

Chapter 7. Permutative Opfibrations from Bipermutative-Indexed Categories

(7.7) A permutative opfibration is a split opfibration with additional properties but no extra structure.

(7.18 (1)) An opcartesian n-linear functor is an n-linear functor with additional properties but no extra structure.
(7.18 (2)) An opcartesian 0-linear functor is an object in the P_0-preimage of $1 \in \mathcal{D}$.
(7.21) In the definition of an opcartesian n-linear functor, axiom 1 follows from axiom 3.
(7.24) An opcartesian 1-linear functor F is precisely a strict symmetric monoidal functor that satisfies $P_1 = P_0 F$ and preserves chosen opcartesian lifts.
(7.25) The multiplicative structure of a small tight bipermutative category defines an opcartesian strong n-linear functor.
(7.30 (1)) An opcartesian n-linear transformation is an n-linear natural transformation with an additional property but no extra structure.
(7.30 (2)) An opcartesian 0-linear transformation is a morphism in the P_0-preimage of $1 \in \mathcal{D}$.
(7.35) $\mathsf{PFib}^{\mathcal{D}}(\langle P \rangle; P_0)$ is a subcategory of $\mathsf{PermCat}^{\mathsf{sg}}(\langle \mathsf{C} \rangle; \mathsf{C}_0)$.
(7.37) Composition in $\mathsf{PFib}^{\mathcal{D}}$ is well defined on opcartesian strong multilinear functors.
(7.42) Composition in $\mathsf{PFib}^{\mathcal{D}}$ is well defined on opcartesian multilinear transformations.
(7.44) If the multiplicative braiding in $\mathcal{D}$ is the identity, then the symmetric group action in $\mathsf{PFib}^{\mathcal{D}}$ is well defined on opcartesian strong multilinear functors. Conversely, if the symmetric group action in $\mathsf{PFib}^{\mathcal{D}}$ is well defined on opcartesian strong multilinear functors, then $\otimes = \otimes^{\mathsf{op}}$ in $\mathcal{D}$.
(7.48) If the multiplicative braiding in $\mathcal{D}$ is the identity, then the symmetric group action in $\mathsf{PFib}^{\mathcal{D}}$ is well defined on opcartesian multilinear transformations. Conversely, if the symmetric group action in $\mathsf{PFib}^{\mathcal{D}}$ is well defined on opcartesian multilinear transformations, then $\otimes = \otimes^{\mathsf{op}}$ in $\mathcal{D}$.
(7.51) $\mathsf{PFib}^{\mathcal{D}}$ is a non-symmetric Cat-multicategory. If the multiplicative braiding in $\mathcal{D}$ is the identity, then $\mathsf{PFib}^{\mathcal{D}}$ is a Cat-multicategory. Conversely, if $\mathsf{PFib}^{\mathcal{D}}$ is a Cat-multicategory, then $\otimes = \otimes^{\mathsf{op}}$ in $\mathcal{D}$.
(7.52) There exists a non-symmetric Cat-multifunctor U from $\mathsf{PFib}^{\mathcal{D}}$ to $\mathsf{PermCat}^{\mathsf{sg}}$. If the multiplicative braiding in $\mathcal{D}$ is the identity, then U is a Cat-multifunctor.
(7.56) The first-factor projection is a small permutative opfibration over $\mathcal{D}$.
(7.61) $\int_{\mathcal{D}} : \mathcal{D}\mathsf{Cat}(\langle\rangle; Z) \longrightarrow \mathsf{PFib}^{\mathcal{D}}(\langle\rangle; U_Z)$ is an isomorphism.
(7.62) $\int_{\mathcal{D}} : \mathcal{D}\mathsf{Cat}(\langle X \rangle; Z) \longrightarrow \mathsf{PFib}^{\mathcal{D}}(\langle U_X \rangle; U_Z)$ is a functor.
(7.64) $\int_{\mathcal{D}} : \mathcal{D}\mathsf{Cat} \longrightarrow \mathsf{PFib}^{\mathcal{D}}$ is a non-symmetric Cat-multifunctor, which is, furthermore, a Cat-multifunctor if the multiplicative braiding in $\mathcal{D}$ is the identity.

Chapter 8. The Grothendieck Construction is a Cat-Multiequivalence

(8.9) $\int_{\mathcal{D}} : \mathcal{D}\mathsf{Cat} \longrightarrow \mathsf{PFib}^{\mathcal{D}}$ is essentially surjective on objects.
(8.20) $\int_{\mathcal{D}} : \mathcal{D}\mathsf{Cat}(\langle X \rangle; Z) \longrightarrow \mathsf{PFib}^{\mathcal{D}}(\langle U_X \rangle; U_Z)$ is injective on objects.
(8.28) $\int_{\mathcal{D}} : \mathcal{D}\mathsf{Cat}(\langle X \rangle; Z) \longrightarrow \mathsf{PFib}^{\mathcal{D}}(\langle U_X \rangle; U_Z)$ is injective on morphism sets.
(8.57) $\int_{\mathcal{D}} : \mathcal{D}\mathsf{Cat}(\langle X \rangle; Z) \longrightarrow \mathsf{PFib}^{\mathcal{D}}(\langle U_X \rangle; U_Z)$ is surjective on objects.

(8.75) $\int_{\mathcal{D}} : \mathcal{D}\mathsf{Cat}(\langle X\rangle ; Z) \longrightarrow \mathsf{PFib}^{\mathcal{D}}(\langle U_X\rangle ; U_Z)$ is surjective on morphism sets.
(8.76) $\int_{\mathcal{D}} : \mathcal{D}\mathsf{Cat} \longrightarrow \mathsf{PFib}^{\mathcal{D}}$ is a non-symmetric Cat-multiequivalence. It is a Cat-multiequivalence if the multiplicative braiding in $\mathcal{D}$ is the identity.

Chapter 9. The Cat-Multifunctor A

(9.7) Cat_* is a complete and cocomplete symmetric monoidal closed category.
(9.11) Γ-Cat is the category of pointed functors $(\mathcal{F}, \underline{0}) \longrightarrow (\mathsf{Cat}, \mathbf{1})$ and pointed natural transformations.
(9.13) Each Γ-category is canonically a pointed functor $(\mathcal{F}, \underline{0}) \longrightarrow (\mathsf{Cat}_*, \mathbf{1})$.
(9.23) Γ-Cat is complete, cocomplete, and symmetric monoidal closed. It is enriched, tensored, and cotensored over Cat_*. It is a Cat_*-multicategory.
(9.24) The monoidal unit diagram J satisfies $J\underline{n} \cong \underline{n}$.
(9.32) The multimorphism categories of Γ-Cat are given by $\mathsf{Map}_*\big(\bigwedge_j X_j\, , Z\big)$.
(9.37 and 9.39) Objects and morphisms in $\Gamma\text{-}\mathsf{Cat}(\langle\,\rangle; Z)$ are determined by their values at $1 \in \underline{1}$.
(9.53) The symmetric group action in Γ-Cat corresponds to permuting the domain factors.
(9.59) Composition in Γ-Cat is given by the pointed Day convolution product and composition in Map_*.
(9.66) Each Γ-category X yields a strictly unital strong symmetric monoidal functor $AX : \mathcal{A} \longrightarrow \mathsf{Cat}$.
(9.73) Each Γ-category morphism F yields a monoidal natural transformation AF.
(9.74) The underlying category of a Cat-multicategory has the same objects and 1-ary 1-cells as morphisms.
(9.79) $A : \Gamma\text{-}\mathsf{Cat} \longrightarrow \mathcal{A}\mathsf{Cat}$ is a functor.
(9.85) $A : \Gamma\text{-}\mathsf{Cat}(\langle\rangle ; Z) \xrightarrow{\cong} \mathcal{A}\mathsf{Cat}(\langle\rangle ; AZ)$ is an isomorphism.
(9.110) $A : \Gamma\text{-}\mathsf{Cat}(\langle X\rangle ; Z) \longrightarrow \mathcal{A}\mathsf{Cat}\big(\langle AX\rangle ; AZ\big)$ is a functor that is injective on objects and morphism sets.
(9.112) A preserves the symmetric group action.
(9.116) A preserves the multicategorical composition.
(9.120) $A : \Gamma\text{-}\mathsf{Cat} \longrightarrow \mathcal{A}\mathsf{Cat}$ is a Cat-multifunctor.

Chapter 10. Inverse K-Theory is a Pseudo Symmetric Cat-Multifunctor

(10.3) The inverse K-theory functor $\mathcal{P}$ is the composite $\int_{\mathcal{A}} \circ A$.
(10.12) $\mathcal{P}$ extends to a pseudo symmetric Cat-multifunctor $\int_{\mathcal{A}} \circ A$ that is not a Cat-multifunctor. As a non-symmetric Cat-multifunctor, $\mathcal{P}$ factors as $\mathsf{U} \circ \int_{\mathcal{A}} \circ A$.
(10.15) $\mathcal{P}$ preserves algebraic structures parametrized by non-symmetric Cat-multifunctors.
(10.17) $\mathcal{P}$ sends monoids in Γ-Cat to small tight ring categories.
(10.19) $\mathcal{P}$ preserves algebraic structures parametrized by pseudo symmetric Cat-multifunctors.
(10.30) $\mathcal{P}(-)^{\sigma}$ and $(\mathcal{P}(-))^{\sigma}$ are different on objects in general.

(10.32) The pseudo symmetry isomorphism $\mathcal{P}_\sigma$ is the identity if and only if σ is the identity permutation.
(10.35) An E_∞-algebra is a Cat-multifunctor from the Barratt-Eccles operad $E\Sigma_*$.
(10.39) An E_∞-algebra in a Cat-multicategory is uniquely determined by an object, a 0-ary 1-cell, a 2-ary 1-cell μ, and a 2-ary 2-cell isomorphism $\mu \xrightarrow{\cong} \mu^{\mathsf{op}}$, subject to several axioms.
(10.45) Each E_∞-algebra has an underlying monoid.
(10.46) E_∞-algebras in Cat, $\mathsf{PermCat}$, and $\mathsf{PermCat}^{\mathsf{sg}}$ are, respectively, small permutative categories and small (tight) bipermutative categories.
(10.53) A pseudo symmetric E_∞-algebra is a pseudo symmetric Cat-multifunctor from $E\Sigma_*$.
(10.54) $\mathcal{P}$ preserves pseudo symmetric E_∞-algebras.
(10.56) A pseudo symmetric E_∞-algebra consists of an object, an n-ary 1-cell μ_σ for each permutation σ, and invertible n-ary 2-cells $\mu_\sigma^\tau : \mu_\sigma \xrightarrow{\cong} \mu_\tau$ and $F_{\sigma;\tau} : \mu_{\tau\sigma} \xrightarrow{\cong} \mu_\tau \cdot \sigma$ for each pair of permutations (σ, τ), subject to several axioms.

Bibliography

[AGV72] M. Artin, A. Grothendieck, and J. L. Verdier, *Séminaire de géométrie algébrique du Bois-Marie 1963–1964. Théorie des topos et cohomologie étale des schémas. (SGA 4). Un séminaire dirigé par M. Artin, A. Grothendieck, J. L. Verdier. Avec la collaboration de N. Bourbaki, P. Deligne, B. Saint-Donat. Tome 1: Théorie des topos. Exposés I à IV. 2e éd.*, Lecture Notes in Mathematics, vol. 269, Springer, Cham, 1972 (French). doi:10.1007/BFb0081551

[BW19] J. Beardsley and L. Z. Wong, *The enriched Grothendieck construction*, Adv. Math. **344** (2019), 234–261. doi:10.1016/j.aim.2018.12.009

[BKP89] R. Blackwell, G. M. Kelly, and A. J. Power, *Two-dimensional monad theory*, J. Pure Appl. Algebra **59** (1989), no. 1, 1–41. doi:10.1016/0022-4049(89)90160-6

[BO20] A. M. Bohmann and A. M. Osorno, *A multiplicative comparison of Segal and Waldhausen K-theory*, Math. Z. **295** (2020), no. 3-4, 1205–1243. doi:10.1007/s00209-019-02394-7

[BF78] A. K. Bousfield and E. M. Friedlander, *Homotopy theory of* Γ-*spaces, spectra, and bisimplicial sets*, Geometric applications of homotopy theory (Proc. Conf., Evanston, Ill., 1977), II, Lecture Notes in Math., vol. 658, Springer, Berlin, 1978, pp. 80–130.

[Day70] B. Day, *On closed categories of functors*, Reports of the Midwest Category Seminar, IV, Lecture Notes in Mathematics, Vol. 137, Springer, Berlin, 1970, pp. 1–38. doi:10.1007/BFb0060437

[Elm∞] A. D. Elmendorf, *Multiplicativity in Mandell's inverse K-theory.* arXiv:2110.07512

[EM06] A. D. Elmendorf and M. A. Mandell, *Rings, modules, and algebras in infinite loop space theory*, Adv. Math. **205** (2006), no. 1, 163–228. doi:10.1016/j.aim.2005.07.007

[EM09] ———, *Permutative categories, multicategories and algebraic K-theory*, Algebr. Geom. Topol. **9** (2009), no. 4, 2391–2441. doi:10.2140/agt.2009.9.2391

[Fre17] B. Fresse, *Homotopy of operads and Grothendieck-Teichmüller groups. Part 1. The algebraic theory and its topological background*, Mathematical Surveys and Monographs, vol. 217, American Mathematical Society, Providence, RI, 2017.

[GJO17] N. Gurski, N. Johnson, and A. M. Osorno, *K-theory for 2-categories*, Adv. Math. **322** (2017), 378–472. doi:`10.1016/j.aim.2017.10.011`

[Her04] C. Hermida, *Fibrations for abstract multicategories*, Galois theory, Hopf algebras, and semiabelian categories, Fields Inst. Commun., vol. 43, Amer. Math. Soc., Providence, RI, 2004, pp. 281–293. doi:`10.1090/fic/043`

[JY∞] N. Johnson and D. Yau, *Bimonoidal Categories, E_n-Monoidal Categories, and Algebraic K-Theory. Volume III: From Categories to Structured Ring Spectra*, available at `https://nilesjohnson.net/`.

[JY21] ———, *2-Dimensional Categories*, Oxford University Press, New York, 2021. doi:`10.1093/oso/9780198871378.001.0001`

[JY22a] ———, *Homotopy equivalent algebraic structures in multicategories and permutative categories*, Theory Appl. Categ. **38** (2022), Paper No. 30, 1156–1208.

[JY22b] ———, *Multifunctorial inverse K-theory*, Ann. K-Theory **7** (2022), no. 3, 507–548. doi:`10.2140/akt.2022.7.507`

[JY22c] ———, *Multifunctorial K-theory is an equivalence of homotopy theories*, J. Homotopy Relat. Struct. **17** (2022), no. 4, 569–592. doi:`10.1007/s40062-022-00317-8`

[JY23] ———, *Multicategories model all connective spectra*, Homology Homotopy Appl. **25** (2023), no. 1, 147–172. doi:`10.4310/hha.2023.v25.n1.a8`

[JS93] A. Joyal and R. Street, *Braided tensor categories*, Adv. Math. **102** (1993), no. 1, 20–78. doi:`10.1006/aima.1993.1055`

[Kel74] G. M. Kelly, *Coherence theorems for lax algebras and for distributive laws*, Category Seminar (Proc. Sem., Sydney, 1972/1973), Lecture Notes in Math., Vol. 420, 1974, pp. 281–375.

[Kel05] ———, *Basic concepts of enriched category theory*, Repr. Theory Appl. Categ. (2005), no. 10, vi+137, Reprint of the 1982 original (Cambridge Univ. Press, Cambridge).

[Lam69] J. Lambek, *Deductive systems and categories II. Standard constructions and closed categories*, Category Theory, Homology Theory and their Applications, I (Battelle Institute Conference, Seattle, Wash., 1968, Vol. One), Springer, Berlin, 1969, pp. 76–122.

[Lap72a] M. L. Laplaza, *Coherence for distributivity*, Coherence in categories, Lectures Notes in Mathematics, Vol. 281, Springer-Verlag, Berlin-Heidelberg, 1972, pp. 29–65.

[Lap72b] ______, *A new result of coherence for distributivity*, Coherence in categories, Lectures Notes in Mathematics, Vol. 281, Springer-Verlag, Berlin-Heidelberg, 1972, pp. 214–235.

[ML69] S. Mac Lane, *One universe as a foundation for category theory*, Reports of the Midwest Category Seminar. III, Lecture Notes in Mathematics, Vol. 106, Springer, Berlin, 1969, pp. 192–200.

[ML98] ______, *Categories for the working mathematician*, second ed., Graduate Texts in Mathematics, vol. 5, Springer-Verlag, New York, 1998.

[Man10] M. A. Mandell, *An inverse K-theory functor*, Doc. Math. **15** (2010), 765–791.

[May78] J. P. May, *The spectra associated to permutative categories*, Topology **17** (1978), no. 3, 225–228. doi:10.1016/0040-9383(78)90027-7

[MV20] J. Moeller and C. Vasilakopoulou, *Monoidal Grothendieck construction*, Theory Appl. Categ. **35** (2020), Paper No. 31, 1159–1207.

[Pow89] A. J. Power, *A general coherence result*, J. Pure Appl. Algebra **57** (1989), no. 2, 165–173. doi:10.1016/0022-4049(89)90113-8

[Qui73] D. G. Quillen, *Higher algebraic K-theory I: Higher K-theories*, Proc. Conf., Battelle Memorial Inst., Seattle, Wash., (1972), Lecture Notes in Math., Vol. 341, 1973, pp. 85–147.

[Seg74] G. Segal, *Categories and cohomology theories*, Topology **13** (1974), 293–312. doi:10.1016/0040-9383(74)90022-6

[Shu08] M. Shulman, *Framed bicategories and monoidal fibrations*, Theory Appl. Categ. **20** (2008), No. 18, 650–738.

[Tho95] R. W. Thomason, *Symmetric monoidal categories model all connective spectra*, Theory Appl. Categ. **1** (1995), No. 5, 78–118.

[Wal85] F. Waldhausen, *Algebraic K-theory of spaces*, Algebraic and geometric topology (New Brunswick, N.J., 1983), Lecture Notes in Math., vol. 1126, Springer, Berlin, 1985, pp. 318–419. doi:10.1007/BFb0074449

[Yau∞I] D. Yau, *Bimonoidal Categories, E_n-Monoidal Categories, and Algebraic K-Theory. Volume I: Symmetric Bimonoidal Categories and Monoidal Bicategories*, available at https://u.osu.edu/yau.22/main/.

[Yau∞II] ———, *Bimonoidal Categories, E_n-Monoidal Categories, and Algebraic K-Theory. Volume II: Braided Bimonoidal Categories with Applications*, available at https://u.osu.edu/yau.22/main/.

[Yau16] ———, *Colored operads*, Graduate Studies in Mathematics, vol. 170, American Mathematical Society, Providence, RI, 2016.

[Yau20] ———, *Involutive category theory*, Lecture Notes in Mathematics, vol. 2279, Springer, Cham, 2020. doi:10.1007/978-3-030-61203-0

[Yau22] ———, *Infinity operads and monoidal categories with group equivariance*, Singapore: World Scientific, 2022. doi:10.1142/12687

Index